STUDENT STUDY GUIDE

JOHN D. CUTNELL
KENNETH W. JOHNSON
Southern Illinois University at Carbondale

MARK J. COMELLA

To accompany

PHYSICS

Fifth Edition

JOHN D. CUTNELL
KENNETH W. JOHNSON
Southern Illinois University at Carbondale

John Wiley & Sons, Inc.

New York • Chichester • Weinheim • Brisbane • Singapore • Toronto

Cover Photo: ©Karl Weatherly / Tony Stone Images, New York.

ISBN 0-471-35582-8

Printed in the United States of America

10 9 8 7 6 5 4 3 2

Printed and bound by Courier Stoughton, Inc.

TO THE STUDENT

This study guide has been written to accompany *Physics*, fifth edition, by John D. Cutnell and Kenneth W. Johnson. It is designed to assist you in understanding the concepts and in developing the problem-solving skills necessary to master a first course in college physics. The Study Guide is intended as a resource to be used after reading the text.

Organization of the Study Guide

Each chapter of the Study Guide is organized in the following manner:

- **Preview** The *Preview* provides a brief synopsis of the chapter contents.

- **Quick Reference** The *Quick Reference* contains a summary of *Important Terms* and a list of *Equations* associated with the main concepts and principles encountered in the chapter.

- **Discussion of Selected Sections** This section provides discussions of most of the sections found in the text with the order of topics following that of the text. The topics discussed are usually the main topics in the chapter, as well as topics that students generally find most difficult. These discussions include a large number of quantitative examples to supplement those found in the text.

- **Practice Problems** The *Practice Problems* are provided to give additional practice at solving "single concept" problems. Detailed solutions are provided in the *Solutions and Answers* section.

In addition, this study guide includes advice on *How to Take a Physics Course*.

Additional Web Resources

The Password that accompanies this Student Study Guide provides access to many additional learning resources at our Web site (**http://www.wiley.com/college/cutnell**). These resources include:

- **Additional Interactive LearningWare Problems** These interactive tutorials take you step-by-step through the solution of 118 problems. The problems are in addition to the 60 Interactive LearningWare problems found in the text itself.

- **Simulation Exercises** The *Simulation Exercises* allow the user to control various aspects of a physical situation and to examine the outcome. They are great for exploring how various parameters affect the outcome.

- **Chapter Quiz** Each *Chapter Quiz* contains an average of thirteen multiple-choice questions that are intended to help you diagnose any weak areas in the concepts of the chapter. Answers for all questions are provided.

- **Sample Examinations** Four multiple-choice *Sample Examinations*, with answers, help you to prepare for examinations. The examinations cover the following chapters: Exam 1, Chapters 1–9; Exam 2, Chapters 10–15; Exam 3, Chapters 16–23; Exam 4, Chapters 24–32.

- **MCAT Review Problems** This section contains problems designed to help students review for the MCAT exam that is required as part of the admission process to medical school. The physical sciences portion of the exam differs radically from many other science tests. Most of the questions are associated with passages. A typical passage describes an experiment, makes an argument, or presents some information, often covering material you have never seen. The questions test whether you can use your physics knowledge to interpret the experiment, settle the argument, or understand the new material. Knowing lots of formulas will *not* help. Instead, you need to understand the *concepts*. Many passages deal with multiple concepts simultaneously. Since these review problems do the same thing, they will also help you prepare for *regular* physics exams, not just the MCAT. In order to test your ability to handle real-life journal articles and medical textbooks, the MCAT contains a few poorly written passages. To prepare you for this ordeal, some of these review problems intentionally present information unclearly or too quickly. Sometimes, the vocabulary and symbols on the exam (and in these review problems) differ slightly from those in your text. Fortunately, by avoiding panic, you can sort through the confusion and answer the questions, most of which do not require you to understand all the details of the passage. Since you may not bring calculators, graphing paper, or other aides to the MCAT, the exam questions do not require difficult calculations. Often, approximations can get you the answer quickly.

- **Links to Other Web Sites** A collection of links to other Web sites that offer physics tutorial help.

HOW TO TAKE A PHYSICS COURSE

Often students entering their first course in physics anticipate having a "hard time". Much of what they have heard from friends and relatives about physics courses leads them to believe that physics can not be learned well by the average person. This is simply not true. The secret to success in a physics course is to learn the material day by day. Trying to "cram" physics the night before an exam is what leads to the "hard time" in the course. Physics is a subject which requires time for the concepts to be absorbed and understood.

Physics is NOT learned by memorizing equations and then trying to find the right numbers to plug into them. Physics is learned by using it day after day to solve problems and by thinking about the concepts and relating them to everyday experience.

How to Learn Physics

Learning physics is a building process much like building a house. You must first build a good foundation and learn to use the proper tools. The foundation in physics consists of a set of laws and principles. The tools are mathematics and problem solving skills. If you build a good foundation, the rest of the structure will be straight and sound. If the foundation is weak, the whole structure may collapse.

After the foundation is laid, everything else rests on it. The same is true of the foundation that you will build in physics. If a principle or concept is not well understood, it will come back to plague you later. Learn everything well as you encounter it. The following will aid you in your effort.

1. **Attend lecture faithfully**. Physics professors are prone to include material which is either not in the text or modified to some extent.

2. **Take GOOD notes**. Good note taking skills are learned by practice in the classroom. Good note taking does not mean writing down everything that is put on the board or appears on the overhead transparencies. Good note taking requires that you are already familiar with the topic by reading the text in advance of the lecture. The material presented in the lecture can then be digested properly and your notes will consist of the digested material in the form of outlines of the major concepts and problem solving methods. Do not write down all of the little facts that you already know unless they add something to your understanding.

3. **Use your notes**. A good way to study is to compile the day's notes and the textbook material into a written discussion of the concepts and problem solving methods. Pretend that you are explaining the material to someone as you write. Also, try out your understanding by teaching the material to a sympathetic friend or fellow physics student. This "teaching" is a very effective way to learn. Indeed, physics professors know physics so well because they teach.

4. **Learn the laws, principles, and terminology of physics**. Avoid memorizing verbatim from the book. Simply knowing the words is useless, you must gain an understanding of the meaning. Restating the laws, principles, and definitions in **your own words** aids in the understanding of their meaning. Memorization is not learning, however, learning sometimes involves memorization. Frequently used equations and constants should be memorized. Actually, most of the equations which you will need, you will automatically remember after doing a lot of homework problems.

5. **Solve problems**. Solving lots of problems from the text and the study guide helps you to develop a deeper understanding of the concepts. In fact, it is often argued, that if you can't solve physics problems, then you don't know physics. After solving several problems of a particular type you will develop "physical intuition". This means that you can analyze a problem and predict the results before you actually solve it.

Problem solving skills are developed by practice and often vary from student to student. There is no cookbook approach to problem solving. However, some general guidelines to problem solving can be written down. Use the following guidelines to help you develop your problem solving technique.

a. Understand the problem. This requires that you carefully read the problem and determine exactly what is asked for in the problem. An invaluable aid to understanding a problem is a labeled diagram of the situation presented in the problem. This can be as simple as a quick sketch of the objects involved and their relationships to one another. Include all of the information given in the problem on your sketch.

b. Determine the physical principles involved in the problem. What kind of a problem is it? Mechanics, electricity, etc. or a combination? What physical principles apply in this problem? Newton's laws, conservation of energy, conservation of momentum, etc.?

c. Try to gain a general understanding of what the results should be like before you do calculations. Does the ball go left or right? Up or down? Does the car speed up or slow down?

d. Write the mathematical expressions of the relevant laws and principles in the notation of your particular problem. Take advantage of any zero quantities to simplify the expressions. DO NOT plug in any other numbers.

e. After you have the mathematical expressions ask yourself, "how can I manipulate these mathematical equations to find what I am asked to find?" Students often become stuck at this point because they look at the equations and realize that they don't know a quantity which appears to be needed, say, the mass of an object. DON'T worry. Solve the equations and obtain an algebraic form for the solution. Often the unknown quantity, such as the mass, will cancel out. If it doesn't, THEN worry about how to find it. In any case, you have a general solution to the problem. All you need now are some specific numbers to "plug" in.

f. Plug in your numbers to get a specific answer for your problem. Since you already have the general solution, you may plug in different numbers to get the specific solution to a new problem without having to do all of the work again.

g. Interpret your solution. Do the units work out right? Does the answer make sense? A simple example of an answer which does not make sense is when your mathematics tells you that a ball rises when dropped and you know the object will fall.

How to Study for Exams

1. If you have **studied** your physics **each day** and **solved** a good number of **problems**, you should need little additional preparation for exams. The night before an exam, you might review the important concepts, refresh your memory on some problem solving techniques, go to bed early and get a good night's sleep. If you don't know the material, then cramming will be of little help. Cramming means trying to memorize far more material than possible in the short time available with very little understanding of any of it. This kind of studying may help a little on questions involving pure recall, but none on questions involving reasoning and deep thought. DON'T CRAM!

2. More preparation is required for the final exam even if you have studied day by day, particularly, if the final is comprehensive. This is because so much time has elapsed since you have studied some of the material that it may have become hazy in your mind. Plan your study to include a review spread out over several days. If you learned the material well in the first place, it will quickly come back to you. Cramming for a final is even less effective than it is for other exams. Again, DON'T do it.

How to Use the Text and the Study Guide

Aside from your instructor, the textbook is the most important source of information available to you. Read the text daily, paying particular attention to the examples worked out in the text. Then try some of the chapter problems on your own. If you need additional help, consult this study guide. It has several worked out example problems and practice problems with the solutions given. It is a good idea to try to solve the practice problems on your own before looking at the solutions. Use the solutions only to check to see if you have done the problem right or to get you started again if you become stuck on a problem.

The text has a large amount of information in the appendices which can be of great help if you know what is there. Spend some time browsing through the appendices. Don't try to memorize all that you find there, but keep in mind what is available. Later, when you need a constant or conversion, etc. you will recall that these are in the appendices and you will not have to spend a great deal of time trying to find them. The Study Guide does not have the information contained in the appendices of the text. The authors of the Study Guide are expecting that you will use both books together.

ACKNOWLEDGMENTS

Our sincerest thanks goes to Georgia Kamvosoulis Mederer who edited the entire manuscript. Her keen eyes caught all kinds of typos, inconsistencies, and even incorrect solutions. We have worked with Georgia before and have always admired her diligence and professionalism. We also wish to acknowledge Kurt Mederer for his invaluable technical assistance in preparing the electronic files.

John D. Cutnell
Kenneth W. Johnson

CONTENTS

Chapter

CHAPTER 1
Introduction and Mathematical Concepts

PREVIEW

In this chapter you will be introduced to the physical units most frequently encountered in physics. After completion of the chapter you will be able to convert between these systems of units and use them as an aid in problem solving.

Also, after a review of trigonometry, you will become acquainted with vectors and the methods of vector addition. Upon completion of this material you should be able to add and subtract vectors graphically, decompose vectors into their components and use the components to reconstruct the vectors.

QUICK REFERENCE

Important terms

Scalar quantity
   A quantity which can be described by a single number.
Vector quantity
   A quantity which can be adequately described by a number (magnitude) and a direction.
Vector components
   Two perpendicular vectors which added together produce the original vector.

Systems of Units

System	Length	Mass	Time
SI	meter (m)	kilogram (kg)	seconds (s)
BE	foot (ft)	slug (sl)	seconds (s)
CGS	centimeters (cm)	gram (g)	seconds (s)

Trigonometry

$$\sin \theta = \frac{h_o}{h} \quad (1.1) \qquad \theta = \sin^{-1}\left(\frac{h_o}{h}\right) \quad (1.4)$$

$$\cos \theta = \frac{h_a}{h} \quad (1.2) \qquad \theta = \cos^{-1}\left(\frac{h_a}{h}\right) \quad (1.5)$$

$$\tan \theta = \frac{h_o}{h_a} \quad (1.3) \qquad \theta = \tan^{-1}\left(\frac{h_o}{h_a}\right) \quad (1.6)$$

$$h^2 = h_o{}^2 + h_a{}^2 \qquad (1.7)$$

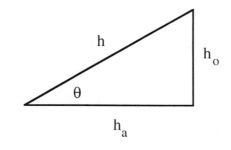

Graphical Addition of Vectors

Vectors may be added (or subtracted) by placing them head to tail.

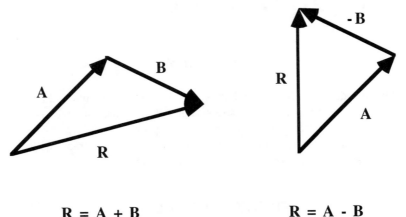

$$R = A + B \qquad\qquad R = A - B$$

Components of a Vector

The following expressions refer to the vector diagram shown below.

$$A_x = A \cos \theta \qquad\qquad A_y = A \sin \theta$$

$$A = \sqrt{A_x^2 + A_y^2} \qquad\qquad \theta = \tan^{-1}\left(\frac{A_y}{A_x}\right)$$

Addition of Vectors Using Components

If $C = A + B$, then the components of C are:

$$C_x = A_x + B_x \qquad\qquad C_y = A_y + B_y$$

and the magnitude and direction of C are:

$$C = \sqrt{C_x^2 + C_y^2} \qquad\qquad \theta = \tan^{-1}\left(\frac{C_y}{C_x}\right)$$

where θ is measured counterclockwise from the + x axis.

DISCUSSION OF SELECTED SECTIONS

1.3 The Role of Units in Problem Solving

Units are very important in the study of physics in that all physical quantities have units. These may be either the *base units* of length (m), mass (kg), and time (s), or *derived units* such as the joule (kg m^2/s^2).

When used in algebraic expressions, the units which accompany the numbers can be used to check not only the accuracy of the calculation, but also the validity of the equation. For this reason, the units will always be displayed along with the numbers in this study guide. You are encouraged to do the same in your solutions to problems. Remember, if the units do not work out, your solution is not right either.

Example 1 *Manipulating Units and Converting Between Systems*
Convert 5.00 mi/h to m/s.

Conversion factors:

$$1 \text{ mi} = 5280 \text{ ft} \qquad\qquad 1 \text{ ft} = 0.305 \text{ m}$$
$$1 \text{ km} = 1000 \text{ m} \qquad\qquad 1 \text{ h} = 3600 \text{ s}$$

Each of the equalities above can be used to form a fraction or *conversion factor* that is equal to unity (i.e., equal to 1). In multiplying by unity we do not change the value of the physical quantity; we are merely expressing the same quantity in a different set of units. One side of the equality will appear in the numerator and the other side will appear in the denominator of the fraction. The specific choice will be made so that the unwanted units cancel and the desired units appear in the final answer.

$$5.00 \text{ mi/h} = 5.00 \, \frac{\text{mi}}{\text{h}} \left(\frac{5280 \text{ ft}}{1 \text{ mi}} \right) \left(\frac{0.305 \text{ m}}{1 \text{ ft}} \right) \left(\frac{1 \text{ h}}{3600 \text{ s}} \right) = 2.24 \, \frac{\text{m}}{\text{s}}.$$

Notice that the first conversion factor changes miles to feet, the second changes feet to meters, and the third conversion factor changes hours to seconds. Again notice that in each conversion factor, the numerator is equal to the denominator and, therefore, the conversion factor is equal to unity.

Example 2 *Another Look at Converting Units*
Convert 2.00 mi^2 to ft^2.

We will start with the fact that 1 mi = 5280 ft. In setting up the conversion factor as a fraction equal to unity, however, we must square the fraction so that it converts square miles (mi^2) to square feet (ft^2).

$$2.00 \text{ mi}^2 = 2.00 \text{ mi}^2 \left(\frac{5280 \text{ ft}}{1 \text{ mi}} \right)^2 = 5.58 \times 10^7 \text{ ft}^2$$

Notice that the conversion factor is still equal to unity ($1^2 = 1$). Also notice that the final answer is expressed in scientific notation to preserve the correct number of significant figures; three in this case.

Example 3 *Using Units to Check Equations*

An equation which may result from the application of the conservation of energy principle is

$$\frac{1}{2} mv^2 = mgh$$

where m is a mass with the units of kg, v is a velocity with the units of m/s, g is an acceleration with the units of m/s^2 and h is a height with the units of m.

This equation may be checked for validity by simply substituting the units into it and manipulating them as if they were numbers. If the units on the right and left side of the equation do not match, the equation is definitely NOT VALID.

$$(kg) \left(\frac{m}{s} \right)^2 = (kg) \left(\frac{m}{s^2} \right)(m)$$

$$(kg) \left(\frac{m^2}{s^2} \right) = (kg) \left(\frac{m^2}{s^2} \right)$$

$$\frac{kg\ m^2}{s^2} = \frac{kg\ m^2}{s^2}$$

In this case, the units on each side of the equation do match and the equation may be a valid one. This procedure does NOT guarantee that the equation has any "true" meaning, however.

By the way, the above combination of units appear so often in physics that they are given the special name of joules (J). This is the unit of ENERGY.

1.4   Trigonometry

Most likely, you have already become acquainted with the basic trigonometric functions in a high school or college course. These can be found in the Quick Reference section of this book.

Many times the use of the "trig" functions in physics involves finding the length of one side of a right triangle when you know one other side and one of the acute angles. This is particularly important in finding the components of a vector which will be studied in section 1.7.

Other variations are, of course, possible and useful. For instance, if two sides of a right triangle are known, then the trigonometric functions can be used to find the angle.

Example 4 *Finding one side of a right triangle if one other side and an angle are known*

An observer, whose eyes are 6.0 ft above the ground, is standing 105 ft away from a tree. The ground is level, and the tree is growing perpendicular to it. The observer's line of sight with the tree top makes an angle of 20.0° above the horizontal. How tall is the tree?

The observer's eye, the tree top, and a point on the trunk are the vertices of a right triangle. The height of the tree is

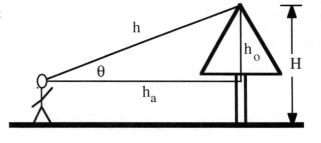

$$H = 6.0 \text{ ft} + h_o$$

Now h_o can be found from the right triangle by using equation (1.3)

$$h_o = h_a \tan \theta = (105 \text{ ft}) \tan 20.0° = 38.2 \text{ ft.}$$
$$H = 6.0 \text{ ft} + 38.2 \text{ ft} = 44.2 \text{ ft.}$$

Another common need in physics is to find one side of a triangle when the other two sides are known but no acute angle is given. Then you should use the Pythagorean theorem (1.7) or one of its variants.

Example 5 *Using the Pythagorean theorem*
In the previous example, it is desired to know the straight line distance from the person's eye to the top of the tree. The Pythagorean theorem gives

$$h = \sqrt{h_o^2 + h_a^2} = \sqrt{(38.2 \text{ ft})^2 + (105 \text{ ft})^2} = 112 \text{ ft.}$$

1.6   Vector Addition and Subtraction

If vectors are colinear, then they may be added or subtracted by simply adding or subtracting their magnitudes. The directions of the vectors are usually specified by calling a vector pointing to the right (or up) positive and a vector pointing to the left (or down) negative. Then the sign of the resultant vector tells which way it points.

Example 6 *Adding and subtracting colinear vectors*
Find the resultant, **C,** of the following vectors (u is an arbitrary unit).

Case 1

$A = +3.0$ u

$B = +2.0$ u

Case 2

$A = +3.0$ u

$B = -2.0$ u

In both cases the vectors are colinear. One could imagine placing the vectors **A** and **B** in a "tail-to-head" fashion. In case 1, both **A** and **B** point in the same direction. The resultant vector will have a magnitude equal to the length of **A** plus the length of **B**. Since both vectors are directed to the right, the resultant will also point to the right. In case 2, the vectors **A** and **B** point in opposite directions. If they are placed in a tail-to-head fashion it will be clear that part of **A** is canceled by **B**. The resultant will have a magnitude that is equal to the magnitude of **A** minus the magnitude of **B**. Since **A** has the greater magnitude, the resultant will point in the direction of **A**. The resultant **C** is shown below for both cases.

Case 1

C = +5.0 u

Case 2

C = +1.0 u

When vectors are not colinear but perpendicular, they may be added by using the Pythagorean theorem and the tangent function. This method yields the magnitude and direction of the resultant vector as a number and an angle.

Example 7 *Adding perpendicular vectors*
Find the resultant, **C**, of the vectors shown where **A** = + 2.0 u and **B** = + 5.0 u.

The resultant is **C** = **A** + **B** as shown in the diagram. Notice that it was constructed by placing the vectors **A** and **B** in a "tail-to-head" fashion and then connecting the "head" of **A** to the "tail" of **B**.

The length of the resultant vector, C, is given by the Pythagorean theorem to be

$$C = \sqrt{A^2 + B^2}$$

$$C = \sqrt{(2.0 \text{ u})^2 + (5.0 \text{ u})^2}$$

$$C = 5.4 \text{ u.}$$

The direction of **C**, as specified by the angle, θ, is given by (1.6) to be

$$\theta = \tan^{-1}\left(\frac{A}{B}\right) = \tan^{-1}\left(\frac{2.0 \text{ u}}{5.0 \text{ u}}\right) = 22°.$$

Now **C** = 5.4 u, 22° counterclockwise from **B**.

1.7 The Components of a Vector

As you might infer from the preceding example, any vector, **C**, may be expressed as the sum of two perpendicular vectors. When these perpendicular vectors are placed along the x and y axes of a Cartesian coordinate system, they are referred to as the x and y components (C_x and C_y) of the vector.

The components of a vector may be found by using a vector diagram similar to the above and equations (1.1) and (1.2). Please keep in mind that if equations (1.1) and (1.2) are used to find the components of a vector, the angle, θ, is defined as being measured from the x axis.

Example 8 *Finding the components of a vector*
Find the components of the vector $\mathbf{A} = 5.0$ u, $\theta = 30$ °.

An application of (1.2) to the vector shown in the figure gives

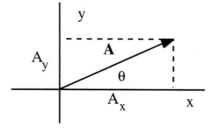

$$A_x = A \cos \theta = (5.0 \text{ u}) \cos 30°$$
$$A_x = 4.3 \text{ u.}$$

A similar application of (1.1) gives

$$A_y = A \sin \theta = (5.0 \text{ u}) \sin 30°$$
$$A_y = 2.5 \text{ u.}$$

Sometimes you will encounter situations where you will need to find the x and y components of a vector, but the angle given, ϕ, is NOT measured from the x axis. In this case, the components may NOT be given by

$$A_x = A \cos \phi \quad \text{and} \quad A_y = A \sin \phi.$$

You now have two choices. You may either find the angle, ϕ, in terms of the angle, θ, or you may try to find a different right triangle in which to apply equations (1.1) and (1.2).

Example 9 *Finding the components of a vector when the angle is not measured from the +x axis*
Find the x and y components of the vector, $\mathbf{A}$, whose magnitude is 10.0 u and which makes an angle of 45° with the +y axis. Refer to the figure shown.

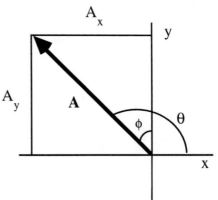

We may use $A_x = A \cos \theta$ and $A_y = A \sin \theta$ if we can find the angle θ.

From the figure it is seen that

$$\theta = 90° + \phi \text{ so that}$$
$$\theta = 90° + 45° = 135°.$$

Now

$$A_x = (10.0 \text{ u}) \cos 135°$$
$$A_x = -7.07 \text{ u}$$
$$A_y = (10.0 \text{ u}) \sin 135°$$
$$A_y = +7.07 \text{ u.}$$

We may also apply equations (1.1) and (1.2) to the right triangle shown in the figure to obtain

$$A_x = - A \sin \phi = - (10.0 \text{ u}) \sin 45° = - 7.07 \text{ u}$$

and

$$A_y = + A \cos \phi = + (10.0 \text{ u}) \cos 45° = + 7.07 \text{ u.}$$

1.8 Addition of Vectors by Means of Components

Since like components of two or more vectors are colinear, they may simply be added, as in example 6, to give the corresponding component of the resultant vector. In this way, all of the components of the resultant vector may be found. The magnitude and direction of the resultant vector may then be determined by equations (1.6) and (1.7).

Example 10 *Adding vectors by the component method*
Add the vectors **A**, **B**, and **C** shown in the figure using the component method. A = 5.0 m, B = 7.0 m and C = 4.0 m.

An application of equations (1.1) and (1.2) to each of the triangles shown in the figure gives

$$A_x = + A \cos 20°$$
$$A_x = + (5.0 \text{ m}) \cos 20°$$
$$A_x = + 4.7 \text{ m.}$$

$$B_x = - B \cos 40°$$
$$B_x = - (7.0 \text{ m}) \cos 40°$$
$$B_x = - 5.4 \text{ m.}$$

$$C_x = + C \sin 25°$$
$$C_x = + (4.0 \text{ m}) \sin 25°$$
$$C_x = + 1.7 \text{ m.}$$

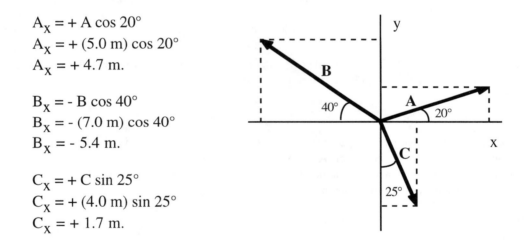

The x component of the resultant vector is

$$R_x = A_x + B_x + C_x$$

$$R_x = 4.7 \text{ m} - 5.4 \text{ m} + 1.7 \text{ m}$$

$$R_x = 1.0 \text{ m.}$$

Repeating the above for the y component gives

$$A_y = + A \sin 20° = + (5.0 \text{ m}) \sin 20° = + 1.7 \text{ m.}$$

$$B_y = + B \sin 40° = + (7.0 \text{ m}) \sin 40° = + 4.5 \text{ m.}$$

$$C_y = - C \cos 25° = - (4.0 \text{ m}) \cos 25° = - 3.6 \text{ m.}$$

The y component of the resultant vector is

$$R_y = A_y + B_y + C_y = 1.7 \text{ m} + 4.5 \text{ m} - 3.6 \text{ m}$$

$$R_y = 2.6 \text{ m}.$$

Now the magnitude of the resultant can be found from the Pythagorean theorem

$$R = \sqrt{R_x^2 + R_y^2}$$

$$R = \sqrt{(1.0 \text{ m})^2 + (2.6 \text{ m})^2} = 2.8 \text{ m}$$

The angle that R makes with the +x axis is

$$\theta = \tan^{-1}(R_y/R_x) = \tan^{-1}(2.6 \text{ m}/1.0 \text{ m}) = 69°.$$

PRACTICE PROBLEMS

The following problems are provided to give you additional practice solving single concept problems. Some work space has been left for you after or to the right of the problem. The solutions to these problems will be found at the end of this chapter.

1. How many significant figures are in the numbers

    a. 2.03

    b. 0.0054

    c. 2.8×10^5

    d. 1500

    e. 0.23×10^{-4}

2. Convert the following into the indicated units.

    a. 5.00 m to ft

    b. 1.2 ft/s to m/s

    c. 60.0 mi/h to km/h

    d. 9.80 m/s^2 to ft/s^2

    e. 535 kg to slugs

3. A basketball coach insists that his players be at least 180.0 cm tall. Would a player of height 5 ft 11.5 in tall qualify for the team?

4. A football field is 100.0 yards long. Express this distance in millimeters.

5. Check the following equations for possible validity.

    a. $x^2 = 1/2 \, gt$ where x is in m, g is in m/s^2, t is in s.

    b. $v^2 = 2 \, ax$ where v is in m/s, a is in m/s^2 and x is in m.

6. A right triangle has a side of length 3.5 m. The angle opposite the 3.5 m side is 25°. Find the length of the other sides.

7. A 500.0 m tall building casts a shadow 800.0 m long over level ground. What is the sun's elevation angle above the horizon?

8. A right triangle has two sides of length 25 ft and 15 ft. Find the length of the hypotenuse and all angles.

9. A bridge 50.0 m long crosses a chasm. If the bridge is inclined at an angle of 20.0° to the horizontal, what is the difference in height between the two ends?

10. A displacement vector of magnitude 5.0 m points in an easterly direction. A second displacement vector points north and has a magnitude of 9.7 m. Find the magnitude and direction of the vector sum.

11. An electric field vector, **E**, has a magnitude of 1.0 newton per coulomb (N/C) and makes an angle of 33° CCW from the +x axis of a Cartesian coordinate system. Find the components of **E**.

12. A magnetic field vector, **B**, is oriented 65° clockwise from the -y axis. It has a magnitude of 0.010 tesla (T). Find the x and y components of **B**.

13. A vector, **A**, has a magnitude of 10.0 u and points 21.0° north of west. A second vector, **B**, has a magnitude of 5.2 u and points 47.0° east of south. Find the magnitude and direction of the sum by the component method.

14. A car drives 2 km west, then 8 km south, and then 10.0 km at an angle 53° north of east. Find the car's final displacement (magnitude and direction).

SOLUTIONS AND ANSWERS

Practice Problems

---

1. a. 2.03 has **THREE** significant figure.
    b. 0.0054 has **TWO** significant figures. Note that it can be written as 5.4×10^{-3}.
    c. 2.8×10^5 has **TWO** significant figures.
    d. 1500 has **TWO** significant figures since both zeros are in doubt.
    e. 0.23×10^{-4} has **TWO** significant figures.

---

2. a. We have
    $$5.00 \text{ m} = (5.00 \text{ m})(3.28 \text{ ft/1 m}) = \textbf{16.4 ft.}$$
    b. Also
    $$1.2 \text{ ft/s} = (1.2 \text{ ft/s})(0.305 \text{ m/1 ft}) = \textbf{0.37 m/s.}$$
    c. We know
    $$60.0 \text{ mi/h} = (60.0 \text{ mi/1 h})(5280 \text{ ft/1 mi})(0.305 \text{ m/1 ft})(1/1000 \text{ km/1 m}) = \textbf{96.6 km/h.}$$
    d. In this case
    $$9.80 \text{ m/s}^2 = (9.80 \text{ m/s}^2)(3.28 \text{ ft/1 m}) = \textbf{32.1 ft/s}^2.$$
    e. Finally
    $$535 \text{ kg} = (535 \text{ kg})(6.85 \times 10^{-2} \text{ sl/1 kg}) = \textbf{36.6 sl.}$$

---

3. The conversion looks like
    $$180.0 \text{ cm} = (180.0 \text{ cm})/(1 \text{ in/2.54 cm}) = 70.9 \text{ in} = 5 \text{ ft } 10.9 \text{ in. } \textbf{YES}$$

---

4. In this case
    $$100.0 \text{ yds} = (100.0 \text{ yds})(3 \text{ ft/yd})(12 \text{ in/ft})(2.54 \text{ cm/in})(10 \text{ mm/cm}) = \textbf{91 440 mm.}$$

---

5. a. Substitute the units into the equation to see if both sides match.
    $$m^2 = (m/s^2)(s) = m/s$$
    The units on each side do NOT match so **the equation is not valid.**

    b. Similarly,
    $$m^2/s^2 = (m/s^2)(m) = m^2/s^2.$$
    The units DO match so the **equation MAY be valid.**

---

6. The hypotenuse is given by (1.1)
    $$h = (3.5 \text{ m})/\sin 25° = \textbf{8.3 m.}$$
    The side adjacent is given by (1.2),
    $$h_a = (8.3 \text{ m})\cos 25° = \textbf{7.5 m.}$$

---

7.   The shadow, the building, and a line drawn from the top of the building to the corresponding point on the end of the shadow form a right triangle. The angle between this latter line and the horizontal is the elevation angle of the sun and is

$$\theta = \tan^{-1} (500.0 \text{ m}/800.0 \text{ m}) = \mathbf{32.01°}.$$

---

8.   The length of the hypotenuse is given by (1.7),

$$h^2 = (25 \text{ ft})^2 + (15 \text{ ft})^2 ,$$

$$h = \mathbf{29 \ ft}.$$

The angle opposite the 25 ft side is given by (1.6),

$$\theta = \tan^{-1} (25/15). \ \theta = \mathbf{59°}.$$

The angle opposite the 15 ft side is

$$\theta = \tan^{-1} (15/25) = \mathbf{31°}.$$

Note that the angles add to 90° as they should.

---

9.   We are given the hypotenuse and one angle in a right triangle. (1.1) gives the side opposite the angle to be

$$h_o = (50.0 \text{ m}) \sin 20.0° = \mathbf{17.1 \ m}.$$

---

10. The vectors are perpendicular; hence the magnitude of the vector sum can be found from the Pythagorean theorem.

$$\text{magnitude} = \mathbf{11 \ m}.$$

The angle the vector makes with the east-west line (+x) is

$$\theta = \tan^{-1} (9.7/5.0) = \mathbf{63°} \text{ N of E}.$$

---

11. The x component is

$$E_x = (1.0 \text{ N/C}) \cos 33° = \mathbf{0.84 \ N/C}.$$

The y component is

$$E_y = (1.0 \text{ N/C}) \sin 33° = \mathbf{0.54 \ N/C}.$$

---

12. The angle as measured from the +x axis is $\theta = 270° - 65° = 205°$.
Then

$$B_x = (0.010 \text{ T}) \cos 205° = \mathbf{- \ 0.0091 \ T}$$

and

$$B_y = (0.010 \text{ T}) \sin 205° = \mathbf{- \ 0.0042 \ T}.$$

---

13. The x components are:

$$A_x = -(10.0 \text{ u}) \cos 21.0° = -9.34 \text{ u}, \ B_x = (5.2 \text{ u}) \sin 47.0° = 3.80 \text{ u}, \ R_x = - 5.54 \text{ u}.$$

The y components are:

$$A_y = (10.0 \text{ u}) \sin 21.0° = 3.58 \text{ u}, \quad B_y = -(5.2 \text{ u}) \cos 47.0° = -3.55 \text{ u}, \quad R_y = 0.03 \text{ u}.$$

The magnitude of the resultant vector is found from the Pythagorean theorem to be

$$R = \sqrt{(-5.54 \text{ u})^2 + (0.03 \text{ u})^2} = \mathbf{5.54 \text{ u}}$$

The angle is

$$\theta = \tan^{-1}(0.03/5.54) = \mathbf{0.310° \text{ N of W.}}$$

---

14. Let **A** be the westerly displacements, **B** the southerly displacement, and **C** be the remaining displacement. Taking east and north to be positive, the east-west (x) components are:

$$A_x = -2 \text{ km}, \quad B_x = 0 \text{ km}, \quad C_x = +(10.0 \text{ km}) \cos 53° = 6 \text{ km}. \quad R_x = +4 \text{ km}.$$

The north-south (y) components are:

$$A_y = 0 \text{ km}, \quad B_y = -8 \text{ km}, \quad C_y = +(10.0 \text{ km}) \sin 53° = +8 \text{ km}. \quad R_y = 0 \text{ km}.$$

The magnitude of the resultant vector is given by the Pythagorean theorem to be

$$R = \sqrt{(4 \text{ km})^2 + (0 \text{ km})^2} = \mathbf{4 \text{ km}}$$

and the direction is

$$\theta = \tan^{-1}(0/4) = 0°. \text{ That is, } \mathbf{EAST}.$$

CHAPTER 2
Kinematics in One Dimension

PREVIEW

In this chapter you will learn how to describe the motion of an object in one dimension, that is, straight line motion. You will be introduced to the concepts of displacement, speed, velocity, and acceleration. Equations for straight line motion with constant acceleration are developed, with applications including freely falling bodies.

QUICK REFERENCE

Important Terms

Mechanics
The branch of physics that deals with the motion of objects and the forces that change it.

Kinematics
A branch of mechanics which describes the motion of an object without explicit reference to the forces that act on the object.

Displacement
A vector representing the change in the position of an object, drawn from the initial to the final position. The most common units for displacement are meters (m), centimeters (cm), or feet (ft).

Average Speed
The distance (a scalar quantity) an object moves in a given amount of time divided by the time. Units are (m/s), (cm/s), or (ft/s).

Average Velocity
The displacement (a vector quantity) of an object divided by the elapsed time. Units are the same as average speed.

Average Acceleration
The change in the velocity (a vector quantity) of an object divided by the elapsed time. Units are (m/s^2), (cm/s^2), or (ft/s^2).

Instantaneous Velocity
The rate of change of displacement. Units are (m/s), (cm/s), or (ft/s).

Instantaneous Speed
The magnitude of the instantaneous velocity vector. Units are the same as instantaneous velocity.

Instantaneous Acceleration
The rate of change of velocity. Units are (m/s^2), (cm/s^2), or (ft/s^2).

Free Fall
An idealized motion, in which air resistance is neglected and the acceleration is nearly constant. This acceleration is due to gravity and taken to be $9.80 \ m/s^2$ or $32.2 \ ft/s^2$ at the surface of the earth.

Kinematic Equations for One Dimensional Motion (Constant Acceleration)

$$v = v_0 + a t \qquad (2.4) \qquad x = \frac{1}{2}(v_0 + v)\, t \qquad (2.7)$$

$$x = v_0 t + \frac{1}{2} a t^2 \quad (2.8) \qquad v^2 = v_0^2 + 2 a x \qquad (2.9)$$

DISCUSSION OF SELECTED SECTIONS

2.1 Displacement

The displacement is a **vector quantity** whose magnitude represents the shortest distance between the initial and final points, and whose direction points from the initial to the final position.

Example 1

A car travels 3.00 km due west, then 4.00 km due south, as shown in the figure. What is the car's displacement? The displacement vector, **x**, is the vector sum of vectors **A** and **B**, that is **x** = **A** + **B**.

$$x = \sqrt{A^2 + B^2}$$

$$x = \sqrt{(3.00 \text{ km})^2 + (4.00 \text{ km})^2}$$

$$x = 5.00 \text{ km}$$

$$\theta = \tan^{-1}\left(\frac{B}{A}\right) = \tan^{-1}\left(\frac{4.00 \text{ km}}{3.00 \text{ km}}\right)$$

$$\theta = 53.1° \text{ South of West}$$

2.2 Speed and Velocity

The **average speed** is a scalar quantity defined as the distance traveled divided by the amount of time taken to travel this distance.

Example 2

If, in Example 1, the car takes 10.0 minutes to travel from its initial to its final position, what is its average speed?

$$\text{Average Speed} = \frac{\text{Distance}}{\text{Elapsed Time}}$$

$$= \left(\frac{7.00 \text{ km}}{10.0 \text{ min}}\right)\left(\frac{1000 \text{ m}}{1 \text{ km}}\right)\left(\frac{1 \text{ min}}{60 \text{ s}}\right) = 11.7 \text{ m/s.}$$

The **average velocity** is a vector quantity defined as the displacement divided by the elapsed time. In general, the average speed is not the same as the magnitude of the average velocity. There is one special case where these two quantities will be equal. When an object travels in a straight line and does *not* change the direction of motion, the average speed of the object over any time interval will be equal to the magnitude of the average velocity over the same time interval.

Example 3
Find the average velocity of the car in Examples 1 and 2.

$$\overline{\mathbf{v}} = \frac{\text{Displacement}}{\text{Elapsed Time}} = \left(\frac{5.00 \text{ km}}{10.0 \text{ min}}\right)\left(\frac{1000 \text{ m}}{1 \text{ km}}\right)\left(\frac{1 \text{ min}}{60 \text{ s}}\right) = 8.33 \text{ m/s at } 53.1^{\circ} \text{ S of W.}$$

Notice that the average speed and the magnitude of the average velocity are **not** the same.

The **instantaneous velocity** indicates both how fast and in what direction an object is moving at a particular instant of time. To determine the instantaneous velocity we measure the time Δt it takes the object to travel a very small displacement $\Delta \mathbf{r}$. The instantaneous velocity is then given by

$$\mathbf{v} = \lim_{\Delta t \to 0} \frac{\Delta \mathbf{x}}{\Delta t} \qquad (2.3)$$

The notation $\Delta t \to 0$ essentially means that we want to isolate an instant of time and find the velocity during that instant. The **instantaneous speed** is simply the magnitude of the instantaneous velocity.

2.3 Acceleration

Acceleration is any change in velocity for an object divided by the time interval over which the change occurs. This means a change in MAGNITUDE and/or DIRECTION. We define the **average acceleration** of an object as

$$\overline{\mathbf{a}} = \frac{\mathbf{v} - \mathbf{v}_0}{t - t_0} = \frac{\Delta \mathbf{v}}{\Delta t} \qquad (2.4)$$

Example 4
A car traveling at 10 m/s in a straight line speeds up to 30 m/s in 5.0 s. What is the average acceleration?

$$\overline{\mathbf{a}} = \frac{\mathbf{v} - \mathbf{v}_0}{\Delta t} = \frac{30 \text{ m/s} - 10 \text{ m/s}}{5.0 \text{ s}} = \frac{+20 \text{ m/s}}{5.0 \text{ s}} = +4 \frac{\text{m/s}}{\text{s}} = +4 \text{ m/s}^2$$

We have assumed that the direction in which the car moves is the + direction, and the + sign in the above result indicates the acceleration is in this same direction.

The **instantaneous acceleration** can be defined by analogy with the instantaneous velocity:

$$\mathbf{a} = \lim_{\Delta t \to 0} \frac{\Delta \mathbf{v}}{\Delta t} \qquad (2.5)$$

2.4 Equations of Kinematics for Constant Acceleration

The motion of an object traveling with constant acceleration along a straight line can be described by equations (2.4), (2.7), (2.8), and (2.9), as listed in the **Quick Reference** section. Note that these equations are valid only for uniform or constant acceleration. If the acceleration is changing, these equations cannot be used. One exception is if the motion can be broken up into segments in which the acceleration is constant for each segment. The kinematic equations can then be applied to each segment separately.

Example 5
Starting from rest, a car accelerates for 8.0 s to a final speed of 16 m/s. Find the acceleration of the car and the distance traveled during this time interval.

Using equation (2.4) with the initial velocity equal to zero we obtain

$$a = (v - v_0)/t = (16 \text{ m/s} - 0 \text{ m/s})/(8.0 \text{ s}) = +2.0 \text{ m/s}^2.$$

To find the distance traveled use equation (2.8)

$$x = v_0 t + (1/2) at^2 = 0 + (1/2) (2.0 \text{ m/s}^2) (8.0 \text{ s})^2 = 64 \text{ m}.$$

Note that this last result can also be obtained using equation (2.7)

$$x = (1/2) (v_0 + v)t = (1/2) (0 \text{ m/s} + 16 \text{ m/s}) (8.0 \text{ s}) = 64 \text{ m}.$$

2.6 Freely Falling Bodies

In the absence of air resistance, it is found that all bodies near the surface of the earth fall vertically with the same acceleration. The acceleration of a freely falling body is called the **acceleration due to gravity**, and its magnitude is given by the symbol "g". This acceleration is always directed downward, toward the center of the earth. Near the surface of the earth g is approximately constant and has a value of

$$g = 9.80 \text{ m/s}^2 \quad \text{or} \quad 32.2 \text{ ft/s}^2.$$

Example 6
A ball is dropped from the top of a 123 m high cliff. Ignoring air resistance find the final velocity of the ball, and the amount of time it takes for it to hit the ground below.

The final velocity can be obtained using equation (2.9) and taking the direction of motion as the y direction.

$$v = \sqrt{v_0^2 + 2ay} = \sqrt{(0 \text{ m/s})^2 + 2(+9.80 \text{ m/s}^2)(123 \text{ m})}$$
$$v = 49.1 \text{ m/s}.$$

The time of flight can be obtained using equation (2.4)

$$v = v_0 + at$$

or

$$t = \frac{v - v_0}{a} = \frac{49.1 \text{ m/s} - 0 \text{ m/s}}{+9.80 \text{ m/s}^2} = 5.0 \text{ s}.$$

2.7 Graphical Analysis of Velocity and Acceleration

Consider an object moving in a straight line at a constant speed. The position, x, is plotted as a function of time, t, in the following graph.

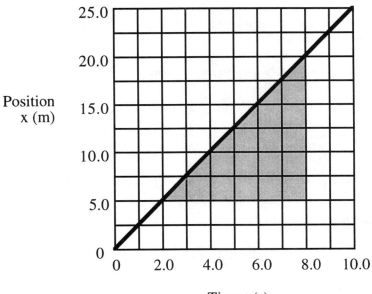

Time t (s)

We can see from the graph that the position is increasing linearly with time, i.e., the graph is a straight line. Between the times 2.0 and 8.0 s note that the object travels from 5.0 to 20.0 m. The change in time is $\Delta t = 6.0$ s, while the change in position is $\Delta x = +15.0$ m. The ratio of $\Delta x/\Delta t$ is called the **slope** of the straight line. That is

$$\text{Slope} = \frac{\Delta x}{\Delta t} = \frac{+15.0 \text{ m}}{6.0 \text{ s}} = +2.5 \text{ m/s}$$

Note that the slope has units of velocity (m/s). In fact, the slope of the position vs. time graph does represent the average velocity of the moving object, as evidenced by the fact that the above equation is identical to equation (2.2). In the real world, however, the situation may not be quite as simple. Consider the following example.

Example 7 *Determining the slope of a position vs. time graph*
Consider the following position vs. time graph and determine the velocity over each segment.

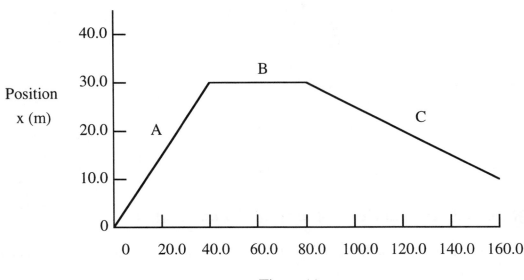

Time t (s)

Segment A: $v = \Delta x/\Delta t = (+30.0\ m)/(40.0\ s) = +0.750\ m/s.$

Segment B: $v = \Delta x/\Delta t = (0\ m)/(40.0\ s) = 0\ m/s.$

Segment C: $v = \Delta x/\Delta t = (-20.0\ m)/(80.0\ s) = -0.250\ m/s.$

In segment A the object is traveling with a positive velocity of 0.75 m/s. Over segment B the object does not change position, so its velocity is zero. Over segment C the slope of the graph is negative and hence the velocity is negative, i.e., the object is now moving in the negative direction.

Example 8 *Accelerated motion, slope of a tangent line*

Using the position vs. time graph shown at the right, we can determine the velocity of the object at say, t = 40 s. We see that the velocity is changing with time. The velocity at any instant of time can be determined by measuring the slope of the tangent to the curve at that point.

This will then provide us with a value of the **instantaneous velocity** at the point in question.

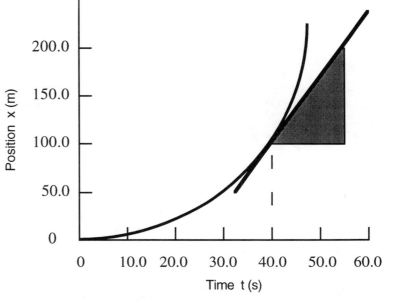

The slope at t = 40.0 s can be determined by referring to the shaded triangle. The slope is

$$v = slope = \Delta x/\Delta t = (200.0\ m - 100.0\ m)/(55.0\ s - 40.0\ s) = (+100.0\ m)/(15.0\ s) = +6.67\ m/s.$$

If we were to plot a graph of velocity vs. time for the data in the drawing for Example 8 we would obtain the graph shown at the right.

Its slope is:

Slope $= \Delta v/\Delta t$
$= (6.7\ m/s - 2.0\ m/s)/(40.0\ s - 12.0\ s)$
$= (+4.7\ m/s)/(28.0\ s) = +0.17\ m/s^2.$

Notice that the slope of the velocity vs. time graph represents the acceleration of the object over this time interval.

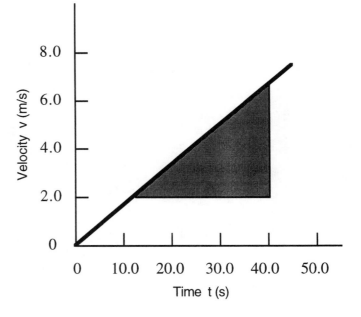

PRACTICE PROBLEMS

1. A car travels 10.0 km due north, then 8.00 km due east, the total trip taking 30.0 minutes. Find (a) the displacement of the car, (b) the average speed (in m/s), and (c) the average velocity (in m/s).

2. A bicycle accelerates to a velocity of 25.0 mi/h in 5.00 s. The average acceleration over this time interval is 4.00 ft/s^2. What was the initial velocity of the bike?

3. A train accelerates from rest to a speed of 25.0 m/s in 10.0 s. Assuming the acceleration to be constant over this interval find (a) the magnitude of the acceleration, and (b) the distance traveled during this interval.

4. A drag racer, starting from rest, travels a quarter mile in 6.0 s. What is the racer's speed (in mi/h) at the end of the race?

5. Suppose a car traveling at 12.0 m/s sees a traffic light turn red. After 0.510 s have elapsed, the driver applies the brakes, and the car decelerates at 6.20 m/s^2. What is the stopping distance of the car, as measured from the point where the driver first notices the red light?

6. A car starts from rest heading due east. It first accelerates at 3.0 m/s^2 for 5.0 s and then continues without further acceleration for 20.0 s. It then brakes for 8.0 s in coming to rest. (a) What is the car's velocity after the first 5.0 seconds? (b) What is the car's acceleration over the last 8.0 s interval? (c) What is the total displacement?

7. A ball is thrown upward with an initial velocity of 20.0 m/s. What maximum height does the ball reach?

8. A stone is thrown vertically downward from a 2.00 x 10^2 m high cliff at an initial velocity of 5.00 m/s. (a) How long does it take for the stone to reach the base of the cliff? (b) What is the stone's final velocity?

9. A sports car, picking up speed, passes between two markers in a time of 4.1 s. The markers are separated by 120 m. All the while, the car accelerates at 1.8 m/s^2. What is its speed at the second marker?

10. A heavy ball is dropped into a lake from a height of 30.0 m above the water. It hits the water with a certain velocity and continues to sink to the bottom of the lake at this same constant velocity. It reaches the bottom of the lake 10.0 s after it was dropped. How deep is the lake?

11. Using the position-time graph shown below, draw the corresponding velocity-time graph.

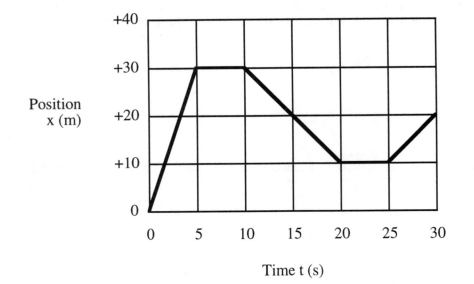

12. A snowmobile moves according to the velocity-time graph shown below. What is the snowmobile's average acceleration during each of the segments labeled A, B, and C?

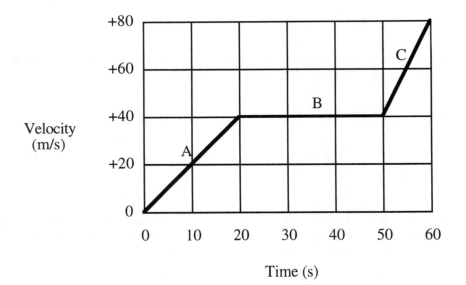

SOLUTIONS AND ANSWERS

Practice Problems

---

1. a. The total displacement is the vector sum of the individual displacements. Since these displacements are perpendicular, the vector sum is given by the Pythagorean theorem.

$$x^2 = (10.0 \text{ km})^2 + (8.00 \text{ km})^2$$
$$x = 12.8 \text{ km}.$$

   The angle is found using
$$\tan \theta = (10.0 \text{ km}/8.00 \text{ km})$$
$$\theta = \textbf{51.3}° \textbf{ N of E.}$$

   b. Avg. Speed = Distance/Time = (18.0 km/30.0 min)(1000 m/1 km)(1 min/60 s) = **10.0 m/s**.
   c. Avg. Velocity = Displacement/Time = (12.8 km/30.0 min)(1000 m/1 km)(1 min/60 s) = **7.11 m/s**.

---

2. The average acceleration is given by equation (2.4),

$$a = \Delta v/\Delta t = (v - v_0)/\Delta t$$

   or
$$v_0 = v - at$$

   where
$$v = (25.0 \text{ mi/h})(5280 \text{ ft/1 mi})(1 \text{ h}/3600 \text{ s}) = 36.7 \text{ ft/s}.$$

   Then
$$v_0 = 36.7 \text{ ft/s} - (4.00 \text{ ft/s}^2)(5.00 \text{ s}) = \textbf{16.7 ft/s.}$$

---

3. a. Using equation (2.4) we have
$$a = (v - v_0)/\Delta t = (+25.0 \text{ m/s})/(10.0 \text{ s}) = \textbf{+2.50 m/s}^2.$$

   b. Using equation (2.7) we have
$$x = (1/2)(v + v_0)t = (1/2)(25.0 \text{ m/s} + 0)(10.0 \text{ s}) = \textbf{125 m.}$$

---

4. Using $v_0 = 0$, x = (0.250 mi)(5280 ft/mi) = 1320 ft, t = 6.0 s and equation (2.7) we solve for v to get

$$v = 2x/t = 2(1320 \text{ ft})/(6.0 \text{ s}) = (440 \text{ ft/s})(1 \text{ mi}/5280 \text{ ft})(3600 \text{ s}/1 \text{ h}) = \textbf{3.0 x 10}^2 \textbf{ mi/h.}$$

---

5. The motion should be divided into two segments; segment 1 is constant velocity and segment 2 is deceleration.

$$x_1 = v_0 t = (12.0 \text{ m/s})(0.510 \text{ s}) = 6.12 \text{ m};$$
$$x_2 = (v^2 - v_0^2)/2a = -(12.0 \text{ m/s})^2/[2(-6.20 \text{ m/s}^2)] = 11.6 \text{ m}.$$

The stopping distance is therefore

$$x = x_1 + x_2 = 6.12 \text{ m} + 11.6 \text{ m} = \textbf{17.7 m.}$$

---

6. a. $a = 3.0$ m/s^2, $t = 5.0$ s, $v_0 = 0$. Equation (2.4) gives

$$v = v_0 + at = (3.0 \text{ m/s}^2)(5.0 \text{ s}) = \mathbf{15 \text{ m/s East.}}$$

   b. $v_0 = 15$ m/s, $v = 0$, $t = 8.0$ s. Equation (2.4) gives

$$a = (v - v_0)/t = (-15 \text{ m/s})/(8.0 \text{ s}) = \mathbf{1.9 \text{ m/s}^2 \text{ West.}}$$

   c. The distance is then

$$x = x_1 + x_2 + x_3 = (1/2)a_1t_1^2 + v_2t_2 + + v_2t_3 + (1/2)a_3t_3^2 = \mathbf{4.0 \times 10^2 \text{ m.}}$$

---

7. $v_0 = 20.0$ m/s, $v = 0$, $a = -9.80$ m/s^2. Equation (2.9) is

$$v^2 = v_0^2 + 2ax$$

Solving for x we obtain

$$x = -v_0^2/2a = -(20.0 \text{ m/s})^2/[2(-9.80 \text{ m/s}^2)] = \mathbf{20.4 \text{ m.}}$$

---

8. a. $v_0 = 5.00$ m/s, $a = 9.80$ m/s^2, $x = 2.00 \times 10^2$ m. Equation (2.8): $x = v_0t + (1/2)at^2$.
   Substituting yields the quadratic equation

$$4.90 \text{ } t^2 + 5.00 \text{ } t - 2.00 \times 10^2 = 0$$

   which has a solution

$$t = \mathbf{5.90 \text{ s.}}$$

   b. Since $\qquad v^2 = v_0^2 + 2ax = (5.00 \text{ m/s})^2 + 2(9.80 \text{ m/s}^2)(2.00 \times 10^2 \text{ m})$
   then

$$v = \mathbf{62.8 \text{ m/s.}}$$

---

9. $a = 1.8$ m/s^2, $x = 120$ m, $t = 4.1$ s. Equation (2.8),

$$x = v_0t + (1/2)at^2$$

   yields

$$v_0 = [x - (1/2)at^2]/t = 25.6 \text{ m/s.}$$

   Equation (2.4), gives

$$v = v_0 + at = 25.6 \text{ m/s} + (1.8 \text{ m/s}^2)(4.1 \text{ s}) = \mathbf{33 \text{ m/s.}}$$

---

10. Find the velocity with which the ball hits the water and the time it takes to do so. Since $v_0 = 0$ we have

$$x = (1/2)at^2$$

    which gives

$$t = \sqrt{(2x/a)} = \sqrt{2(30.0\text{m})/(9.80\text{m}/\text{s}^2)} = 2.47 \text{ s.}$$

    Now

$$v = at = (9.80 \text{ m/s}^2)(2.47 \text{ s}) = 24.2 \text{ m/s.}$$

    The time it takes to sink is

$$t' = 10.0 \text{ s} - 2.47 \text{ s} = 7.5 \text{ s.}$$

    The depth of the lake is therefore found from

$$\text{Depth} = vt' = (24.2 \text{ m/s})(7.5 \text{ s}) = \mathbf{180 \text{ m.}}$$

11.

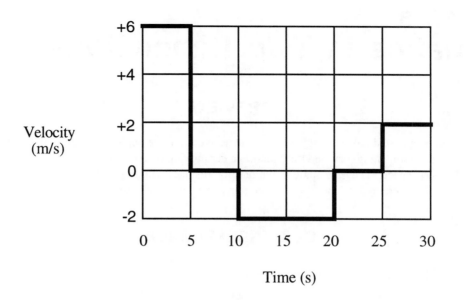

Velocity
(m/s)

Time (s)

---

12. During segment:

A: a = (40 m/s)/(20 s) = **2 m/s^2**.

B: a = (0 m/s)/(30 s) = **0 m/s^2**.

C: a = (40 m/s)/(10 s) = **4 m/s^2**.

CHAPTER 3
Kinematics in Two Dimensions

PREVIEW

In this chapter the concepts of displacement, velocity, and acceleration are applied to motion in two dimensions. You will learn to describe motion along a curved line, such as a car moving on a race track, a ball flying through the air, or a spacecraft in deep space. The kinematic equations for displacement, velocity, and acceleration developed in Chapter 2 will be applied here to describe projectile motion and relative velocity.

QUICK REFERENCE

Important Terms

Projectile Motion
An object moving in two dimensions which experiences only the acceleration due to gravity. Air resistance is assumed to be negligible and there is no acceleration in the x direction, i.e., $a_x = 0$. In the y direction on earth we have $a_y = -9.80$ m/s^2. A baseball or football thrown from one person to another is a good example.

Relative Velocity
The velocity of a moving object can take on different values depending on the point of view of the observer. That is, the value of the velocity must be measured 'relative' to a particular observer.

Kinematic Equations for Two Dimensional Motion (Constant Acceleration)

$$v_x = v_{0x} + a_x t \qquad (3.3a) \qquad\qquad v_y = v_{0y} + a_y t \qquad (3.3b)$$

$$x = \frac{1}{2}(v_{0x} + v_x)t \quad (3.4a) \qquad\qquad y = \frac{1}{2}(v_{0y} + v_y)t \quad (3.4b)$$

$$x = v_{0x}t + \frac{1}{2}a_x t^2 \quad (3.5a) \qquad\qquad y = v_{0y}t + \frac{1}{2}a_y t^2 \quad (3.5b)$$

$$v_x^2 = v_{0x}^2 + 2a_x x \qquad (3.6a) \qquad\qquad v_y^2 = v_{0y}^2 + 2a_y y \qquad (3.6b)$$

DISCUSSION OF SELECTED SECTIONS

3.2 Equations of Kinematics in Two Dimensions

In Chapter 2 we emphasized that displacement, velocity, and acceleration were vector quantities. In one-dimensional motion the direction of these vectors was simply denoted by a positive or negative sign. In the more complex realm of two dimensions we must be careful to treat both the x and y motion independently, with the understanding that they will add together as vectors.

In treating two-dimensional motion we must be careful to view the x and y components of the motion as separate but related quantities. We will see that the x part of the motion occurs exactly as it would if the y part did not occur at all. Similarly, the y part of the motion occurs exactly as it would if the x part of the motion did not exist. Therefore, two sets of kinematic equations are needed to describe the full two dimensional motion.

Example 1 *Components of the velocity vector*
A ball is traveling at a constant velocity of 20.0 m/s at an angle of 30.0 degrees with respect to the x axis. How far does the ball travel in the x and y directions in 1 minute? See the diagram shown below.

First find the x and y components of the velocity.

$$v_x = v \cos \theta = (20.0 \text{ m/s})\cos 30.0^\circ = 17.3 \text{ m/s.}$$

$$v_y = v \sin \theta = (20.0 \text{ m/s})\sin 30.0^\circ = 10.0 \text{ m/s.}$$

Since there is no acceleration, the x and y displacements are given by equations (3.5a) and (3.5b) with $v_{0x} = v_x$ and $v_{0y} = v_y$. We therefore have,

$$x = v_x t = (17.3 \text{ m/s}) (60 \text{ s}) = 1040 \text{ m.}$$
$$y = v_y t = (10.0 \text{ m/s}) (60 \text{ s}) = 600 \text{ m.}$$

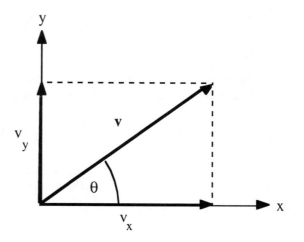

3.3   Projectile Motion

When we throw a ball, or fire a bullet from a gun, the motion is one in which only the vertical component of the velocity changes with time. The horizontal velocity component remains constant (assuming that air resistance is negligible). We refer to this type of motion as **projectile motion**. The equations describing the trajectory of such a projectile are equations (3.3) through (3.6). Consider the following example.

Example 2
A football is kicked with an initial velocity of 82 ft/s at an angle of 53° with respect to the horizontal. How high does the football go and how far down the field does it land?

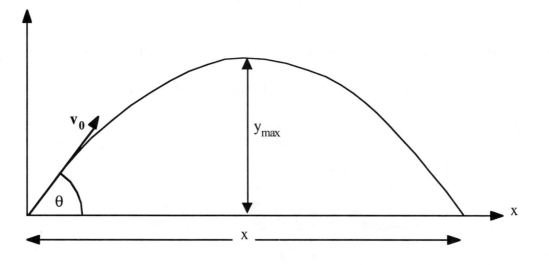

The x and y motions can be treated separately. Begin by looking at the motion in the y direction. We have

$$v_{0y} = v_0 \sin \theta = (82 \text{ ft/s}) \sin 53^\circ = 65 \text{ ft/s;} \quad a_y = -32.2 \text{ ft/s}^2; \quad v_y = 0.$$

Notice we have chosen the final velocity to be zero. In effect, we are looking at the first half of the motion in the y direction, i.e., the portion of the motion in which the ball is rising. When the ball reaches its maximum height, $v_y = 0$ and $y = y_{max}$. Using equation (3.6b) we can solve for this maximum height y,

$$y_{max} = y = \frac{-v_{0y}^2}{2a_y} = \frac{-(65 \text{ ft/s})^2}{2(-32.2 \text{ ft/s}^2)} = 66 \text{ ft}$$

To find the distance the football travels in the x direction, we need to know the time of flight for the ball. We can find the time it takes for the ball to rise using equation (3.3b), i.e.,

$$t = -v_{0y}/a_y = -(65 \text{ ft/s})/(-32.2 \text{ ft/s}^2) = 2.0 \text{ s.}$$

The total time of flight is twice this value, or 4.0 seconds. Now we can find the x distance using equation (3.5a) and noting that $a_x = 0$ and $v_{0x} = v_0 \cos \theta = (82 \text{ ft/s}) \cos 53° = 49 \text{ ft/s}$. We have

$$x = v_{0x}t = (49 \text{ ft/s}) (4.0 \text{ s}) = 2.0 \times 10^2 \text{ ft.}$$

So the ball rose 66 ft in the air, stayed up in the air 4.0 s (hang time), and traveled 2.0×10^2 ft.

Example 3

A jet fighter is traveling horizontally with a speed of 111 m/s at an altitude of 3.00×10^2 m, when the pilot accidentally releases an outboard fuel tank. (a) How much time elapses before the tank hits the ground? (b) What is the speed of the tank just before it hits the ground? (c) What is the horizontal distance traveled by the tank?

(a) The y direction data is: $y = -3.00 \times 10^2$ m, $v_{0y} = 0$, $a_y = -9.80$ m/s^2. Using equation (3.5b) and solving for t,

$$t = \sqrt{\frac{2y}{a_y}} = \sqrt{\frac{2(-3.00 \times 10^2 \text{ m})}{(-9.80 \text{ m/s}^2)}} = 7.82 \text{ s.}$$

(b) The y component of the velocity as it strikes the ground can be found using equation (3.3b),

$$v_y = v_{0y} + a_y t = 0 + (-9.80 \text{ m}/\text{s}^2)(7.82 \text{ s}) = -76.6 \text{ m}/\text{s.}$$

The speed as it hits the ground can be obtained by calculating the vector sum of the x and y components. The x component of the velocity is constant at 111 m/s. Therefore,

$$v_{tot} = \sqrt{v_x^2 + v_y^2} = \sqrt{(111 \text{ m/s})^2 + (-76.6 \text{ m/s})^2} = 135 \text{ m/s.}$$

(c) The horizontal distance can be obtained using equation (3.5a) with $a_x = 0$ and $v_{0x} = 111$ m/s,

$$x = v_{0x} t = (111 \text{ m/s}) (7.82 \text{ s}) = 868 \text{ m.}$$

Note that the initial velocity, v_{0x}, plays no role in determining the time required for the fuel tank to hit the ground. Even if the jet were moving faster, the tank would still take 7.82 s to reach the ground. However, the speed on impact and the horizontal distance would be affected by v_{0x}.

3.4 Relative Velocity

The velocity measured for a particular object depends on the observer who is making the measurement. For example, suppose a ground based observer is measuring the velocity of two cars, each moving in the same direction. Car A is moving at 45 km/h relative to the ground, while car B is moving at 55 km/h relative to the ground. The observer in car A, however, measures the velocity of car B as 10 km/h relative to himself. Car B measures A's velocity as -10 km/h in his reference frame. So each observer obtains different values for the velocities because each is making measurements from different points of reference.

Example 4 *Relative velocity in one dimension*
A passenger in a moving train is walking towards the dining car in the front of the train. The train is moving at a speed of 21 ft/s and the passenger is walking at 4 ft/s relative to the train. How far will the passenger move in 10.0 seconds relative to (a) the train, and (b) a ground based observer.

Use the following symbols to represent the different relative velocities:

$$v_{TG} = \text{velocity of the train relative to the ground.}$$
$$v_{PT} = \text{velocity of the passenger relative to the train.}$$
$$v_{PG} = \text{velocity of the passenger relative to the ground.}$$

We can see that the three velocities are related in the following manner:

$$v_{PG} = v_{PT} + v_{TG} = 4 \text{ ft/s} + 21 \text{ ft/s} = 25 \text{ ft/s.}$$

The distances measured in each case are therefore,

(a) $s_{PT} = v_{PT}t = (4 \text{ ft/s}) (10.0 \text{ s}) = 40 \text{ ft.}$

(b) $s_{PG} = v_{PG}t = (25 \text{ ft/s}) (10.0 \text{ s}) = 250 \text{ ft.}$

When dealing with relative velocities in two dimensions it is important to emphasize the vector nature of the velocities. For example, if a boat is sailing across a stream, and the stream is flowing, both the boat's and the stream's velocity must be considered. Similarly, the velocity of an airplane in flight is effected by the velocity of the wind. The following example illustrates this effect.

Example 5 *Relative velocity in two dimensions*
A boat whose speed is 10.0 mi/h in still water travels across a river that is 2.00 miles wide. The river current is 5.00 mi/h directed parallel to the river bank. How long does it take for the boat to cross the river and how far downstream will the boat arrive?

First find the velocity of the boat relative to the shore. From the diagram we can see that

$$v_{BS} = v_{BW} + v_{WS}$$

Since the vectors v_{BW} and v_{WS} are perpendicular to each other, the magnitude and direction of v_{BS} can be found from the Pythagorean theorem.

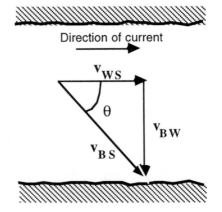

Direction of current

$$v_{BS} = \sqrt{v_{BW}^2 + v_{WS}^2}$$

$$= \sqrt{(10.0 \text{ mi/h})^2 + (5.00 \text{ mi/h})^2}$$

$$= 11.2 \text{ mi/h.}$$

To find the amount of time and downstream distance we need to know the angle θ shown in Figure A. We have

$$\theta = \tan^{-1}\left(\frac{v_{BW}}{v_{WS}}\right) = \tan^{-1}\left(\frac{10.0 \text{ mi/h}}{5.00 \text{ mi/h}}\right) = 63.4^{\circ}.$$

In order to find the time taken for the trip we need the total distance traveled. We see that this distance is

$$r = \frac{2.00 \text{ mi}}{\sin \theta} = \frac{2.00 \text{ mi}}{\sin 63.4^{\circ}} = 2.24 \text{ mi.}$$

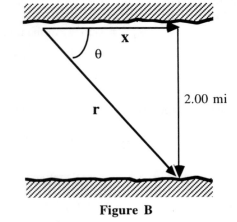

Note that the angle θ that $\mathbf{v_{BS}}$ makes with $\mathbf{v_{WS}}$ (in Figure A) is the same angle θ that $\mathbf{r}$ makes with $\mathbf{x}$ (in Figure B).

Figure B

Therefore, the time taken for the boat to cross the river can be obtained using

$$t = \frac{r}{v_{BS}} = \frac{2.24 \text{ mi}}{11.2 \text{ mi/h}} = 0.20 \text{ h.}$$

The distance the boat lands downstream can be found from

$$x = \frac{2.00 \text{ mi}}{\tan \theta} = \frac{2.00 \text{ mi}}{\tan 63.4^{\circ}} = 1.00 \text{ mi.}$$

PRACTICE PROBLEMS

1. A pitcher throws a baseball at a speed of 90.0 mi/h towards home plate, which is located 60.0 ft away. How far has the ball dropped when it reaches the plate? Assume the ball leaves the pitcher's hand traveling horizontally.

2. An arrow is shot at an initial velocity of 25.0 m/s at an angle of 30.0 degrees with respect to the horizontal. (a) What is the maximum height reached by the arrow? (b) How far does the arrow travel in the horizontal direction in returning to the same level from which it was shot?

3. A plane is traveling at a velocity of 2.0×10^2 m/s at an angle of 45 degrees below the horizontal. It releases a bomb at an altitude of 1.0×10^3 m, which strikes a target on the ground. (a) How far was the plane from the target when the bomb was released? (b) How long did it take the bomb to strike the target after being released? (c) What was the bomb's speed at the moment of impact?

4. A diver springs upward from a three-meter board. At the instant she contacts the water her speed is 8.90 m/s and her body makes an angle of 75.0° with respect to the surface of the water. (a) Determine her initial velocity, both magnitude and direction. (b) How much time does she spend in the air?

5. A ball rolls off the top of a stairway with a horizontal velocity of 2.0 m/s. The steps are each 20.0 cm high and 20.0 cm wide. Which step will the ball hit first?

6. When a cannon is aimed at an angle of 45° above the horizontal, a cannon ball lands 1.0×10^2 m down range. What was the muzzle velocity of the cannon ball?

7. The compass of an aircraft indicates it is headed due north, and its airspeed indicator shows that it is moving through the air at 120 mi/h. If a wind of 50.0 mi/h blowing from west to east suddenly arises, what is the velocity of the aircraft relative to the earth?

8. An airplane pilot wishes to fly due south. A wind of 25 km/h is blowing toward the west. If the flying speed of the plane is 300.0 km/h (its speed in still air), in what direction should the pilot head?

9. Two trains approach a railroad station, one from the north at 30.0 m/s, the other from the east at 40.0 m/s. Find the velocities of (a) the railroad station and (b) the south-bound train; relative to the west-bound train.

10. A boat can travel at a speed of 20.0 km/h relative to the water. The boat sails across a river (perpendicular to the river bank) and reaches a point 5.00 km directly across from its launch point. The water is flowing at a speed of 8.00 km/h parallel to the river bank. (a) At what angle must the boat steer to reach its destination? (b) How much time is required for the boat to make the trip?

SOLUTIONS AND ANSWERS

Practice Problems

---

1. The horizontal speed is
$$v_x = (90.0 \text{ mi/h})(1\text{h}/3600 \text{ s})(5280 \text{ ft}/1 \text{ mi}) = 132 \text{ ft/s}.$$
The time to travel 60.0 ft is
$$t = x/v_x = (60.0 \text{ ft})/(132 \text{ ft/s}) = 0.455 \text{ s}.$$

$$y = v_{0y}t + (1/2)a_y t^2 = 0 + (1/2)(32.2 \text{ ft/s}^2)(0.455 \text{ s})^2 = \textbf{3.33 ft.}$$

---

2. a. The initial velocity is
$$v_{0y} = v_0 \sin 30.0° = (25.0 \text{ m/s})(0.500) = 12.5 \text{ m/s};$$

Use equation (3.6b) and solve for y

$$y = -v_0^2/2a_y = - (12.5 \text{ m/s})^2/[2(-9.80 \text{ m/s}^2)] = \textbf{7.97 m.}$$

b. Find the time from $v_y = v_{0y} + a_y t$

$$t = -v_{0y}/a_y = - (12.5 \text{ m/s})/(-9.80 \text{ m/s}^2) = 1.28 \text{ s}.$$
Finally, we have
$$x = v_{0x}t = v_0 \cos 30.0° \ t = (25.0 \text{ m/s}) \cos 30.0° \ (2.56 \text{ s}) = \textbf{55.4 m.}$$

---

3. Using equation (3.6b) we have

$$v_y^2 = v_{0y}^2 + 2a_y y = [(2.0 \times 10^2 \text{ m/s}) \sin 45°]^2 + 2(-9.80 \text{ m/s}^2)(-1.0 \times 10^3 \text{m}),$$

which yields $v_y = 2.0 \times 10^2$ m/s. Now, $v_x = v_0 \cos 45° = 141$ m/s. The time can be found from equation (3.3b),

$$t = (v_y - v_{0y})/ a_y = (2.0 \times 10^2 \text{ m/s} - 1.4 \times 10^2 \text{ m/s})/(9.80 \text{ m/s}^2) = 6.1 \text{ s}.$$

a. $x = v_{0x}t = (141 \text{ m/s})(6.1 \text{ s}) = 860 \text{ m}, \ y = 1.0 \times 10^3 \text{ m},$

$$s^2 = x^2 + y^2 = (860 \text{ m})^2 + (1.0 \times 10^3 \text{ m})^2, \ \text{so that } s = \textbf{1300 m.}$$

b. We have already seen that t = **6.1 s.**

c. Finally,
$$v^2 = v_x^2 + v_y^2 = (141 \text{ m/s})^2 + (2.0 \times 10^2 \text{ m/s})^2 \text{ so } v = \textbf{240 m/s.}$$

---

4. a. At a height of 3.00 m above the water find v_0. We know

$$v_y = v \sin 75.0° = (8.90 \text{ m/s})\sin 75.0° = 8.60 \text{ m/s}.$$

From equation (3.6b) we have

$$v_{0y}^2 = v_y^2 - 2a_y y = (8.60 \text{ m/s})^2 - 2(9.80 \text{ m/s}^2)(3.00 \text{ m}) \rightarrow v_{0y} = 3.89 \text{ m/s.}$$

Also,

$$v_{0x} = v_x = v \cos 75.0° = (8.90 \text{ m/s}) \cos 75.0° = 2.30 \text{ m/s} \rightarrow v_0^2 = v_{0x}^2 + v_{0y}^2 \text{ so that}$$

$$v_0 = \textbf{4.52 m/s.}$$

The direction is obtained from

$$\theta = \tan^{-1}(v_{0y}/v_{0x}) = \tan^{-1}[(3.89 \text{ m/s})/(2.30 \text{ m/s})] = \textbf{59.4°.}$$

b. The time to go up is $t_{up} = -v_{0y}/a_y = 0.397$ s; the time down is $t_{down} = v_y/a_y = 0.878$ s. So total time is

$$t = t_{up} + t_{down} = 0.40 \text{ s} + 0.88 \text{ s} = \textbf{1.28 s.}$$

---

5. In order to determine which step the ball lands on, the ball must fall at least a distance equal to its horizontal displacement. For example, at x = 0.2 m,

$$t = x/v_x = (0.2 \text{ m})/(2.0 \text{ m/s}) = 0.1 \text{ s,}$$
$$y = (1/2)a_y t^2 = 0.05 \text{ m.}$$

So the ball goes past the first step. At x' = 0.8 m (4th step), t' = 0.4 s and y' = 0.78 m, so it goes past the 4th step. However, at x" = 1.0 m (5th step), t" = 0.5 s, y" = 1.23 m. So the ball lands on the **fifth step.**

---

6. The time taken to travel half the distance in the x direction, using equation (3.5a), is

$$t = x/(v_0 \cos 45°) = (50.0 \text{ m})/(v_0 \cos 45°).$$

The time it takes for the ball to reach its maximum height is the also this value of t, and using equation (3.3b) with $v_y = 0$,

$$t = v_{0y}/a_y = v_0 \sin 45°/(9.8 \text{ m/s}^2).$$

Equating the two expressions for the time, and noting that cos 45° = sin 45° yields,

$$(v_0 \cos 45°)^2 = (50.0 \text{ m})(9.8 \text{ m/s}^2).$$

Thus,

$$v_0^2 = (50.0 \text{ m})(9.8 \text{ m/s}^2)/(\cos^2 45°)$$
$$v_0 = \textbf{31 m/s.}$$

---

7. A vector diagram shows that

$$v^2 = v_P^2 + v_W^2 = (120 \text{ mi/h})^2 + (50.0 \text{ mi/h})^2$$
$$v = \textbf{130 mi/h.}$$

The direction is

$$\theta = \tan^{-1}(v_P/v_W) = \tan^{-1}[(120 \text{ mi/h})/(50.0 \text{ mi/h})] = \textbf{67° North of East.}$$

---

8. Since the final direction of motion and the direction of the wind are perpendicular, we can find the angle between the plane and wind, i.e.,

$$\theta = \cos^{-1}(v_W/v_P) = \cos^{-1}[(25 \text{ km/h})/(300.0 \text{ km/h})] = \textbf{85° S of E.}$$

---

9. a. We can write $\textbf{v}_{RW} = \textbf{v}_{RG} + \textbf{v}_{GW}$ where R $\rightarrow$ railroad station, W $\rightarrow$ west bound train, G $\rightarrow$ ground. So

$$v_{RW} = 0 + 40 \text{ m/s} = \textbf{40 m/s East.}$$

b. We have $\textbf{v}_{SW} = \textbf{v}_{SG} + \textbf{v}_{GW}$. Since $\textbf{v}_{SG}$ and $\textbf{v}_{GW}$ are perpendicular, we can write $v_{SW}^2 = v_{SG}^2 + v_{GW}^2$. So,

$$v_{SW}^2 = (30.0 \text{ m/s})^2 + (40.0 \text{ m/s})^2$$
$$v_{SW} = \textbf{50.0 m/s.}$$

Also,

$$\theta = \tan^{-1}(v_{SG}/v_{GW}) = \textbf{36.9° S of E.}$$

---

10. a. A vector diagram shows that the angle at which the boat must steer is $\theta = \cos^{-1}(v_S/v_B)$, where the velocity v_S is that of the stream and v_B is that of the boat. So,

$$\theta = \cos^{-1}[(8.00 \text{ km/h})/(20.0 \text{ km/h})] = \textbf{66.4°.}$$

b. The velocity across the river is

$$v^2 = v_B^2 - v_S^2 = (20.0 \text{ km/h})^2 - (8.00 \text{ km/h})^2$$
$$v = 18.3 \text{ km/h.}$$

Since it's 5.00 km across the river, the time to get across is

$$t = x/v = (5.00 \text{ km})/(18.3 \text{ km/h}) = \textbf{0.273 h.}$$

CHAPTER 4
Forces and Newton's Laws of Motion

PREVIEW

In this chapter you will begin the study of **dynamics**, that branch of physics which explains why objects accelerate. You will be introduced to the concept of **force**, and study Newton's laws of motion, which apply to all forces that occur in nature. You will learn how to construct free-body diagrams, and use them to analyze systems subject to such forces as gravity and friction. You will also apply Newton's laws of motion to solve a number of different problems. The applications will include both equilibrium and non-equilibrium problems.

QUICK REFERENCE

Important Terms

Force
> The push or pull required to change the state of motion of an object, as defined by Newton's second law. It is a vector quantity with units of newtons (N), dynes (dyn), or pounds (lb).

Inertia
> The natural tendency of an object to remain at rest or in uniform motion at a constant speed in a straight line.

Mass
> A quantitative measure of inertia. Units are kilograms (kg), grams (g), or slugs (sl).

Inertial Reference Frame
> A reference frame in which Newton's law of inertia is valid.

Free-Body Diagram
> A vector diagram that represents all of the forces acting on an object.

Gravitational Force
> The force of attraction that every particle of mass in the universe exerts on every other particle.

Weight
> The gravitational force exerted by the earth (or some other large astronomical body) on an object.

Normal Force
> One component of the force that a surface exerts on an object with which it is in contact. This component is directed normal, or perpendicular, to the surface.

Friction
> The force that an object encounters when it moves or attempts to move along a surface. It is always directed parallel to the surface in question.

Tension
> The tendency of a rope (or similar object) to be pulled apart due to the forces that are applied at either end.

Equilibrium
> The state an object is in if it has zero acceleration. Mathematically, equilibrium means $\sum \mathbf{F} = 0$.

Apparent Weight
> The force that an object exerts on the platform of a scale. It may be larger or smaller than the true weight, depending on the acceleration of the object and the scale.

Newton's Laws of Motion

First Law
>   An object continues in a state of rest or in a state of motion at a constant speed along a straight line, unless compelled to change that state by a net force. By "net" force we mean the **vector sum** of all of the forces acting on an object.

For example, consider a spaceship in deep space, isolated from any other object or force. If the ship is stationary, it will remain so. But if the ship is moving (its rocket engines are shut down) it will continue to move in a straight line with a constant speed. So if the ship were traveling into deep space at say, 100 000 mi/h, it would continue to move at this speed in a straight line, even without the rocket engines firing, until an outside force acted to stop or change its motion.

Second Law
>   When a net force ΣF acts on an object of mass m, the acceleration **a** that results is directly proportional to the net force and has a magnitude that is inversely proportional to the mass. The direction of the acceleration is the same as the direction of the net force. This statement is usually written as

$$\Sigma \mathbf{F} = m\mathbf{a} \qquad \text{or} \qquad \mathbf{a} = \Sigma \mathbf{F}/m \qquad (4.1)$$

The symbol $\Sigma \mathbf{F}$ represents the net force, that is, the **vector sum** of all the forces acting on an object. This means that the components of the forces must be examined. For example, in two dimensions, equation (4.1) becomes

$$\Sigma F_x = ma_x \qquad (4.2a)$$
$$\Sigma F_y = ma_y \qquad (4.2b)$$

Equations (4.1) and (4.2) can be used to determine the **units** of force. They are

SI	$kg\ m/s^2$ = newton (N)
CGS	$g\ cm/s^2$ = dyne (dyn)
BE	slug ft/s^2 = pound (lb)

Third Law
>   Whenever one body exerts a force on a second body, the second body exerts an oppositely directed force of equal magnitude on the first body.

This is sometimes referred to as the "action-reaction" law, i.e., "for every action there is an equal, but opposite, reaction". An example is a spaceship firing its rocket engines. The engines eject hot gases out the rear of the rocket at very high velocity, this is the "action". The "reaction" is that the rocket accelerates in the forward direction.

Universal Gravitation

Every particle in the universe exerts an attractive force on every other particle in the universe. The force acting between two particles of masses m_1 and m_2 separated by a distance r has a magnitude given by

$$F = G \frac{m_1 m_2}{r^2} \qquad (4.3)$$

where G is the universal gravitational constant, whose value (obtained experimentally) is given by

$$G = 6.673 \times 10^{-11}\ N\ m^2/kg^2.$$

The **weight** of an object on the earth is the gravitational force that the earth exerts on an object. This force is always directed downward, toward the center of the earth. For the earth (mass = M_E, radius = R_E) we can write

$$W = G \frac{M_E \, m}{R_E^2}$$

where the gravitational acceleration is $g = GM_E/R_E^2 = 9.80 \text{ m/s}^2$. The weight of an object can therefore be written

$$W = m \, g \qquad\qquad (4.5)$$

DISCUSSION OF SELECTED SECTIONS

4.3/4.4   Newton's Second Law of Motion and the Vector Nature of Newton's Second Law of Motion

Newton's second law, as expressed in equations (4.1) and (4.2), contains vectors. The equation $\Sigma \mathbf{F} = \mathbf{ma}$ means that we need to determine the net force acting on an object. The terminology, $\Sigma \mathbf{F}$, means that you must determine the vector sum (Σ = summation sign) of **all** the forces acting on an object. You should begin by drawing a diagram that represents the object and all the forces acting on it, a so-called **free-body diagram**.

Example 1 *A Free-Body Diagram*
A 2.0 kg object is subjected to a 6.0 N force acting in the x direction, and an 8.0 N force acting in the y direction. What is the acceleration (magnitude and direction) of this object?

The first diagram shows the mass m and forces acting on it. The first diagram is the free-body diagram where the mass is placed at the origin of the x-y coordinate system and the forces are shown acting from this point.

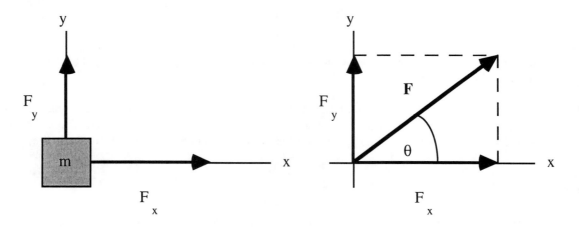

In order to find the acceleration of the object, **a**, we need to find the net force, **F**, acting on the object. Since the forces acting on m are perpendicular, we can find the net force using

$$F = \sqrt{F_x^2 + F_y^2} = \sqrt{(6.0 \text{ N})^2 + (8.0 \text{ N})^2} = 10 \text{ N}.$$

$$\theta = \tan^{-1}\left(\frac{F_y}{F_x}\right) = \tan^{-1}\left(\frac{8.0 \text{ N}}{6.0 \text{ N}}\right) = 53^\circ.$$

Since the acceleration is in the same direction as the net force, **a** is directed 53° up from the +x axis. Its magnitude is

$$a = F/m = (10 \text{ N})/(2.0 \text{ kg}) = 5 \text{ m/s}^2.$$

In most cases, the forces acting on an object can have any orientation in space. The individual forces must be resolved into x and y components before the net force, and hence the acceleration, can be found. The next example illustrates this.

Example 2 *Finding the net force and acceleration*
Suppose a 1.8 kg object is subjected to forces as illustrated in the following diagram. The magnitudes of the forces are F_1 = 20.0 N, F_2 = 14.0 N, F_3 = 15.0 N. What is the net force and acceleration acting on the object?

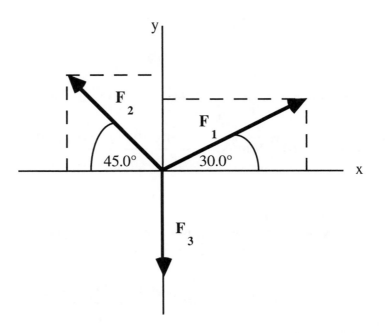

To find $\Sigma\mathbf{F}$ and the accelerations, we need to look at the components of the forces and use equations (4.2a) and (4.2b). That is, we need to sum the force components in the x direction (ΣF_x) and y direction (ΣF_y). We have

Force	x component
$\mathbf{F_1}$	+(20.0 N) cos 30.0° = +17.3 N
$\mathbf{F_2}$	- (14.0 N) cos 45.0° = - 9.9 N
$\mathbf{F_3}$	0

ΣF_x = +17.3 N - 9.9 N = +7.4 N

Force	y component
$\mathbf{F_1}$	+(20.0 N) sin 30.0° = +10.0 N
$\mathbf{F_2}$	+(14.0 N) sin 45.0° = + 9.9 N
$\mathbf{F_3}$	-15.0 N

ΣF_y = +10.0 N + 9.9 N - 15.0 N = +4.9 N

The accelerations in the x and y directions can now be obtained using ΣF_x and ΣF_y found above:

$$a_x = \frac{\Sigma F_x}{m} = \frac{+7.4 \text{ N}}{1.8 \text{ kg}} = 4.1 \text{ m/s}^2 ; \quad a_y = \frac{\Sigma F_y}{m} = \frac{+4.9 \text{ N}}{1.8 \text{ kg}} = 2.7 \text{ m/s}^2 .$$

4.7   The Gravitational Force

Consider two particles of masses m_1 and m_2 separated by a distance r. Each mass exerts a force of attraction on the other mass. The force is directed along the line joining the particles and the force has a magnitude given by

$$F = G \frac{m_1 m_2}{r^2} \tag{4.3}$$

Example 3
Find the force between objects of mass $m_1 = 1.00 \times 10^3$ kg, $m_2 = 2.00 \times 10^3$ kg, separated by r = 3.00 m.

$$F = G \frac{m_1 m_2}{r^2} = (6.67 \times 10^{-11} \text{ N m}^2/\text{kg}^2)\frac{(1.00 \times 10^3 \text{ kg})(2.00 \times 10^3 \text{ kg})}{(3.00 \text{ m})^2} = 1.48 \times 10^{-5} \text{ N}$$

Example 4 *The earth's gravity and weight*
The weight of an object on earth is defined as the gravitational force exerted by the earth on the object. Find the weight on earth of a 50.0 kg mass. Use $M_E = 5.98 \times 10^{24}$ kg, $R_E = 6.38 \times 10^6$ m for the earth's mass and radius.

$$W = G \frac{M_E m}{R_E^2} = (6.67 \times 10^{-11} \text{ N m}^2/\text{kg}^2)\frac{(5.98 \times 10^{24} \text{ kg})(50.0 \text{ kg})}{(6.38 \times 10^6 \text{ m})^2} = 4.90 \times 10^2 \text{ N}$$

Note that according to Newton's second law, F = ma = mg, since g is the acceleration of gravity. Therefore,

$$g = \frac{GM_E}{R_E^2} = \frac{(6.67 \times 10^{-11} \text{ N m}^2/\text{kg}^2)(5.98 \times 10^{24} \text{ kg})}{(6.38 \times 10^6 \text{ m})^2} = 9.80 \text{ m/s}^2 .$$

So the weight of an object on earth can be expressed as W = mg. This expression will prove to be quite useful.

Weight

The weight of an object is not the same as its mass. The mass of an object is the same regardless of where the object is located in the universe. The weight, being simply the force that gravity exerts on an object, depends on how close the object is to another massive object such as the earth. Close to the surface of the earth, an object of mass, m, has a weight of mg, where "g" is the acceleration due to gravity. The acceleration, g, close to the surface of the earth is

$$9.80 \text{ m/s}^2 \text{ or } 32.2 \text{ ft/s}^2 .$$

4.8/4.9   The Normal Force and Static and Kinetic Frictional Forces

The normal force, $\mathbf{F_N}$, is the perpendicular force that a surface exerts on an object with which it is in contact. For an object at equilibrium on a flat surface (see figure below) we can see that $F_N = W$. The normal force is simply the surface's reaction to the forces pushing the object against the surface.

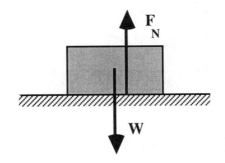

If an object is at equilibrium on an inclined plane, the magnitude of the normal force will be equal to the magnitude of a component of the object's weight, as is shown in the figure.

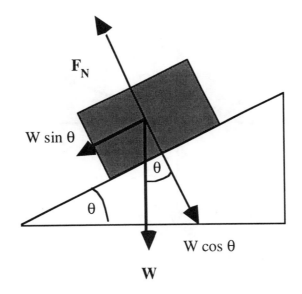

The magnitude of the normal force is $F_N = W \cos \theta$, which is the case even for a flat surface, i.e., for $\theta = 0$ since $\cos \theta = 1$, therefore $F_N = W$.

When an object is in contact with a surface, there is another force in addition to the normal force and gravity acting on it. This additional force is **friction**. When the object slides over the surface, the surface exerts a frictional force on the object. This is known as **kinetic friction**. However, a frictional force exists even if two objects are not in motion. This force is **static friction**.

An important characteristic of the frictional force is that its magnitude is proportional to the magnitude of the normal force. For example, an object sliding on a rough surface is subject to the force of kinetic friction, $\mathbf{f_k}$, according to

$$f_k = \mu_k F_N \qquad (4.8)$$

where μ_k is the **coefficient of kinetic friction**. The coefficient of kinetic friction can have values which are usually $0 \leq \mu_k \leq 1$.

Example 5

A 5.0 kg object is pulled over a rough horizontal surface by a 100.0 N force directed at an angle of 14.0° with respect to the surface (see figure). If the coefficient of kinetic friction is 0.40, find the acceleration of the object.

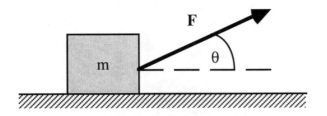

We begin by drawing a free-body diagram for this example. The forces which must be included are the applied force, **F**, the frictional force, $\mathbf{f_k}$, the normal force, $\mathbf{F_N}$, and the weight, **W** = m**g**. If the object is moving in the +x direction, we have

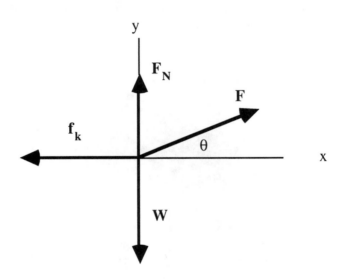

Look at the forces in the x and y directions:

y direction: $F_N + F_y - W = 0$

since there is no acceleration in the y direction, so

$$F_N = W - F_y = mg - F \sin \theta$$

$$F_N = (5.0 \text{ kg})(9.80 \text{ m/s}^2) - (100.0 \text{ N}) \sin 14.0° = 25 \text{ N}.$$

x direction: $F_x - f_k = ma_x$

since the object is accelerating in the x direction, so

$$F \cos \theta - \mu_k F_N = ma_x$$

$$(100.0 \text{ N}) \cos 14.0° - (0.40)(25 \text{ N}) = 87 \text{ N} = ma_x$$

Therefore

$$a_x = (87\ \text{N})/m = (87\ \text{N})/(5.0\ \text{kg}) = 17\ \text{m}/\text{s}^2.$$

Suppose an object is resting on a flat surface. If we apply a small horizontal force, friction resists this force and the object remains at rest. This is the force of static friction, $\mathbf{f_s}$. As we gradually increase the applied force, the object will eventually begin to move once we have overcome the maximum frictional force, $\mathbf{f_s}^{\mathbf{max}}$. If the coefficient of static friction is μ_s, the magnitude of the maximum static frictional force can be written

$$f_s{}^{max} = \mu_s\ F_N \qquad\qquad (4.7)$$

Example 6 *Static Friction*
A 10.0 kg block sits on a flat surface whose coefficient of static friction is $\mu_s = 0.60$ and whose coefficient of kinetic friction is $\mu_k = 0.40$. (a) What horizontal force is required to get the block to move? (b) If we continue to apply the same force as in part (a), what will the block's acceleration be?

(a) We can use equation (4.7) to find the force F necessary to overcome friction, that is,

$$F = f_s{}^{max}$$
$$F = \mu_s\ F_N = \mu_s\ mg$$
$$F = (0.60)(10.0\ \text{kg})(9.80\ \text{m/s}^2) = 59\ \text{N}.$$

(b) Using the above force, Newton's second law, $\Sigma F = ma$, and equation (4.8) for kinetic friction, we have

$$\Sigma F = F - f_k$$
$$\Sigma F = F - \mu_k\ F_N$$
$$\Sigma F = F - \mu_k mg = ma$$

Now

$$a = \frac{\Sigma F}{m} = \frac{F - \mu_k mg}{m} = \frac{59\ \text{N} - (0.40)(10.0\ \text{kg})(9.80\ \text{m/s}^2)}{10.0\ \text{kg}} = 2.0\ \text{m/s}^2$$

Example 7
A box weighing 147 N sits on a horizontal surface. The coefficient of static friction between the box and the surface is 0.70. If a force **P** is exerted on the box at an angle directed 37° below the horizontal, what must the magnitude of **P** be to get the box moving? The box and the associated free body diagram are shown below.

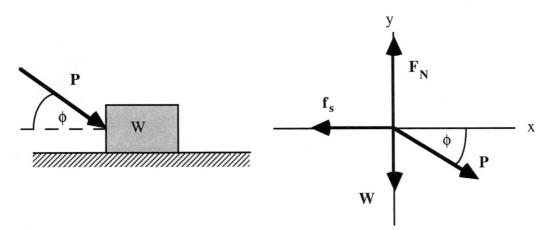

Use the component equations of Newton's second law, equations (4.2b) and (4.2a).

In the y direction

$$\Sigma F_y = ma_y$$

The box has no acceleration in the vertical direction so

$$F_N - P_y - W = 0$$

which gives

$$F_N = P \sin \phi + W \qquad (1)$$

In the x direction

$$\Sigma F_x = ma_x$$

The box has no acceleration in the horizontal direction so

$$P_x - f_s = 0$$

or

$$P \cos \phi - \mu_s F_N = 0$$

or

$$P \cos \phi - \mu_s(P \sin \phi + W) = 0 \qquad (2)$$

Solving (2) for P yields:

$$P = \frac{\mu_s W}{\cos \phi - \mu_s \sin \phi} = \frac{(0.70)\,(147\ N)}{\cos 37° - (0.70)\sin 37°} = 270\ N.$$

4.10   The Tension Force

Forces are sometimes applied by ropes, cables, or wires that pull on an object. We usually assume that the rope is massless and that the magnitude of the tension, **T,** is the same throughout the rope.

Example 8 *The tension force*
A 12.0 kg object is pulled upward by a massless rope with an acceleration of 3.00 m/s^2. What is the tension in the rope?

The tension force is directed upward (+) while the weight of the object (mg) is directed downward (-). Since the object is being accelerated upward (+) we can write $\Sigma \mathbf{F} = m\mathbf{a}$ as

$$T - mg = ma$$

so

$$T = m\,(g + a)$$
$$T = (12.0\ kg)(9.80\ m/s^2 + 3.00\ m/s^2)$$
$$T = \ 154\ N.$$

Example 9

Suppose the object in the previous example were accelerating downward at 3.00 m/s^2 instead of upward. What would the tension in the rope be in this case?

The only difference in this case would be that the acceleration is downward, so take a = -3.00 m/s^2. We have,

$$T - mg = ma,$$

so that

$$T = m (g + a)$$
$$T = (12.0 \text{ kg}) (9.80 \text{ m/s}^2 - 3.00 \text{ m/s}^2)$$
$$T = 82 \text{ N}.$$

The situations depicted in examples 8 and 9 can be observed when you're in an elevator. Suppose you're on the bottom floor and the elevator starts accelerating upward. You feel as if you are being pushed downward, and the tension in the elevator cable increases. When you're on the top floor and the elevator begins accelerating downward, you feel as if you are being lifted off the floor and there is a corresponding decrease in the elevator cable tension. When the elevator is at rest, or moving with a steady speed, you feel "normal" and the cable tension would just be equal to the weight of you and the elevator.

4.11   Equilibrium Applications of Newton's Laws of Motion

An object that is at rest or traveling with a constant velocity has zero acceleration and is said to be in "equilibrium". This implies that the net force acting on the object is zero. In two dimensions we can write

$$\Sigma F_x = 0 \qquad\qquad (4.9a)$$

$$\Sigma F_y = 0 \qquad\qquad (4.9b)$$

There are five steps that can be followed in solving equilibrium problems. They are:

Step 1.   Select an object, the "system", about which the most information is known. If two or more objects are connected, it may be necessary to treat each individually.

Step 2.   Draw a "free-body" diagram for the system. As discussed previously, this is a drawing that represents the object and ALL the forces acting on it.

Step 3.   Choose a convenient set of x, y axes for the object and resolve all the forces into components that point along these axes. By "convenient" we mean that you should choose your axes so that as many forces as possible point directly along the x or y axis. This will minimize the number of calculations needed.

Step 4.   Apply equations (4.9a) and (4.9b) by setting the sum of the components of the forces in the x direction equal to zero and the sum of the components in the y direction equal to zero.

Step 5.   Solve the two equations obtained in step 4 for the desired unknown quantities. In many cases the equations must be solved simultaneously for the two unknowns.

The following are a number of examples employing the method just outlined.

Example 10 *A weight suspended by cables*
A 280 lb block is suspended by two cables, as shown below. Find the tension in each cable.

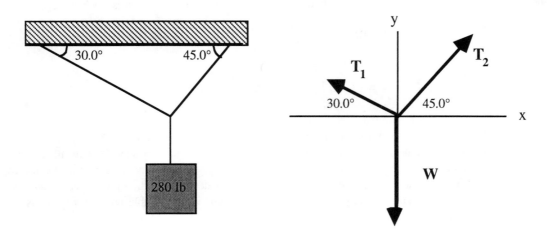

Choosing the block as our object, we have drawn the free-body diagram and chosen x and y axes, as shown. We now apply equations (4.9a) and (4.9b) using the following table

Force	x component	y component
T_1	$-T_1 \cos 30.0$	$+T_1 \sin 30.0°$
T_2	$+T_2 \cos 45.0°$	$+T_2 \sin 45.0°$
W	0	$- 280$ lb

Use Newton's second law

$$\sum F_x = -T_1 \cos 30.0° + T_2 \cos 45.0° = 0 \qquad (1)$$

$$\sum F_y = +T_1 \sin 30.0° + T_2 \sin 45.0° - 280 \text{ lb} = 0 \qquad (2)$$

Solve for T_1 in equation (1),

$$T_1 = T_2 \left(\frac{\cos 45.0°}{\cos 30.0°} \right) = 0.816 \, T_2 \qquad (3)$$

Substitute the result into equation (2),

$$(0.816) \, T_2 \sin 30.0° + T_2 \sin 45.0° = 280 \text{ lb}$$

Solving for T_2 then yields,

$$T_2 = \frac{280 \text{ lb}}{(0.816 \sin 30.0° + \sin 45.0°)} = 250 \text{ lb}$$

And finally, using this in equation (3) yields

$$T_1 = (0.816) \, T_2 = 200 \text{ lb}.$$

Example 11

Consider the following system in equilibrium. Determine the value of the weight, **W**, and the tension in the left wire.

The free body diagram can be drawn in two different ways, as shown below. In (a), we can make the y axis vertical and draw in the forces. Perhaps a better choice would be to align the coordinate axes as shown in diagram (b). This is more convenient because two of the three forces now lie along the axes, and are thus resolved into x and y components automatically.

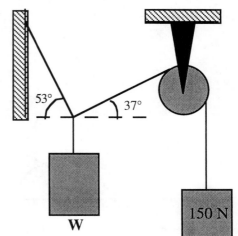

The equilibrium equations, (4.9a) and (4.9b) can now be written as

$$\Sigma F_x = 150 \text{ N} - W \sin 37° = 0$$

which gives

$$W = (150 \text{ N})/(\sin 37°) = 250 \text{ N}$$

and

$$\Sigma F_y = T - W \cos 37° = 0$$

which gives

$$T = W \cos 37° = 200 \text{ N}.$$

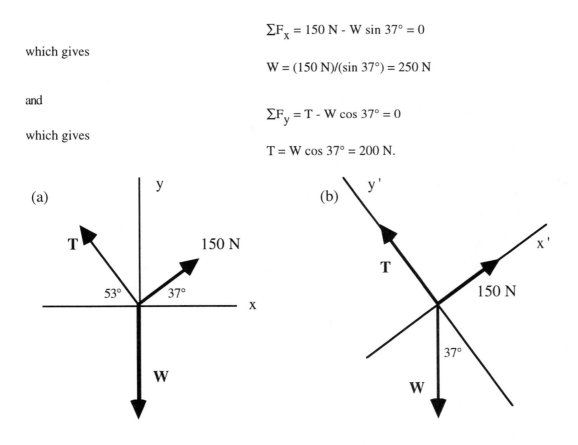

Notice that the choice of x and y axes in (b) made the equilibrium equations rather easy to solve for W and T. If we had chosen the axes as in drawing (a), the equations would have been a bit more difficult to solve, i.e.,

$$\Sigma F_x = (150 \text{ N}) \cos 37° - T \cos 53° = 0$$

$$\Sigma F_y = (150 \text{ N}) \sin 37° + T \sin 53° - W = 0.$$

Thus, the appropriate choice of coordinate axes can facilitate the solution quite a bit in most cases.

Example 12 *Object on an inclined plane*
Consider a mass M = 25.0 kg sitting on an inclined plane whose angle of inclination is θ = 30.0°. A massless cord connected to M passes over a frictionless pulley and is attached to a mass m = 15.0 kg. The mass M slides up the plane at a constant speed. Find the coefficient of kinetic friction between the mass M and the plane.

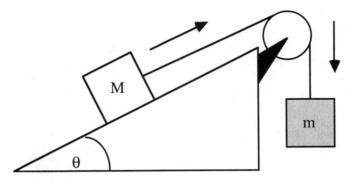

Since the masses move with constant velocity (a = 0), the system is in equilibrium. We need to look at each mass individually in order to determine the necessary quantities. The free-body diagrams for M and m look like

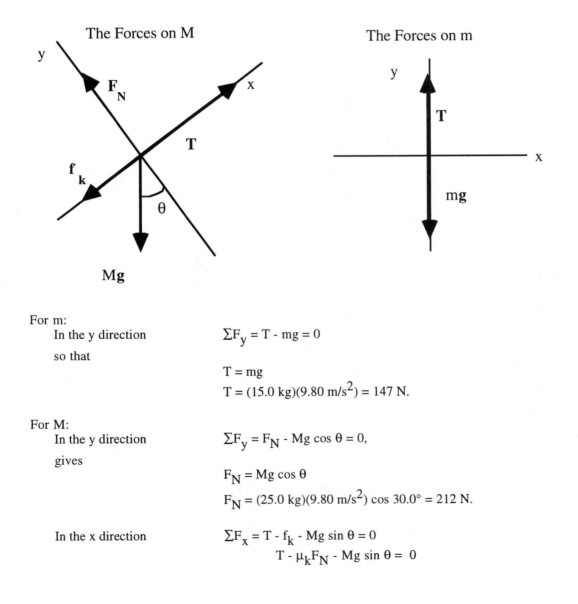

For m:
    In the y direction

$$\Sigma F_y = T - mg = 0$$

    so that

$$T = mg$$
$$T = (15.0 \text{ kg})(9.80 \text{ m/s}^2) = 147 \text{ N}.$$

For M:
    In the y direction

$$\Sigma F_y = F_N - Mg \cos \theta = 0,$$

    gives

$$F_N = Mg \cos \theta$$
$$F_N = (25.0 \text{ kg})(9.80 \text{ m/s}^2) \cos 30.0° = 212 \text{ N}.$$

    In the x direction

$$\Sigma F_x = T - f_k - Mg \sin \theta = 0$$
$$T - \mu_k F_N - Mg \sin \theta = 0$$

Solving for μ_k yields

$$\mu_k = \frac{T - Mg \sin \theta}{F_N} = \frac{147 \text{ N} - (25.0 \text{ kg})(9.80 \text{ m/s}^2) \sin 30.0°}{212 \text{ N}} = 0.12.$$

4.12   Nonequilibrium Applications of Newton's Laws of Motion

Nonequilibrium occurs when the forces acting on an object are not balanced. In this case the object accelerates and Newton's second law is

$$\Sigma F_x = ma_x \qquad (4.2a) \qquad\qquad \Sigma F_y = ma_y \qquad (4.2b)$$

Example 13 *Atwood's Machine*
Two masses, m_1 and m_2 (assume $m_2 > m_1$), are connected by a massless string that passes over a massless, frictionfree pulley, as shown in the figure. Find the acceleration of the masses and the tension in the string.

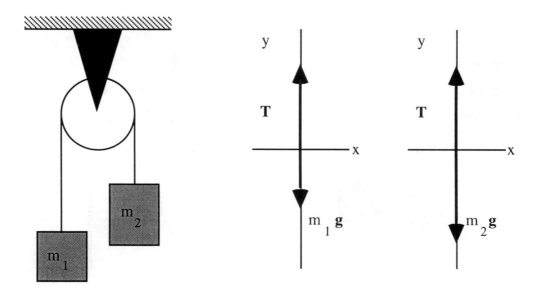

We need to look at each mass individually in order to obtain the two equations necessary to solve for the unknown quantities **a** and **T**. The free-body diagram for each mass is shown above. Since both masses are connected by the string, the magnitudes of their accelerations are the same. Since $m_2 > m_1$, m_1 has an upward acceleration and m_2 has a downward acceleration. We also assume that the tension in the string on each side of the pulley is the same.

For m_1 the acceleration is upward, so $T > m_1 g$ and we can therefore write,

$$T - m_1 g = m_1 a.$$

For m_2 the acceleration is down, so $m_2 g > T$. Therefore, we can write,

$$m_2 g - T = m_2 a.$$

Solving for T in the first equation and substituting into the second gives,

$$m_2 g - (m_1 g + m_1 a) = m_2 a.$$

Finally, solving for the acceleration yields,

$$a = \frac{(m_2 - m_1)}{(m_1 + m_2)} g$$

Substituting this expression for the acceleration into either of the mass equations, we can solve for the tension.

$$T = \frac{2 m_1 m_2}{(m_1 + m_2)} g$$

Example 14
Consider the following arrangement with M_1 = 15.0 kg, M_2 = 10.0 kg, and μ_k = 0.30. Find the acceleration of the system and the tension in the connecting cord. Free-body diagrams accompany the drawing.

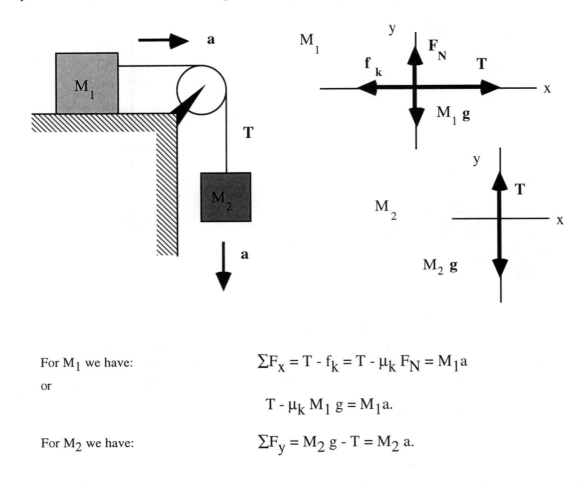

For M_1 we have:

$$\Sigma F_x = T - f_k = T - \mu_k F_N = M_1 a$$

or

$$T - \mu_k M_1 g = M_1 a.$$

For M_2 we have:

$$\Sigma F_y = M_2 g - T = M_2 a.$$

Solving for T in the second equation and substituting into the first yields:

$$M_2 g - M_2 a - \mu_k M_1 g = M_1 a$$

or

$$(M_1 + M_2)\, a = (M_2 - \mu_k M_1)\, g$$

The acceleration is therefore,

$$a = \frac{(M_2 - \mu_k M_1)g}{(M_1 + M_2)} = \frac{[10.0 \text{ kg} - (0.30)(15.0 \text{ kg})](9.80 \text{ m/s}^2)}{(15.0 \text{ kg} + 10.0 \text{ kg})} = 2.2 \text{ m/s}^2.$$

The tension can now be found by substituting this value of a into the equation for M_2.

$$T = M_2\, (g - a)$$
$$T = (10.0 \text{ kg})\, (9.80 \text{ m/s}^2 - 2.2 \text{ m/s}^2) = 76 \text{ N}.$$

PRACTICE PROBLEMS

1. A block of mass 4.0 kg rests on a horizontal surface. What horizontal force is required to accelerate the block at 5.0 m/s^2 if (a) there is no friction, and (b) if the coefficient of kinetic friction is 0.25?

2. An object of mass 10.0 kg is subjected to the following forces: 25.0 N due east, 50.0 N at an angle of 53.0° north of east, and 40.0 N due west. What is the acceleration of the object?

3. At what altitude above the earth's surface would the acceleration of gravity be 7.00 m/s^2? Use the values of the earth's mass and radius given in example 4.

4. Two 75 kg masses are separated by a distance of 1.0 m. A 5.0 kg object is placed 0.25 m from one of the masses (0.75 m from the other) along the line joining the masses. What is the gravitational force on the object?

5. Two blocks are connected by a cord on a horizontal surface. A force F pulls on the blocks, as shown below. Find the acceleration of the blocks and the tension in the connecting cord if (a) the surface is frictionless, and (b) the coefficient of kinetic friction is 0.20.

6. Traveling at a speed of 58.0 km/h, the driver of an automobile suddenly locks the wheels by slamming on the brakes. The coefficient of kinetic friction between the tires and the road is 0.720. How far does the car skid before coming to a halt? Ignore effects of air resistance.

7. You are standing on a scale in an elevator, the scale reads 176 lb when the elevator is at rest. The elevator begins to accelerate upward at 12.0 ft/s^2. (a) What is the reading on the scale? (b) Suppose the elevator were accelerating downward at 12.0 ft/s^2. What does the scale read now?

8. A 250 lb block rests on a horizontal table. The coefficient of kinetic friction between the block and table is 0.30. What force is required to pull the block at constant speed if (a) a horizontal rope is attached, or if (b) the rope makes an angle of 25° above the horizontal?

9. A space traveler weighs 130 lb on Earth. What will the traveler weigh on another planet whose radius is three times that of the earth and whose mass is twice that of the earth?

10. An object rests on an inclined plane. One end of the plane is lifted until the object begins to slide. If the angle at which the object begins to slide is 35°, what is the coefficient of static friction?

11. Find the tensions T_1 and T_2 in the following diagram.

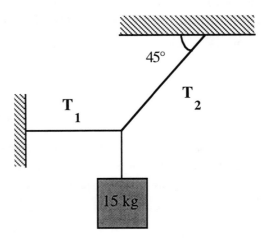

12. A fisherman is fishing from a bridge and using "10-lb test line". In other words, the line will sustain a maximum force of 10.0 lb without breaking. (a) What is the heaviest fish that can be pulled up vertically through the air when the line is reeled in at a constant speed? (b) Repeat part (a), assuming that the line is given an upward acceleration of 6.50 ft/s^2.

13. Find the acceleration of the system and the tension in the cord.
    Take m = 2.0 slugs and M = 3.5 slugs.

14. A 55.0 kg person is riding in a hot air balloon, and a scale shows this person's weight to be 549 N. The acceleration of gravity at the location of the balloon is known to be 9.79 m/s^2. Determine the magnitude and direction of the vertical component of the balloon's acceleration.

15. Find the acceleration of the system if (a) there is no friction, and (b) if the coefficient of kinetic friction is 0.33. Take M = 7.5 kg, and m = 2.5 kg.

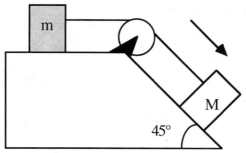

16. Starting from rest, a piece of ice slides down a 45.0° incline in twice the time it takes to slide down a frictionless 45.0° incline. What is the coefficient of kinetic friction between the ice and the incline?

17. What is the magnitude of the force **F** (assumed horizontal) required to push the block up the plane at a constant speed? Take the coefficient of kinetic friction to be 0.25.

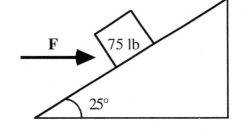

18. Block 1 weighs 96 lb. The coefficient of static friction between block 1 and the table is 0.50. Find the maximum weight of block 2 for which the system will remain in equilibrium.

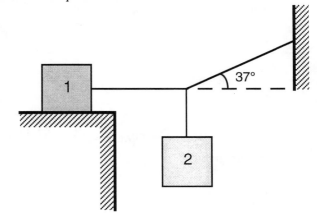

SOLUTIONS AND ANSWERS

Practice Problems

1. (a) Using equation (4.1),
$$F = ma = (4.0 \text{ kg})(5.0 \text{ m/s}^2) = \textbf{2.0 x 10}^1 \textbf{ N}.$$

    (b) We now have two forces,
$$F - f_k = ma,$$
    where
$$f_k = \mu_k F_N = \mu_k mg.$$
    Therefore,
$$F = \mu_k mg + ma,$$
    or
$$F = m(\mu_k g + a) = (4.0 \text{ kg})[(0.25)(9.80 \text{ m/s}^2) + 5.0 \text{ m/s}^2] = \textbf{3.0 x 10}^1 \textbf{ N}.$$

2. Use equations (4.2a) and (4.2b) to find the accelerations in the x and y directions. We have,
$$\Sigma F_x = ma_x$$
    or
$$25.0 \text{ N} + (50.0 \text{ N})\cos 53.0° - 40.0 \text{ N} = (10.0 \text{ kg}) a_x$$
    so
$$a_x = (15.1 \text{ N})/(10.0 \text{ kg}) = \textbf{1.51 m/s}^2.$$
    Also
$$\Sigma F_y = ma_y$$
    or
$$(50.0 \text{ N})\sin 53.0° = 39.9 \text{ N} = (10.0 \text{ kg}) a_y$$
    then
$$a_y = (39.9 \text{ N})/(10.0 \text{ kg}) = \textbf{3.99 m/s}^2.$$

    The acceleration **a** has magnitude a and direction θ given by
$$a^2 = a_x{}^2 + a_y{}^2$$
    or a = **4.27 m/s²**. Also,
$$\theta = \tan^{-1} (a_y/a_x) = \textbf{69.3° N of E}.$$

3. The acceleration is
$$a = GM_E/r^2,$$
    so
$$r^2 = GM_E/a = (6.67 \text{ x } 10^{-11} \text{ N m}^2/\text{kg}^2)(5.98 \text{ x } 10^{24} \text{ kg})/(7.00 \text{ m/s}^2).$$
    Solving gives
$$r = 7.55 \text{ x } 10^6 \text{ m}.$$
    Now
$$r = R_E + h$$
    Hence,
$$h = r - R_E = 7.55 \text{ x } 10^6 \text{ m} - 6.38 \text{ x } 10^6 \text{ m} = \textbf{1.17 x 10}^6 \textbf{ m}.$$

4.   We can use equation (4.3),

$$F_1 = Gm_1M/r_1^2 = G(75 \text{ kg})(5.0 \text{ kg})/(0.25 \text{ m})^2 = 4.0 \times 10^{-7} \text{ N.}$$

The second mass gives

$$F_2 = G(75m \text{ kg})(5.0 \text{ kg})/(0.75 \text{ m})^2 = 4.4 \times 10^{-8} \text{ N.}$$

The resultant is

$$F = F_1 - F_2 = \mathbf{3.6 \times 10^{-7} \text{ N.}}$$

---

5.   (a) $$F = ma \text{ (where } m = m_1 + m_2 = 40 \text{ kg)}$$

so

$$a = F/m = (120 \text{ N})/(40 \text{ kg}) = \mathbf{3.0 \text{ m/s}^2.}$$

The tension is

$$T = m_1 a = (25 \text{ kg})(3.0 \text{ m/s}^2) = \mathbf{75 \text{ N.}}$$

(b) Newton's second law requires

$$F - f_k = ma$$

or

$$F - \mu_k mg = ma$$

Then

$$a = (F - \mu_k mg)/m = \mathbf{1.0 \text{ m/s}^2.}$$

The tension is found from

$$T - f_k = m_1 a$$

or

$$T = m_1 (a + \mu_k g) = \mathbf{74 \text{ N.}}$$

---

6.   The initial velocity is

$$v_0 = (58 \text{ km/h})(1 \text{ h}/3600 \text{ s})(1000 \text{ m}/1 \text{ km}) = 16.1 \text{ m/s.}$$

Only the friction force acts, so Newton's second law

$$f_k = ma$$

gives

$$a = -f_k/m = -\mu_k mg/m = -\mu_k g = -(0.720)(9.80 \text{ m/s}^2) = -7.06 \text{ m/s}^2.$$

Using equation (2.9) we have

$$v^2 = v_0^2 + 2ax$$

or

$$s = -v_0^2/2a = -(16.1 \text{ m/s})^2/[2(-7.06 \text{ m/s}^2)] = \mathbf{18.4 \text{ m.}}$$

---

7.   (a) The reading on the scale will be the same as the normal force exerted by the platform of the scale. We have

$$F_N - W = ma,$$

so that

$$F_N = W + ma = 176 \text{ lb} + [(176 \text{ lb})/(32.2 \text{ ft/s}^2)](12.0 \text{ ft/s}^2) = \mathbf{242 \text{ lb.}}$$

(b) $W - F_N = ma$

gives

$$F_N = W - ma = 176 \text{ lb} - (5.5 \text{ sl})(12.0 \text{ ft/s}^2) = \textbf{110 lb}.$$

---

8.   (a) Since the acceleration is zero we have

$$F - f_k = 0$$

or

$$F = \mu_k F_N = (0.30)(250 \text{ lb}) = \textbf{75 lb}.$$

(b) The normal force is not equal to the weight in this case, instead

$$F_N + F_y = W$$

so that

$$F_N = W - F_y = W - F \sin 25°.$$

Therefore,

$$F \cos 25° = \mu_k (W - F \sin 25°).$$

Solving for F we obtain

$$F = \mu_k W/(\cos 25° + \mu_k \sin 25°)$$
$$F = [(0.30)(250 \text{ lb})]/[\cos 25° + (0.30) \sin 25°)] = \textbf{73 lb}.$$

---

9.   Using equation (4.4) we can find the ratio of the weights on the planet (P) and earth (E).

$$\frac{W_P}{W_E} = \frac{GM_P m/R_P^2}{GM_E m/R_E^2} = \frac{M_P R_E^2}{M_E R_P^2} = \frac{2M_E R_E^2}{M_E (3R_E)^2} = \frac{2}{9}$$

The weight of the traveler on the planet is therefore,

$$W_P = (2/9) \, W_E = (2/9) \, (130 \text{ lb}) = \textbf{29 lb}.$$

---

10. Refer to the following figure

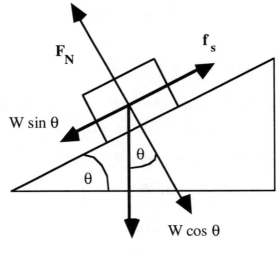

The force pulling the object down the plane is

$$F = W \sin \theta.$$

The frictional force is

$$f_s = \mu_s F_N = \mu_s (W \cos \theta).$$

Just as the block begins to slide we see that

$$F = f_s$$

so that

$$W \sin \theta = \mu_s W \cos \theta$$

Then

$$\mu_s = \sin \theta / \cos \theta = \tan \theta = \tan (35°) = \mathbf{0.70}.$$

---

11. Refer to the diagram in the statement of the problem.
    In the y direction:

$$\Sigma F_y = T_{2y} - mg = 0$$

which gives

$$T_2 \sin 45° = mg$$

or

$$T_2 = mg / \sin 45° = (15 \text{ kg})(9.8 \text{ m/s}^2)/(\sin 45°) = \mathbf{210 \ N}.$$

In the x direction:

$$\Sigma F_x = T_{2x} - T_1 = 0$$

which gives

$$T_1 = T_2 \cos 45°$$

$$T_1 = (210 \text{ N}) \cos 45° = \mathbf{150 \ N}.$$

---

12. (a) Since there is no acceleration, Newton's second law requires that

$$T - W = 0$$

or

$$W = T = \mathbf{10.0 \ lb}.$$

(b) Newton's second law with a non-zero acceleration, is

$$T - mg = ma$$

or

$$T = m(g+a).$$

Solving for m yields

$$m = T/(g + a)$$
$$m = (10.0 \text{ lb})/(32.2 \text{ ft/s}^2 + 6.50 \text{ ft/s}^2) = 0.258 \text{ sl}.$$

Therefore, the weight of the fish is

$$W = mg = (0.258 \text{ sl})(32.2 \text{ ft/s}^2) = \mathbf{8.32 \ lb}.$$

13. Using the results from example 13, we have

$$a = (M - m)g/(M + m)$$
$$a = (1.5 \text{ sl})(32 \text{ ft/s}^2)/(5.5 \text{ sl}) = \textbf{8.7 ft/s}^2.$$

The tension is then

$$T = 2Mmg/(M + m)$$
$$T = 2(3.5 \text{ sl})(2.0 \text{ sl})(32 \text{ ft/s}^2)/(5.5 \text{ sl}) = \textbf{81 N}.$$

---

14. Note that $F_N = 549 \text{ N} > mg = (55.0 \text{ kg})(9.79 \text{ m/s}^2) = 539 \text{ N}$, so the acceleration is **UPWARD**.
Using equation (4.6),

$$F_N = mg + ma$$

we have

$$a = (F_N - mg)/m = (549 \text{ N} - 539 \text{ N})/(55.0 \text{ kg}) = \textbf{0.18 m/s}^2.$$

---

15. (a) Newton's second law gives

$$\Sigma F = (M+m)a$$

where the weight is the only force that is pulling M down the plane. Therefore,

$$\Sigma F = Mg \sin 45° = (M+m)a$$

or

$$a = Mg \sin 45°/(M+m)$$
$$a = (7.5 \text{ kg})(9.8 \text{ m/s}^2) \sin 45°/(10.0 \text{ kg}) = \textbf{5.2 m/s}^2.$$

(b) With friction (on both surfaces),

$$\Sigma F = Mg \sin 45° - \mu_k(Mg \cos 45° + mg) = (M+m)a$$

which gives

$$a = \textbf{2.7 m/s}^2.$$

---

16. Frictionless plane:

$$a = g \sin 45.0°$$
$$a = (9.80 \text{ m/s}^2) \sin 45.0° = 6.93 \text{ m/s}^2;$$

and

$$x = (1/2)at^2.$$

With friction:

$$a' = g \sin 45.0° - \mu_k g \cos 45.0°$$

which gives

$$x = (1/2)a't'^2 = (1/2)a'(2t)^2$$
$$x = 2a't^2$$

or

$$a' = (1/4)a = 1.73 \text{ m/s}^2.$$

Now

$$\mu_k = (g \sin 45.0° - a')/(g \cos 45.0°)$$
$$\mu_k = [(9.80 \text{ m/s}^2) \sin 45.0° - 1.73 \text{ m/s}^2]/[(9.80 \text{ m/s}^2)(\cos 45.0°)] = \textbf{0.75}.$$

17. The free-body diagram is shown below. We have:

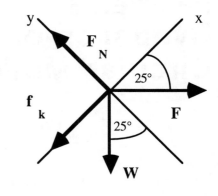

$$\Sigma F_y = F_N - W \cos 25° - F \sin 25° = 0.$$
$$\Sigma F_x = F \cos 25° - f_k - W \sin 25° = 0.$$

Since
$$f_k = \mu_k F_N = \mu_k(W \cos 25° + F \sin 25°)$$
we have:
$$F \cos 25° - \mu_k(W \cos 25° + F \sin 25°) - W \sin 25° = 0.$$

Solving for F we obtain:

$$F = W(\sin 25° + \mu_k \cos 25°)/(\cos 25° - \mu_k \sin 25°) = \textbf{61 lb}.$$

---

18. The free-body diagram for block 2 is shown below.

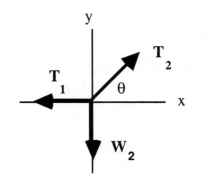

Block 2: $\Sigma F_x = T_{2x} - T_1 = T_2 \cos 37° - T_1 = 0$
so that,
$$T_2 = T_1/\cos 37°.$$

$\Sigma F_y = T_{2y} - W_2 = T_2 \sin 37° - W_2 = 0$
so that,
$$W_2 = T_2 \sin 37° = T_1 \sin 37°/\cos 37°.$$

Similarly for block 1: $T_1 - f_s = 0$
which gives
$$T_1 = \mu_s F_N$$
$$T_1 = \mu_s W_1 = (0.50)(96 \text{ lb}) = 48 \text{ lb}$$
Therefore,
$$W_2 = (48 \text{ lb})\sin 37°/\cos 37° = \textbf{36 lb}.$$

CHAPTER 5
Dynamics of Uniform Circular Motion

PREVIEW

In this chapter you will be introduced to the physics of an object traveling in a circular path. After completion of the chapter, you will be able to find the speed and acceleration of an object traveling in a circle at constant speed along with the period of the motion and the force necessary to keep the object moving in the circle. Applications which will be explored are the motion of a satellite in a circular orbit and artificial gravity.

QUICK REFERENCE

Important Terms

Uniform circular motion
   The motion of an object traveling with constant speed in a circular path.
Period of uniform circular motion
   The time necessary for an object to complete one revolution around the circle.
Centripetal acceleration
   The acceleration that an object must have in order to stay in a circular path. The direction of this acceleration is always toward the center of the circle.
Centripetal force
   The net force that must act on an object to keep it in a circular path. This net force is in the direction of the centripetal acceleration, that is, toward the center of the circle.

Equations of Uniform Circular Motion

The **speed** of an object moving in a circular path of radius, r, with period, T, is

$$v = \frac{2\pi r}{T} \qquad (5.1)$$

The magnitude of the **centripetal acceleration** of an object moving with uniform circular motion is

$$a_c = \frac{v^2}{r} \qquad (5.2)$$

The magnitude of the **centripetal force** acting on an object moving in uniform circular motion is

$$F_c = \frac{mv^2}{r} \qquad (5.3)$$

Banked Curves

The following applies to an object moving without friction around a curve of radius, r, and an angle of bank, θ.

$$\tan \theta = \frac{v^2}{gr} \tag{5.4}$$

Satellites in Circular Orbits

The **speed of a satellite** moving in a circular orbit of radius, r, about the earth is

$$v = \sqrt{\frac{GM_E}{r}} \tag{5.5}$$

The **period of a satellite** in circular orbit about the earth is

$$T = \frac{2\pi r^{3/2}}{\sqrt{GM_E}} \tag{5.6}$$

DISCUSSION OF SELECTED SECTIONS

5.1 Uniform Circular Motion

Example 1 *Finding the speed of a planet*
The planet Neptune travels in a nearly circular orbit of radius, r = 4.5 X 10^9 km, about the sun. It takes Neptune 165 y to make a complete trip around the sun. How fast (in km/h) does Neptune travel in its orbit?

Equation (5.1) is

$$v = \frac{2\pi r}{T} = \frac{2(3.14)(4.5 \times 10^9 \text{ km})}{(165 \text{ y})(365 \text{ d} / 1 \text{ y})(24 \text{ h} / 1 \text{ d})} = 2.0 \times 10^4 \text{ km} / \text{h}$$

Example 2 *Finding the period*
A satellite travels in a circular orbit 3.59 X 10^4 km above the earth with a speed of 1.11 X 10^4 km/h. What is the period of the satellite's motion? The radius of the earth is 6.38 X 10^3 km.

Rearranging equation (5.1) yields

$$T = \frac{2\pi r}{v} = \frac{2(3.14)(3.59 \text{ X } 10^4 \text{ km} + 6.38 \text{ X } 10^3 \text{ km})}{1.11 \text{ X } 10^4 \text{ km/h}} = 23.9 \text{ h}$$

5.2 Centripetal Acceleration

Example 3

Using the data from example 1, find the centripetal acceleration of Neptune in m/s^2.

Equation (5.2) gives

$$a_c = \frac{v^2}{r} = \frac{\left[(2.0 \times 10^4 \, km/h)(10^3 \, m/km)(1 \, h/3600 \, s)\right]^2}{(4.5 \times 10^9 \, km)(10^3 \, m/1 \, km)} = 6.9 \times 10^{-6} \, m/s^2$$

The direction of the centripetal acceleration of Neptune is toward the center of its orbit, that is, toward the sun.

5.3 Centripetal Force

The "centripetal force" is a generic term pertaining to the net force which holds an object in uniform circular motion. Regardless of the identity of the physical force, equation (5.3) must be satisfied if the object is to remain in uniform circular motion. Some possible contributors to the centripetal force are: friction, tension in a rope or string, the gravitational force, the electrical force and the magnetic force. You are familiar with some of these already, and the others will be introduced later in the course.

Example 4

What is the minimum coefficient of static friction necessary to allow a penny to rotate along with a 33 1/3 rpm record (diameter = 0.300 m), when the penny is placed on the outer edge of the record?

The static frictional force acting on the penny is $f_s^{max} = \mu_s F_N = \mu_s mg$. This frictional force plays the role of the centripetal force necessary to hold the penny in its circular path as it revolves with the record so,

$$\mu_s mg = \frac{mv^2}{r}$$

$$\mu_s = \frac{v^2}{rg}$$

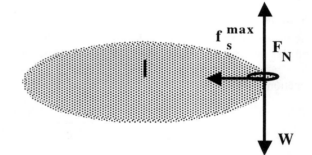

We now need the speed of the penny. Equation (5.1) is $v = 2\pi r/T$, where $T = (60 \, s)/(33 \, 1/3)$. Therefore,

$$\mu_s = \frac{4\pi^2 r}{gT^2} = \frac{4\pi^2 \, (0.150 \, m)}{(9.80 \, m/s^2)\left[(60 \, s)/(33 \, 1/3)\right]^2} = 0.187$$

5.5 Satellites in Circular Orbits

Example 5
At what height above the surface of Mars must a satellite be placed to be in a synchronous orbit? The radius of Mars is 6.79×10^6 m, its mass is 6.40×10^{23} kg and its period of rotation is 1.03 d.

Equation (5.6) written for Mars is

$$T = \frac{2\pi r^{3/2}}{\sqrt{GM_M}}$$

Squaring

$$T^2 = \frac{4\pi^2 r^3}{GM_M}$$

Rearranging and taking the cube root

$$r = \sqrt[3]{\frac{GM_M T^2}{4\pi^2}} = \sqrt[3]{\frac{(6.67 \times 10^{-11}\ \text{Nm}^2/\text{kg}^2)(6.40 \times 10^{23}\ \text{kg})\left[(1.03\ \text{d})(8.64 \times 10^4\ \text{s}/1\ \text{d})\right]^2}{4\pi^2}}$$

$$r = 2.05 \times 10^7\ \text{m}$$

The height of the satellite above the surface of Mars is

$$h = 2.05 \times 10^7\ \text{m} - 6.79 \times 10^6\ \text{m} = 1.37 \times 10^7\ \text{m}.$$

PRACTICE PROBLEMS

1. A grinding wheel (radius 7.6 cm) is rotating at 1750 rpm.
    a. What is the speed of a point on the outer edge of the wheel?

    b. What is the centripetal acceleration of the point?

2. An airplane is flying in a horizontal circle of radius 1.5 km with a speed of 450 km/h. What is the magnitude of the centripetal acceleration of the plane?

3. An airplane is flying in a horizontal circle of radius 1.0 km. What must be the speed of the plane if the pilot is to experience a centripetal acceleration three times that of gravity?

4. If the pilot in problem 3 has a mass of 75 kg, what centripetal force acts on him?

5. A trick motorcyclist moves in a horizontal circle inside a large vertical drum which has a radius of 10.0 m. If the coefficient of static friction between the tires and the drum is 0.60, what is the smallest possible speed that the motorcycle and rider can have?

6. A car can barely negotiate a 50.0 m unbanked curve when the coefficient of static friction between the tires and road is 0.80. How much bank would the curve require if the car is to safely go around the curve without relying on friction?

7. A satellite is placed in a circular polar orbit. What must be the height above the surface of the earth of its orbit if it is to pass over the same point on the equator twice per day?

8. A spacecraft orbits Mars (mass = 6.40×10^{23} kg) in a circular orbit of radius 8.01×10^5 km. What is the period of the spacecraft?

9. How many revolutions per minute must a rotating space station (r = 1200 m) turn to provide an artificial gravity of 0.50 g?

10. An airplane pulls out of a dive in a vertical circle of radius 1.0 km traveling with a speed of 550 km/h. How many times greater is the apparent weight of the pilot than his true weight?

11. The moon of a planet is observed to have a nearly circular orbit, r = 4.00 X 10^5 km, and an orbital period of 21.5 days. What is the mass of the planet?

12. A 1.5 kg toy motorcycle is moving on the inside of a vertical circular track (r = 1.0 m). It arrives at the top of the track with a speed of 5.0 m/s. What force does the track exert on the motorcycle?

SOLUTIONS AND ANSWERS

Practice Problems

---

1. a. Equation (5.1) gives $v = 2\pi r/T$. The period of the motion is

$$T = (1/1750)(60 \text{ s}) = 3.43 \times 10^{-2} \text{ s}.$$

so

$$v = 2\pi(7.6 \times 10^{-2} \text{ m})/(3.43 \times 10^{-2} \text{ s}) = \textbf{14 m/s}.$$

b. Equation (5.2) gives

$$a_c = v^2/r = (14 \text{ m/s})^2/(7.6 \times 10^{-2} \text{ m}) = \textbf{2.6} \times \textbf{10}^3 \text{ m/s}^2.$$

---

2. Equation (5.2) gives

$$a_c = v^2/r = [(450 \times 10^3 \text{ m/h})(1/3600 \text{ h/s})]^2/(1.5 \times 10^3 \text{ m}) = \textbf{1.0} \times \textbf{10}^1 \text{ m/s}^2.$$

---

3. Equation (5.2) gives $v^2 = a_c r$, but $a_c = 3g$ so

$$v^2 = 3gr = 3(9.80 \text{ m/s}^2)(1.0 \times 10^3 \text{ m})$$

or

$$v = \textbf{170 m/s} = \textbf{620 km/h}.$$

---

4. Equation (5.3) is $F_c = mv^2/r$. Now $v^2 = 3gr$ from problem 3, so

$$F_c = 3mg = 3(75 \text{ kg})(9.80 \text{ m/s}^2) = \textbf{2200 N}.$$

---

5. An application of Newton's second law in the vertical direction gives $F_f = \mu_s F_N = mg$. The centripetal force is provided entirely by the normal force acting on the motorcycle, so $mv^2/r = mg/\mu_s$ or

$$v^2 = rg/\mu_s = (10.0 \text{ m})(9.80 \text{ m/s}^2)/(0.60).$$

Hence,

$$v = \textbf{13 m/s}.$$

---

6. When the car goes around the unbanked curve the centripetal force is provided entirely by friction, so

$$mv^2/r = \mu_s mg$$

or

$$v^2 = \mu_s rg.$$

Using this result in equation (5.4) which applies to a frictionless banked curve, we have

$$\theta = \tan^{-1} \mu_s = \tan^{-1}(0.80) = \textbf{39}°.$$

7. The satellite must have a period of 24 h to pass over the same point on the equator twice in one day. Equation (5.6) gives after some rearrangement

$$r^3 = GM_E T^2/(4\pi^2) = (6.67 \times 10^{-11} \text{ N m}^2/\text{kg}^2)(5.98 \times 10^{24} \text{ kg})(8.64 \times 10^4 \text{ s})^2/(4\pi^2),$$

so $r = 4.23 \times 10^7$ m. The height of the orbit above the earth is

$$H = 4.23 \times 10^7 \text{ m} - 0.64 \times 10^7 \text{ m} = \mathbf{3.59 \times 10^7 \text{ m} = 22\ 300 \text{ mi}}.$$

---

8. Equation (5.6) gives the period of the orbit.

$$T = \frac{2\pi r^{3/2}}{\sqrt{GM_M}} = \frac{2\pi(8.01 \times 10^8 \text{ m})^{3/2}}{\sqrt{(6.67 \times 10^{-11} \text{ N m}^2/\text{kg}^2)(6.40 \times 10^{23} \text{ kg})}} = 2.18 \times 10^7 \text{ s} = \mathbf{6.06 \times 10^3 \text{ h}}.$$

---

9. The centripetal acceleration of the space station is $a_c = v^2/r$ where $v = 2\pi r/T$ so $a_c = 4\pi^2 r/T^2$. We want $a_c = 0.50g$, so

$$T^2 = 4\pi^2 r/(0.50g) = 4\pi^2(1200 \text{ m})/(4.90 \text{ m/s}^2).$$

Thus,

$$T = 98 \text{ s} = 98/60 \text{ min}.$$

The desired result is

$$1/T = \mathbf{0.61 \text{ rev/min}}.$$

---

10. The apparent weight of the pilot is $W' = F_c + mg = mv^2/r + mg$. Now

$$W'/W = 1 + v^2/(rg) = 1 + [(550 \times 10^3 \text{ m/h})(1/3600 \text{ h/s})]^2/[(1.0 \times 10^3 \text{ m})(9.80 \text{ m/s}^2)] = \mathbf{3.4}.$$

---

11. Equation (5.6) applied to the planet and its moon gives $M = 4\pi^2 r^3/(GT^2)$.
    The period of the moon is

$$T = (21.5 \text{ d})(8.64 \times 10^4 \text{ s/d}).$$

Evaluation gives

$$M = \mathbf{1.10 \times 10^{25} \text{ kg}}.$$

---

12. Newton's second law applied to the vertical forces acting on the motorcycle at the top of the track gives

$$F_N + mg = mv^2/r$$

or

$$F_N = m(v^2/r - g) = (1.5 \text{ kg})[(5.0 \text{ m/s})^2/(1.0 \text{ m}) - 9.80 \text{ m/s}^2] = \mathbf{23 \text{ N}}.$$

CHAPTER 6
Work and Energy

PREVIEW

In this chapter you will be introduced to the concept of work and the related concepts of kinetic energy and potential energy. After completion of the chapter you will be able to calculate the work done by constant forces, the kinetic energy of a moving object, the gravitational potential energy of an object, and the power developed by a force.

In addition, you will be able to use the principle of conservation of mechanical energy to solve a variety of problems in a much simpler way than you have done previously.

QUICK REFERENCE

Important Terms

Work
The work done by a CONSTANT force acting on an object is the component of the force along the displacement of the object times the magnitude of the displacement.
Kinetic energy
The energy of an object due to its motion.
Gravitational potential energy
The energy of an object due to its position relative to the earth.
Total mechanical energy
The sum of the kinetic and potential energies of an object.
Conservative force
A force which does work on an object which is independent of the path taken by the object between its starting point and its ending point.
Average power
The work done by a force on an object divided by the time taken to do the work.

Work

The **work done on an object** by a force, F, is

$$W = (F \cos \theta) \, s \qquad\qquad (6.1)$$

where s is the magnitude of the displacement of the object and θ is the angle between the force and the displacement.

The **work done by gravity** on an object is

$$W_{gravity} = - mg \, (h_f - h_0) \qquad\qquad (6.4)$$

where m is the mass of the object, h_0 is the initial height of the object, and h_f is the final height of the object.

The **work done by nonconservative forces acting** on an object is

$$W_{nc} = (1/2 \, mv_f^2 - 1/2 \, mv_0^2) + (mgh_f - mgh_0) \qquad (6.6)$$
$$W_{nc} = (KE_f - KE_0) + (PE_f - PE_0) \qquad\qquad (6.7a)$$
$$W_{nc} = \Delta KE + \Delta PE \qquad\qquad (6.7b)$$

Energy

The **kinetic energy** of an object of mass, m, and speed, v, is

$$KE = 1/2 \ mv^2 \tag{6.2}$$

The **gravitational potential energy** of an object a height, h, above the reference level is

$$PE = mgh \tag{6.5}$$

The **total mechanical energy** of an object is the sum of its kinetic and potential energies.

$$E = KE + PE$$

The work done by nonconservative forces produces a change in the total mechanical energy of the object.

$$W_{nc} = E_f - E_0 \tag{6.8}$$

Power

The average power developed by a force which has done work, W, in time, t, is

$$\overline{P} = \frac{Work}{Time} = \frac{W}{t} \tag{6.10a}$$

The average power developed by a force of magnitude F, moving an object with an average speed, v is

$$\overline{P} = F \ \overline{v} \tag{6.11}$$

Theorems and Principles

Work-Energy Theorem - the total work done by all forces acting on an object is

$$W = KE_f - KE_0 = 1/2 \ mv_f^2 - 1/2 \ mv_0^2 \tag{6.3}$$

Principle of Conservation of Mechanical Energy - the total mechanical energy of an object remains constant as the object moves, provided that no net work is done by nonconservative forces.

$$E_f = E_0 \tag{6.9a}$$

DISCUSSION OF SELECTED SECTIONS

6.1 Work Done by a Constant Force

The concept of work plays an important role in physics since it connects, via the work-energy theorem, Newton's second law to the important scalar quantities of kinetic energy and potential energy.

Although work is a scalar quantity, it comes in two varieties-positive and negative. POSITIVE work indicates that work is done ON the object, whereas NEGATIVE work indicates that the work is done BY the object.

Example 1 *Positive and negative work*
A constant force of 40.0 N is needed to keep a car traveling at constant speed as it moves 5.0 km down the road. How much work is done? Is the work done on or by the car?

The force needed to keep the car moving is in the direction of the car's displacement. Equation (6.1) gives the work.

$$W = (F \cos 0°) \, s = + (40.0 \text{ N})(5.0 \text{ X } 10^3 \text{ m}) = + 2.0 \text{ X } 10^5 \text{ J}$$

The work is POSITIVE so the work is done ON the car.

Example 2
There must be a retarding force acting on the car in example 1 if it is to remain traveling at constant speed. What is this force? How much work does the force do? Is this work done on or by the car?

Newton's second law applied along the direction of motion of the car gives

$$F - F_r = ma = 0$$

or

$$F_r = F = 40.0 \text{ N}$$

The work done by the retarding force is then given by (6.1) to be

$$W_r = (F_r \cos 180°) \, s = - (40.0 \text{ N})(5.0 \text{ X } 10^3 \text{ m}) = - 2.0 \text{ X } 10^5 \text{ J}$$

The work is NEGATIVE so the work is done BY the car.

Example 3
A 2.40 X 10^2 N force is pulling an 85.0 kg block across a horizontal surface. The force acts at an angle of 20.0° above the surface. The coefficient of kinetic friction is 0.200, and the block moves a distance of 8.00 m. Find (a) the work done by the pulling force, (b) the work done by the kinetic frictional force, (c) the total work done by all the forces.

a. The work done by the pulling force is

$$W_F = (F \cos 20.0°) \, s$$
$$W_F = (2.40 \text{ X } 10^2 \text{ N})(\cos 20.0°)(8.00 \text{ m})$$
$$W_F = + 1.80 \text{ X } 10^3 \text{ J}$$

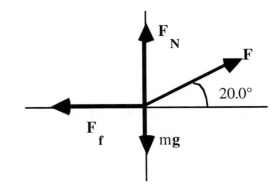

b. The magnitude of the frictional force, $F_f = \mu F_N$, acting on the block is found by applying Newton's second law to the free body diagram.
In the vertical direction

$$F_N + F \sin 20.0° - mg = 0$$

then

$$F_f = \mu(mg - F \sin 20.0°)$$

The work done by this force is

$$W_f = \mu(mg - F \sin 20.0°)(\cos 180°) \, s$$

$$W_f = - (0.200)\{(85.0 \text{ kg})(9.80 \text{ m/s}^2) - (2.40 \times 10^2 \text{ N})(\sin 20.0°)\}(8.00 \text{ m})$$

$$W_f = - 1.20 \times 10^3 \text{ J}$$

c. The total work done by all forces is

$$W = W_F + W_f + W_N + W_g = 1.80 \times 10^3 \text{ J} - 1.20 \times 10^3 \text{ J} + 0 \text{ J} + 0 \text{ J}$$

$$W = 6.0 \times 10^2 \text{ J}$$

Note that the work due to the gravitational force and the normal force is zero since these forces act at a 90° angle to the motion of the block.

6.2 The Work-Energy Theorem and Kinetic Energy

Net work done on an object always produces a change in the kinetic energy of the object according to the work-energy theorem (6.3). The theorem is often very useful in finding the speed of an object if the net work done by the forces acting on it are known or can be calculated.

Example 4 *Using the work-energy theorem to find the speed of an object*
If the block in example 3 starts from rest, what is the speed of the block after it has traveled 8.00 m?

The work-energy theorem gives

$$W = 1/2 \, mv_f^2 - 1/2 \, mv_0^2$$

$$W = 1/2 \, mv_f^2$$

Solving for v_f gives

$$v_f = \sqrt{\frac{2W}{m}} = \sqrt{\frac{2(6.0 \times 10^2 \text{ J})}{85.0 \text{ kg}}} = 3.8 \text{ m/s}$$

Another common use of the work-energy theorem is finding information about the forces acting on an object when information about the object's speed changes is known.

Example 5 *Finding an average force when the speeds are known*
A baseball pitcher can throw a 9.00 ounce (1.75×10^{-2} sl) baseball with speed measured by a radar gun of 90.0 miles per hour (132 ft/s). Assuming that the force exerted by the pitcher on the ball acts over a distance of two arm lengths, each 28.0 inches, what is the average force exerted by the pitcher on the ball?

The work-energy theorem gives $W = 1/2 \, mv_f^2$ if the ball starts at rest. The work done is $W = Fs$, so

$$F = \frac{mv_f^2}{2s} = \frac{(1.75 \times 10^{-2} \text{ sl})(132 \text{ ft/s})^2}{2(4.67 \text{ ft})} = 32.6 \text{ lb}$$

6.3 Gravitational Potential Energy

The gravitational force is one of a class of "special" forces called conservative forces. What makes the gravitational force special is that the work done on an object by gravity depends only on the initial and final positions of the object. The work does NOT depend on how the object got from the initial to the final position.

Example 6 *Work done on an object by gravity*
How much work is done by gravity when a 5.0 kg object moves 1.5 m horizontally and then moves 3.0 m vertically upward? Vertically downward?

The work done by gravity during the horizontal motion is

$$W = (mg \cos 90°)s = 0 \text{ J}$$

The work done by gravity during the vertical motion is

$$W = (mg \cos 180°)s = - (5.0 \text{ kg})(9.80 \text{ m/s}^2)(3.0 \text{ m}) = - 150 \text{ J}$$

The work is negative so the work is done BY the object.

If the object is moving vertically downward, the work done during the vertical motion is

$$W = (mg \cos 0°)s = + (5.0 \text{ kg})(9.80 \text{ m/s}^2)(3.0 \text{ m}) = + 150 \text{ J}.$$

In this case the work is done ON the object since the work is positive.

As the previous example shows, the magnitude of the work done by gravity on the object when it is going up is the same as the magnitude of the work done by gravity on the object when it is going down. The negative of the work done by gravity is called the potential energy of the object. The work-energy theorem then gives rise to equation (6.6).

Example 7 *Using gravitational potential energy*
A 3.00 kg model rocket is launched vertically upward with sufficient initial speed to reach a height of 1.00×10^2 m, even though air resistance (a nonconservative force) performs $- 8.00 \times 10^2$ J of work on the rocket. How high would the rocket have gone if there were no air resistance?

The equation (6.7b) written for the actual motion of the rocket gives

$$\Delta KE = W - \Delta PE$$
$$\Delta KE = W - mgh$$
$$\Delta KE = - 8.00 \times 10^2 \text{ J} - (3.00 \text{ kg})(9.80 \text{ m/s}^2)(1.00 \times 10^2 \text{ m})$$
$$\Delta KE = - 3740 \text{ J}$$

If the rocket were launched with the same initial kinetic energy and no air resistance acts, then (6.7b) is

$$\Delta KE = - \Delta PE = - mgh$$

or

$$h = - \frac{\Delta KE}{mg} = - \frac{(-3740 \text{ J})}{(3.00 \text{ kg})(9.80 \text{ m}/\text{s}^2)}$$
$$h = 127 \text{ m}$$

6.5 The Conservation of Mechanical Energy

If no friction or other nonconservative force acts on an object then its total mechanical energy remains constant. This principle is very useful in solving problems involving the motion of an object. It is often easier to use than Newton's laws or the kinematic equations since it involves scalars rather than vectors. Also, it only requires that you have knowledge about the motion at the beginning and end.

Example 8
A truck is descending a winding mountain road. When the truck is 1480 m above sea level and traveling 15 m/s, its brakes fail. What is the maximum possible speed of the truck at the foot of the mountain 550 m above sea level?

The truck would have the maximum possible speed if friction were ignored. In this case, the total mechanical energy of the truck is conserved as it moves down the mountain road.

$$1/2 \ mv_f^2 + mgh_f = 1/2 \ mv_0^2 + mgh_0$$

Taking the zero of potential energy at sea level we have

$$v_f = \sqrt{v_0^2 + 2g(h_0 - h_f)}$$
$$v_f = \sqrt{(15 \ m/s)^2 + 2(9.80 \ m/s^2)(1480 \ m - 550 \ m)}$$
$$v_f = 140 \ m/s$$

6.7 Power

The average power is the rate at which work is done as given by equation (6.10a).

Example 9
A 73 kg sprinter, starting from rest, reaches a speed of 7.0 m/s in 1.8 s with negligible effect due to air resistance. The sprinter then runs the remainder of the race at a steady speed of 7.0 m/s under the influence of a 35 N force due to air resistance. Calculate the average power needed (a) to accelerate the runner and (b) to sustain the steady speed at which most of the race is run.

(a) In order to find the power, we need to find the work needed to accelerate the runner. The work-energy theorem gives $W = \Delta KE = \frac{1}{2} \ mv_f^2$ so

$$\overline{P} = \frac{1/2 \ mv_f^2}{t} = \frac{(1/2)(73 \ kg)(7.0 \ m/s)^2}{1.8 \ s}$$
$$\overline{P} = 990 \ W$$

$$\overline{P} = \frac{990 \ W}{(746 \ W)/(1 \ hp)} = 1.3 \ hp$$

(b) The force that the sprinter must exert to run at constant speed is equal in magnitude and opposite in direction to the force exerted on him by air resistance. The power needed to run the race at a constant speed is then

$$\overline{P} = Fv = (35 \text{ N})(7.0 \text{ m/s})$$

$$\overline{P} = 250 \text{ W}$$

$$\overline{P} = \frac{250 \text{ W}}{(746 \text{ W})/(1 \text{ hp})} = 0.34 \text{ hp}$$

6.9 Work Done by a Variable Force

The work done by a variable force in moving an object is equal to the area under the (F cos θ) versus s curve.

Example 10

The graph below shows the component (F cos θ) of the net force that acts on a 2.0 kg block as it moves along a flat horizontal surface. Find (a) the net work done on the block, and (b) the final speed of the block if it starts from rest at s = 0.

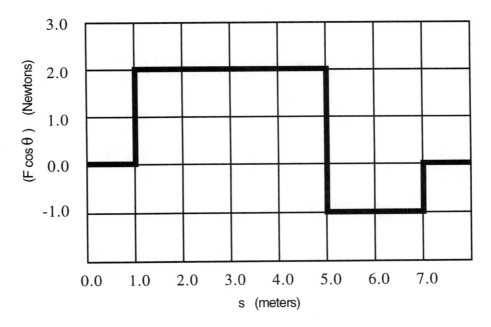

(a) The net work done on the block is equal to the area under the curve.

From 0.0 to 1.0 m, the net force is zero, so no work is done.

From 1.0 m to 5.0 m, there are 8 blocks under the curve. Each block has an area of (1.0 N)(1.0 m) = 1.0 J so

$$W_{1\text{-}5} = 8 \text{ blocks} \left(\frac{1.0 \text{ J}}{1 \text{ block}} \right) = 8.0 \text{ J}$$

From 5.0 to 7.0 m, there are 2 blocks under the curve. Note that the net force is *negative* for this area, so, each block has an area of - 1.0 J.

$$W_{5\text{-}7} = 2 \text{ blocks} \left(\frac{-1.0 \text{ J}}{1 \text{ block}} \right) = -2.0 \text{ J}$$

The net work done on the block is therefore,

$$W_{net} = W_{1\text{-}5} + W_{5\text{-}7} = (+8.0 \text{ J}) + (-2.0 \text{ J}) = 6.0 \text{ J}.$$

(b) The final speed of the block can be found from the work-energy theorem. Since the block starts from rest, we have $W = \Delta KE = \frac{1}{2} mv_f^2 - 0$ so

$$W = \frac{1}{2} mv_f^2.$$

Solving for v_f gives

$$v_f = \sqrt{\frac{2W}{m}} = \sqrt{\frac{2(6.0 \text{ J})}{2.0 \text{ kg}}} = 2.45 \text{ m/s}.$$

PRACTICE PROBLEMS

1. The driver of a 1500 kg car slams on the brakes locking the wheels. A total retarding force of 1800 N acts to stop the car in a distance of 70.0 m. How much work is done in bringing the car to a halt? Is the work done on or by the car?

2. A 2.0 ton truck descending a 5.0° hill is brought to a stop in 250 ft. The driver applies the brakes so that the wheels lock. (a) If the coefficient of kinetic friction between the truck tires and the road is 0.60, how much work is done by friction in stopping the truck? (b) How much work is done on the truck by gravity?

3. In problem 1, how fast was the car traveling initially?

4. In problem 2, how fast was the truck traveling immediately before the brakes were applied?

5. A 2100 lb airplane requires a power of 100.0 hp at the propeller to climb at an angle of 20.0° to the horizontal at constant speed. How much altitude could it gain in 10.0 minutes if air resistance is ignored?

6. A bicyclist tops the crest of a 20.0 m high hill moving with a speed of 1.0 m/s. Ignoring friction find the speed of the bicycle at the bottom of the hill.

7. A 0.60 kg ball thrown vertically upward with an initial speed of 12.0 m/s rises to a height of 5.1 m. What is the average force exerted by the air on the ball? How high would the ball have gone if no air resistance acted on it?

8. A swing is observed to rise to a maximum height of 1.5 m above its lowest point. How fast is it going at its lowest point? Ignore friction and air resistance.

9. A 1500 kg car accelerates from rest to 75 km/h in 45 s. How much power is supplied by the engine to accelerate the car?

10. A locomotive engine can supply 2.0×10^3 hp to accelerate a train from rest to 20.0 m/s in 9.0 min. Find the mass of the train. (Ignore friction.)

SOLUTIONS AND ANSWERS

Practice Problems

1. The work done by the retarding force is

$$W = (F \cos 180°) \, s$$
$$W = -(1800 \text{ N})(70.0 \text{ m}) = -1.3 \text{ X } 10^5 \text{ J.}$$

   This work is negative so the work was actually **done by the car**.

2. a. The magnitude of the frictional force acting on the truck is

$$F_f = \mu F_N = \mu mg \cos 5.0°.$$

   The work done by friction is then

$$W_f = (F_f \cos 180°) \, s = -\mu mgs \cos 5.0°$$
$$W_f = -(0.60)(4.0 \text{ X } 10^3 \text{ lb})(250 \text{ ft}) \cos 5.0° = -6.0 \text{ X } 10^5 \text{ ft-lb.}$$

   b. The work done by gravity is

$$W_{gravity} = (mg \cos 85.0°) \, s$$
$$W_{gravity} = (4.0 \text{ X } 10^3 \text{ lb})(\cos 85.0°)(250 \text{ ft}) = +8.7 \text{ X } 10^4 \text{ ft-lb.}$$

3. The work-energy theorem applied to the car's motion gives

$$W = -1/2 \, mv_0^2$$

   since the final speed of the car is zero. Solving yields

$$v_0^2 = -2W/m = -2(-1.3 \text{ X } 10^5 \text{ J})/(1500 \text{ kg})$$

   so

$$v_0 = 13 \text{ m/s.}$$

4. The work-energy theorem gives for the truck
$$W = -1/2 \, mv_0^2$$
   so
$$v_0^2 = -2W/m = -2(-6.0 \text{ X } 10^5 \text{ ft-lb} + 8.7 \text{ X } 10^4 \text{ ft-lb})(32.2 \text{ ft/s}^2)/(4.0 \text{ X } 10^3 \text{ lb})$$
   or
$$v_0 = 91 \text{ ft/s.}$$

5. The work done by the propeller in 10.0 min is

$$W_{nc} = Pt = (100.0 \text{ hp})(550 \text{ ft-lb/hp})(10.0 \text{ min})(60 \text{ s/min}) = 3.3 \text{ X } 10^7 \text{ ft-lb.}$$

Equation (6.6) written for constant speed motion is

$$W_{nc} = mg(h_f - h_0)$$

then

$$h_f - h_0 = W_{nc}/mg = (3.3 \times 10^7 \text{ ft-lb})/(2.1 \times 10^3 \text{ lb}) = \mathbf{1.6 \times 10^4 \text{ ft.}}$$

---

6. The conservation of energy applied between the top and bottom of the hill with the bottom of the hill chosen as the zero of potential energy gives

$$1/2 \ mv_f^2 + mgh_f = 1/2 \ mv_0^2$$

or

$$v_0^2 = v_f^2 + 2gh_f = (1.0 \text{ m/s})^2 + 2(9.80 \text{ m/s}^2)(20.0 \text{ m})$$

so

$$v_0 = \mathbf{2.0 \times 10^1 \text{ m/s.}}$$

---

7. If the initial position of the ball is taken as the zero of potential energy, then equation (6.6) gives

$$W_{nc} = -1/2 \ mv_0^2 + mgh_f.$$

The work done by the force of air resistance is

$$W_{nc} = (F \cos 180°) \ h_f.$$

Then

$$F = m(v_0^2/2h_f - g) = (0.60 \text{ kg})\{(12.0 \text{ m/s})^2/(10.2 \text{ m}) - 9.80 \text{ m/s}^2\}$$
$$F = \mathbf{2.6 \ N.}$$

In the absence of air resistance the total mechanical energy of the ball is conserved, hence

$$mgh_f = 1/2 \ mv_0^2$$

or

$$h_f = v_0^2/2g = (12.0 \text{ m/s})^2/(19.6 \text{ m/s}^2) = \mathbf{7.35 \ m.}$$

---

8. Let the lowest point of the swing be the level of zero potential energy. The conservation of energy applied between the lowest and highest level of the swing's motion is

$$1/2 \ mv_0^2 = mgh_f$$

or

$$v_0^2 = 2gh_f = 2(9.80 \text{ m/s}^2)(1.5 \text{ m})$$

which gives

$$v_0 = \mathbf{5.4 \ m/s.}$$

---

9. The work done on the car by the engine is, according to the work-energy theorem,

$$W = 1/2 \ mv_f^2 = 1/2(1500 \text{ kg})\{(75 \times 10^3 \text{ m/h})(1/3600 \text{ h/s})\}^2 = 3.3 \times 10^5 \text{ J.}$$

The average power supplied by the engine is

$$P = W/t = (3.3 \times 10^5 \text{ J})/(45 \text{ s}) = \mathbf{7.3 \times 10^3 \ W.}$$

10. The work done by the engine on the train is given by the work-energy theorem to be

$$W = 1/2 \, mv_f^2.$$

Equation (6.10a) gives the work in terms of the power so

$$m = 2W/v_f^2 = 2Pt/v_f^2$$

$$m = 2(2.0 \times 10^3 \text{ hp})(746 \text{ W/1 hp})(9.0 \text{ min})(60 \text{ s/1 min})/(20.0 \text{ m/s})^2 = \textbf{4.0} \times \textbf{10}^\textbf{6} \textbf{ kg}.$$

CHAPTER 7
Impulse and Momentum

PREVIEW

In this chapter you will learn about the concepts of impulse and momentum. In addition, the important principle of conservation of linear momentum is discussed. This concept is then applied to analysis of collisions in one and two dimensions.

QUICK REFERENCE

Important Terms

Impulse (of a force)
   The product of the average force and the time interval during which the force acts. Impulse is a vector quantity.
Linear momentum
   The mass times velocity of an object. Linear momentum is a vector quantity.
Total linear momentum of a system of objects
   The vector sum of the individual momenta of the objects.
Internal forces
   Forces which act between the objects of a system.
External forces
   Forces which act on the objects of a system which come from the outside.
Inelastic collision
   A collision in which the total kinetic energy of the system of objects is NOT constant.
Elastic collision
   A collision in which the total kinetic energy of the system is constant.

Equations Involving Impulse and Momentum

If $\mathbf{F}$ is the average force and Δt is the time over which the force acts, then

$$\textbf{Impulse} = \overline{\mathbf{F}} \, \Delta t \qquad (7.1)$$

The **linear momentum** of an object is

$$\mathbf{p} = m\mathbf{v} \qquad (7.2)$$

where m is the mass of the object and $\mathbf{v}$ is the velocity of the object.

Newton's second law written in terms of momentum is

$$\overline{\mathbf{F}} = \frac{m\mathbf{v}_f - m\mathbf{v}_0}{\Delta t} \qquad (7.3)$$

For a **system of objects** where $\mathbf{P}_f$ is the final total momentum of the system and $\mathbf{P}_0$ is the initial total momentum of the system

$$(\textbf{Sum of external forces}) \, \Delta t = \mathbf{P}_f - \mathbf{P}_0 \qquad (7.6)$$

The final velocities of two objects participating in a **head-on elastic collision** in which ball 2 is initially at rest are:

$$v_{f1} = \left(\frac{m_1 - m_2}{m_1 + m_2}\right) v_{01} \tag{7.8a}$$

$$v_{f2} = \left(\frac{2m_1}{m_1 + m_2}\right) v_{01} \tag{7.8b}$$

The distance x_{cm} of the center-of-mass point from the origin is

$$x_{cm} = \frac{m_1 x_1 + m_2 x_2}{m_1 + m_2} \tag{7.10}$$

Theorems and Principles

Impulse-Momentum Theorem

$$\overline{F} \Delta t = mv_f - mv_0 \tag{7.4}$$

The Principle of Conservation of Linear Momentum

If no external forces act on a system of objects, then the total linear momentum of the system remains constant in time. That is,

$$P_0 = P_f \tag{7.7}$$

DISCUSSION OF SELECTED SECTIONS

7.1 The Impulse-Momentum Theorem

Example 1
A 0.25 kg ball is dropped onto a hard surface. Immediately before the ball hits the surface it is traveling with a speed of 2.5 m/s. As it rebounds it travels with a reduced speed of 2.0 m/s. What is the impulse of the force that the surface exerts on the ball?

Taking up as the positive direction, the impulse-momentum theorem gives

impulse $= \overline{F} \Delta t = mv_f - mv_0$
impulse $= (0.25 \text{ kg})(2.0 \text{ m/s} + 2.5 \text{ m/s}) = +1.1 \text{ kg m/s}$

The direction of the impulse is UP.

Example 2
If the ball in example 1 is in contact with the floor for 15 ms, what is the average force exerted on it by the surface?

$$\overline{F} = \frac{mv_f - mv_0}{\Delta t} = \frac{(0.25 \text{ kg})(2.0 \text{ m/s} + 2.5 \text{ m/s})}{15 \times 10^{-3} \text{ s}} = +75 \text{ N}$$

The direction of the force on the ball is UP.

7.2 The Principle of Conservation of Linear Momentum

The conservation of linear momentum principle can often be used to analyze the motion of systems of colliding objects. The only criterion for its validity is that no EXTERNAL forces act on the system of objects during the collision. Practically, collisions are often of such short duration that external forces such as gravity and friction have little time to influence the motion and may be neglected. In these cases, the conservation of linear momentum principle may be applied to the motion IMMEDIATELY BEFORE and IMMEDIATELY AFTER the collision.

Example 3

Two identical spaceships perform a docking maneuver in which one spaceship catches up with the other and the two lock together. If the faster spaceship travels with a speed of 9015 m/s and the slower travels with 9005 m/s before the docking maneuver, what is the speed of the spaceships after the docking is complete?

The total momentum of the two spaceship system is constant since no external forces act during the docking maneuver. Of course, internal forces do act during the maneuver to lock the spaceships together, but they cannot affect the overall motion of the system.

$$P_0 = P_f$$

$$mv_{01} + mv_{02} = 2mv_f$$

Taking the initial direction of travel of the two spaceships as the positive direction, we have

$$v_f = \frac{v_{01} + v_{02}}{2} = \frac{9015 \text{ m/s} + 9005 \text{ m/s}}{2} = +9010 \text{ m/s}$$

The spaceships move together with a speed of 9010 m/s in the direction of the initial motion.

Example 4

A 1550 kg cannon, initially at rest, fires a 60.0 kg shell horizontally with a speed of 525 m/s. What is the maximum possible recoil velocity of the cannon?

The maximum possible recoil velocity of the cannon corresponds to no friction acting on the cannon to impede its motion. In this case there are no external forces acting in the directions of motion of the cannon and shell, so the conservation of momentum principle can be applied in the horizontal direction. Note that gravity does not affect the motion of the system since it acts vertically and is balanced by the normal force of the ground.

$$P_f = P_0$$

$$m_1 v_{f1} + m_2 v_{f2} = m_1 v_{01} + m_2 v_{02}$$

The initial velocities of both the cannon and shell are zero, that is $v_{01} = v_{02} = 0$ m/s. Assigning the initial velocity of the shell to be positive, we have for the final velocity of the cannon

$$v_{f1} = -\frac{m_2}{m_1} v_{f2} = -\left(\frac{60.0 \text{ kg}}{1550 \text{ kg}}\right)(+525 \text{ m/s}) = -20.3 \text{ m/s}$$

The cannon moves in the negative direction opposite to the motion of the shell with a speed of 20.3 m/s.

Example 5

Two 45 kg boys are playing pitch and catch in a 105 kg boat. The boys and boat are initially at rest; one boy stands in the front of the boat and the other in the back of the boat. One of the boys throws the 0.50 kg ball with a horizontal speed of 26 m/s. What is the speed of the boat immediately after the ball is thrown? What is the speed of the boat after the second boy catches the ball? Ignore friction.

If friction can be ignored then the conservation of linear momentum principle can be applied to the horizontal motion of the system of boat, boys, and ball.

$$P_f = P_0$$

$$m_1 v_{f1} + m_2 v_{f2} = m_1 v_{01} + m_2 v_{02}$$

The initial velocity of the boys and boat, v_{01}, is zero as is the initial velocity of the ball, v_{02}. Taking the direction in which the ball is thrown as positive, the final velocity of the boat is

$$v_{f1} = - \left(\frac{m_2}{m_1} \right) v_{f2} = - \left[\frac{0.50 \text{ kg}}{2(45 \text{ kg}) + 105 \text{ kg}} \right] (+ 26 \text{ m/s}) = - 0.067 \text{ m/s}.$$

7.3 Collisions in One Dimension

When objects collide in the absence of external forces, the total momentum of the system is conserved. The conservation of momentum principle then yields an equation connecting the masses and initial and final velocities of the objects. This equation can then be used to solve for one quantity if all the others are known.

If the collision also happens to be elastic, then the conservation of kinetic energy will provide a second equation containing the masses and initial and final speeds of the objects. The two equations allow you to simultaneously find TWO quantities.

Example 6 *Collision of two objects of equal mass traveling in opposite directions*
Two identical pool balls make a perfectly elastic head-on collision on a frictionless table. The speeds of the balls before the collision are 2.0 m/s and 3.0 m/s. What are the speeds and directions of motion of the balls after the collision?

Before the collision, the velocities of the balls are as shown in the figure. The conservation of momentum principle gives

BEFORE

$$P_f = P_0$$

$V_{01} = + 3.0 \text{ m/s}$ $V_{02} = - 2.0 \text{ m/s}$

$$m v_{f1} + m v_{f2} = m v_{01} + m v_{02}$$

AFTER

$$v_{f1} + v_{f2} = v_{01} + v_{02}$$

V_{f1} V_{f2}

$$v_{f1} + v_{f2} = + 3.0 \text{ m/s} - 2.0 \text{ m/s}$$

$$v_{f1} + v_{f2} = + 1.0 \text{ m/s} \qquad (1)$$

The conservation of kinetic energy gives

$$1/2 \, m v_{f1}^2 + 1/2 \, m v_{f2}^2 = 1/2 \, m v_{01}^2 + 1/2 \, m v_{02}^2$$

$$v_{f1}^2 + v_{f2}^2 = 13.0 \text{ m}^2/\text{s}^2 \qquad (2)$$

The equations (1) and (2) may be solved simultaneously for the velocities by solving (1) for v_{f1}, substituting into (2) and solving the resulting quadratic equation for v_{f2}. In this calculation we omit the units for convenience.

Equation (1) gives

$$v_{f1} = 1.0 - v_{f2}$$

Substitution of this into equation (2) yields

$$(1.0 - v_{f2})^2 + v_{f2}^2 = 13.0$$

or

$$2v_{f2}^2 - 2.0v_{f2} - 12.0 = 0$$

The quadratic formula gives

$$v_{f2} = \frac{-(-2.0) \pm \sqrt{(-2.0)^2 - 4(2)(-12.0)}}{4}$$

$$v_{f2} = \frac{2.0 \pm 10.0}{4} = -2.0 \text{ m/s OR} + 3.00 \text{ m/s}$$

The final velocity of ball one is found by substituting these values into equation (1)

$$v_{f1} = 1.0 \text{ m/s} - 3.00 \text{ m/s} = -2.0 \text{ m/s}$$

$$v_{f1} = 1.0 \text{ m/s} - (-2.0 \text{ m/s}) = +3.0 \text{ m/s}$$

The result, $v_{f1} = +3.0$ m/s and $v_{f2} = -2.0$ m/s, indicates that the balls have passed through each other without colliding and can be ruled out as being physically unrealistic. Hence, after the collision ball 1 has reversed direction and is moving to the LEFT with a speed of 2.0 m/s, while ball 2 has also reversed direction and is moving to the RIGHT with a speed of 3.0 m/s. Note that the balls simply exchanged speeds during the collision. This is true for all one dimensional elastic collisions of two objects with EQUAL MASSES.

Example 7 *Collision of two objects with equal mass traveling in the same direction*
Repeat example 6 assuming that both balls are traveling in the same direction with the 3.0 m/s ball overtaking the 2.0 m/s ball.

The analysis in example 6 applies to this problem with the exception that the velocity $v_{02} = +2.0$ m/s so that equation (1) becomes

$$v_{f1} + v_{f2} = 5.0 \text{ m/s} \qquad (1')$$

Following the same procedure except using (1') instead of (1) results in the quadratic equation

$$2v_{f2}^2 - 10.0v_{f2} + 12.0 = 0$$

with the solutions $v_{f2} = +3.0$ m/s and $+2.0$ m/s.

Equation (1') now gives

$$v_{f1} = 5.0 \text{ m/s} - 3.0 \text{ m/s} = +2.0 \text{ m/s}$$

and

$$v_{f1} = 5.0 \text{ m/s} - 2.0 \text{ m/s} = +3.0 \text{ m/s}$$

The case where $v_{f1} = +3.0$ m/s and $v_{f2} = +2.0$ m/s corresponds to no collision at all. Again, the balls have exchanged their speeds, but their directions of motion after the collision are the same as before the collision.

Example 8 *Elastic collision of two objects with different masses when one object is initially at rest*
A 1.0 kg ball initially traveling with a speed of 2.5 m/s strikes a stationary 0.50 kg ball. Determine the final velocities of the balls. How would the result differ if the 0.50 kg ball were moving with 2.5 m/s and the 1.0 kg ball were at rest?

Equations (7.8a,b) apply to this motion. Note that these were derived by an application of the conservation of kinetic energy and the conservation of momentum principle.

$$v_{f1} = \left(\frac{m_1 - m_2}{m_1 + m_2} \right) v_{01} = \left(\frac{1.0 \text{ kg} - 0.50 \text{ kg}}{1.0 \text{ kg} + 0.50 \text{ kg}} \right)(2.5 \text{ m/s}) = 0.8 \text{ m/s}$$

$$v_{f2} = \left(\frac{2m_1}{m_1 + m_2} \right) v_{01} = \left(\frac{2(1.0 \text{ kg})}{1.0 \text{ kg} + 0.50 \text{ kg}} \right)(2.5 \text{ m/s}) = 3.3 \text{ m/s}$$

In the second case, where the 0.50 kg ball is moving, it becomes ball 1 and the 1.0 kg ball becomes ball 2.

$$v_{f1} = \left(\frac{0.50 \text{ kg} - 1.0 \text{ kg}}{0.50 \text{ kg} + 1.0 \text{ kg}} \right)(2.5 \text{ m/s}) = -0.8 \text{ m/s}$$

$$v_{f2} = \left(\frac{2(0.50 \text{ kg})}{0.50 \text{ kg} + 1.0 \text{ kg}} \right)(2.5 \text{ m/s}) = +1.7 \text{ m/s}$$

These results indicate that the direction of the final velocity of the striking ball depends on whether or not it is more massive than the struck ball. If the striking ball is more massive than the struck ball, then it will continue in its original direction of travel. If the striking ball is less massive than the struck ball then it will rebound in a direction opposite its original direction of travel. This is true independently of the initial speed of the striking ball as can be seen from an examination of equation (7.8a).

The conservation of momentum principle may still apply even if kinetic energy is not conserved during the collision. This is true if the frictional forces which take energy from the system are INTERNAL rather than EXTERNAL forces.

Example 9 *A completely inelastic collision*
A 1500 kg automobile traveling with a speed of 55 km/h makes a head-on collision with an 1800 kg automobile traveling with a speed of 35 km/h in the opposite direction. The automobiles lock together during the collision and move together with a final speed, v. Find v.

The collision is completely inelastic since the automobiles lock together. If the friction between the automobile's tires and the road and air resistance can be neglected during the collision, then no net external force acts on the cars and the total linear momentum of the system is conserved.

$$\mathbf{P}_f = \mathbf{P}_0$$

$$m_1\mathbf{v}_{f1} + m_2\mathbf{v}_{f2} = m_1\mathbf{v}_{01} + m_2\mathbf{v}_{02}$$

The final velocities of the automobiles are equal so

$$v = \frac{m_1v_{01} + m_2v_{02}}{m_1 + m_2} = \frac{(1500 \text{ kg})(55 \text{ km/h}) + (1800 \text{ kg})(-35 \text{ km/h})}{1500 \text{ kg} + 1800 \text{ kg}} = +5.9 \text{ km/h}$$

The direction the velocity of the cars after the collision is the same as the initial direction of the 1500 kg automobile.

7.4 Collisions in Two Dimensions

Example 10
A 1200 kg automobile traveling with a velocity of 65 km/h, 35° north of east makes a completely inelastic collision with an identical automobile traveling with a velocity of 32 km/h, 45° south of west. What is the magnitude and direction of the velocity of the cars after the collision?

If friction between the tires and the road and air resistance can be ignored, then no net external force acts on the cars and the total VECTOR momentum of the system is constant during the collision.

In the x direction:

$$P_{fx} = P_{0x}$$

$$(m_1 + m_2)v_f \cos \theta = m_1v_{01} \cos 35° - m_2v_{02} \cos 45°$$

$$(2400 \text{ kg})v_f \cos \theta = (1200 \text{ kg})(65 \text{ km/h}) \cos 35° - (1200 \text{ kg})(32 \text{ km/h}) \cos 45°$$

$$v_f \cos \theta = 15.3 \text{ km/h} \qquad (1)$$

One extra figure has been retained in the answer.

In the y direction:

$$P_{fy} = P_{0y}$$

$$(m_1 + m_2)v_f \sin\theta = m_1 v_{01} \sin 35° - m_2 v_{02} \sin 45°$$

$$v_f \sin\theta = 7.33 \text{ km/h} \tag{2}$$

Again, one extra figure has been retained.

Division of (2) by (1) yields

$$\theta = \tan^{-1}(7.33 \text{ km/h}/15.3 \text{ km/h}) = 25.6°$$

Substitution of this angle into (2) gives the final speed of the cars to be $v_f = 17$ km/h. The angle that the final velocity makes with the x axis rounded to proper significant figures is $\theta = 26°$.

PRACTICE PROBLEMS

1. A batter hits a 0.14 kg ball which is initially traveling at a speed of 40.0 m/s. The ball loses 25% of its kinetic energy during the impact with the bat. What is the impulse of the force that the bat exerts on the ball?

2. A large ball has a mass which is ten times larger than a small ball. The small ball has speed which is fifteen times larger than the large ball's speed. Compare the momenta of the two balls. Which ball would be harder to stop in a given time period? Why?

3. Two 75 kg campers are in the middle of a creek in a 45 kg boat without a paddle. All they have on board the boat is a quantity of 0.500 kg cans of beans. One of the campers is able to throw a can with a speed of 26.8 m/s. How many cans of beans must the camper throw to give the boat a speed of 1.0 m/s? Ignore friction.

4. A 45 kg boy is standing on a 2.0 kg skateboard which is at rest on the ground. The boy jumps off the front of the skateboard with a horizontal speed of 0.50 m/s. What is the maximum possible speed of the skateboard? Which way does it move?

5. A 35 kg child runs with a speed of 4.5 m/s and "belly flops" onto a 7.0 kg snow sled. What is the speed of the sled and child immediately after the child comes to rest on the sled? Assume that no friction acts between the sled runners and the snow.

6. A 0.50 kg wooden block is dropped from a height of 15.0 m. At what height must a 5.0 g bullet traveling upward strike the block in order to momentarily stop its fall? Assume that the time required for the bullet to come to rest in the block is negligibly short and that the bullet travels with a speed of 320 m/s immediately before striking the block.

7. A 0.60 kg brick is thrown into a stationary 20.0 kg wagon with a velocity of 15.0 m/s directed 20.0° below the horizontal. What is the speed of the wagon and brick after the brick comes to rest in the wagon?

8. A 5.0 kg ball traveling with a speed of 5.0 m/s makes an elastic collision with a 3.0 kg ball which is traveling in the same direction with a speed of 1.5 m/s. What are the velocities of the balls immediately after the collision?

9. A 1500 kg car is traveling east with a speed of 85 km/h when it collides with a 1200 kg car traveling north with a speed of 65 km/h. The cars lock together during the collision. What is the velocity of the cars after the collision?

10. A pool ball moving with a speed of 2.5 m/s makes an elastic head-on collision with an identical ball traveling in the opposite direction with a speed of 5.9 m/s. Find the velocities of the balls after the collision.

SOLUTIONS AND ANSWERS

Practice Problems

1. The impulse of the force is equal to the change of momentum of the ball. In order to find the change in momentum of the ball we need to find its final speed.

$$1/2 \, mv_f^2 = (0.75)(1/2)mv_0^2,$$

gives the final speed of the ball to be

$$v_f = \sqrt{0.75} \; v_0.$$

If the direction of the ball's initial velocity is taken as positive, then the impulse of the force is

$$\textbf{Impulse} = m\textbf{v}_f - m\textbf{v}_0 = -mv_0(\sqrt{0.75} + 1) = -(0.14 \text{ kg})(40.0 \text{ m / s})(\sqrt{0.75} + 1) = \textbf{ -- 1.0} \times \textbf{10}^1 \textbf{ kg m / s.}$$

The negative sign indicates that the impulse is in a direction opposite the initial direction of travel of the ball.

---

2. The ball with the larger mass has a momentum $p_L = m_L v_L$ and the ball with the smaller mass has a momentum $p_S = m_S v_S$. Now

$$m_L = 10 \, m_S \text{ and } v_L = (1/15)v_S$$

so

$$\textbf{p}_L = \textbf{2/3 } \textbf{m}_S \textbf{v}_S = \textbf{2/3 } \textbf{p}_S.$$

The ball with the larger mass has the smaller momentum. **The smaller mass ball will be the hardest to stop** since the impulse-momentum theorem requires that more force is needed to produce a larger momentum change in the same time.

---

3. If friction is ignored, no external force acts on the boat in the horizontal direction, so the total momentum of the boat plus contents is conserved in that direction. After N cans of beans have been thrown with a horizontal speed of v_c the boat will have a speed of v_B. The conservation of momentum requires

$$(m - Nm_c)v_B - Nm_c v_c = 0$$

or

$$N = \frac{mv_B}{m_c(v_B + v_c)} = \frac{(195 \text{ kg})(1.0 \text{ m/s})}{(0.500 \text{ kg})(1.0 \text{ m/s} + 26.8 \text{ m/s})} = \textbf{ 14.}$$

---

4. If friction is ignored, no external forces act on the boy and skateboard in the horizontal direction. Total momentum is conserved in that direction. The conservation of momentum principle gives $m_B v_B + m_S v_S = 0$ or

$$v_S = -(m_B/m_S)v_B = -(45/2.0)(0.50 \text{ m/s}) = \textbf{ -- 11 m/s.}$$

The negative sign indicates that the skateboard moves in a direction opposite the boy.

---

5. The child and sled make a completely inelastic collision. The total momentum is conserved in the horizontal direction if friction is ignored. Then

$$(m_c + m_s)v = m_c v_{0c} \text{ or } (42 \text{ kg})v = (35 \text{ kg})(4.5 \text{ m/s}) \text{ so } v = \textbf{3.8 m/s.}$$

6. The mechanical energy of the block (object 1) is conserved as the block falls so

$$1/2\ m_1 v_1{}^2 + m_1 gh = m_1 g(15.0\ m)$$

or

$$h = 15.0\ m - 1/2\ v_1{}^2/g.$$

If the collision time is short, then the total momentum of the bullet (object 2) and block system will be approximately conserved since gravity cannot change it very much in a short time. Then

$$m_2 v_2 - m_1 v_1 = 0 \text{ or } v_1 = (m_2/m_1)v_2.$$

Using this result in the conservation of energy expression above yields

$$h = 15.0\ m - (1/2)(m_2/m_1)^2 v_2{}^2/g = 15.0\ m - (1/2)(5.0\ X\ 10^{-3}\ kg/0.50\ kg)^2(320\ m/s)^2/(9.80\ m/s^2) = \textbf{14.5 m}.$$

---

7. The brick and wagon make a totally inelastic collision. If friction between the wagon and ground is ignored, then no net force acts on the system in the horizontal direction so the total momentum is conserved in that direction.

$$(m_B + m_W)v = m_B v_B \cos 20.0°$$

or

$$(20.6\ kg)v = (0.60\ kg)(15.0\ m/s) \cos 20.0°$$

or

$$v = \textbf{0.41 m/s.}$$

---

8. Both the conservation of momentum and the conservation of kinetic energy apply. The former gives

$$(5.0\ kg)v_{f1} + (3.0\ kg)v_{f2} = (5.0\ kg)(5.0\ m/s) + (3.0\ kg)(1.5\ m/s)$$

$$5.0 v_{f1} + 3.0 v_{f2} = 29.5 \qquad (1)$$

The conservation of kinetic energy gives

$$1/2\ (5.0\ kg)v_{f1}{}^2 + 1/2\ (3.0\ kg)v_{f2}{}^2 = 1/2\ (5.0\ kg)(5.0\ m/s)^2 + 1/2\ (3.0\ kg)(1.5\ m/s)^2$$

$$2.5 v_{f1}{}^2 + 1.5 v_{f2}{}^2 = 65.9 \qquad (2)$$

Solving (1) for v_{f1} and substituting the result into (2) and rearranging gives

$$2.4 v_{f2}{}^2 - 17.7 v_{f2} + 21.1 = 0$$

The quadratic formula yields solutions of $v_{f2} = +1.5$ m/s and $\textbf{+ 5.9 m/s}$.
Substituting these values into (1) results in $v_{f1} = + 5.0$ m/s and $\textbf{+ 2.4 m/s}$.
The first of these values correspond to no collision at all so they are rejected.

---

9. The cars make a completely inelastic collision in which the total momentum is conserved if friction between the tires and the road is ignored. In the north-south (y) direction

$$m_1 v_{01y} = (m_1 + m_2) v_{fy}$$

or

$$v_{fy} = (1200/2700)(65 \text{ km/h}) = 29 \text{ km/h}$$

and in the east-west (x) direction

$$m_2 v_{02x} = (m_1 + m_2) v_{fx}.$$

or

$$v_{fx} = (1500/2700)(85 \text{ km/h}) = 47 \text{ km/h}.$$

The speed of the cars after the collision is found from the Pythagorean theorem,

$$v_f^2 = v_{fx}^2 + v_{fy}^2,$$

to be

$$v_f = \textbf{55 km/h}.$$

The direction that the velocity vector makes with east is

$$\theta = \tan^{-1}(29/47) = \textbf{32° north of east.}$$

---

10. The balls have the same mass, so we immediately know that **they will exchange speeds**. The directions of travel must be determined. Before the collision, the total momentum of the system is

$$P_0 = m(v_{01} + v_{02}) = m[+2.5 \text{ m/s} + (-5.9 \text{ m/s})] = -(3.4 \text{ m/s})m$$

which is in the direction of motion of the 5.9 m/s ball, say to the left. After the collision, the total momentum must be the same in magnitude and still to the left.

$$P_f = m(v_{f1} + v_{f2}) = -(3.4 \text{ m/s})m \text{ so } v_{f1} + v_{f2} = -3.4 \text{ m/s}$$

where the magnitude of v_{f1} is 5.9 m/s and the magnitude of v_{f2} is 2.5 m/s. The only way for these to be true is for $v_{f1} = \textbf{- 5.9 m/s}$ and $v_{f2} = \textbf{+ 2.5 m/s}$. That is, **the balls also exchange directions of motion.**

CHAPTER 8
Rotational Kinematics

PREVIEW

In this chapter we will discuss rotational motion. Basic quantities such as angular displacement, angular velocity and angular acceleration are introduced along with the kinematic equations of rotation. Centripetal and tangential accelerations are also studied in detail.

QUICK REFERENCE

Important Terms

Angular displacement
    The angle through which a rigid object rotates about a fixed axis. By convention, this quantity is positive if the rotation is counterclockwise, negative if it is clockwise. Units are radians (rad).

Radian
    An SI unit of angular displacement defined as the circular arc length s traveled by a point on a rotating body divided by the radial distance r of the point from the axis: $\theta = s/r$. Also, 2π radians = 1 revolution = $360°$.

Average angular velocity
    The angular displacement of an object divided by the elapsed time. Units are (rad/s).

Instantaneous angular velocity
    The angular velocity of an object at a given instant of time.

Average angular acceleration
    The change in angular velocity divided by the elapsed time. Units are (rad/s^2).

Tangential velocity
    The linear velocity of a point on a rotating body. It represents how fast the point is moving along the arc of the circle. Units are (m/s).

Tangential acceleration
    The linear acceleration of a point on a rotating body, measured in the tangential direction. Units are (m/s^2).

Uniform circular motion
    The motion of an object moving at constant tangential speed in a circular path.

Angular Quantities

Angular Displacement - the angle θ swept out when a rigid body rotates about a fixed axis. If s is the arc length swept out and r the radius, as shown in the diagram, we have

$$\theta \text{ (radians)} = \frac{\text{arc length}}{\text{radius}} = \frac{s}{r} \qquad (8.1)$$

The angle θ can be measured in revolutions, degrees, or radians. The number of radians that corresponds to one revolution or 360° is

$$\theta = \frac{s}{r} = \frac{2\pi r}{r} = 2\pi \text{ rad}$$

Average Angular Velocity - the angular displacement divided by the elapsed time. That is

$$\overline{\omega} = \frac{\theta - \theta_0}{t - t_0} = \frac{\Delta\theta}{\Delta t} \qquad (8.2)$$

Instantaneous Angular Velocity - the angular velocity at a given instant of time. That is

$$\omega = \lim_{\Delta t \to 0} \overline{\omega} = \lim_{\Delta t \to 0} \frac{\Delta\theta}{\Delta t} \qquad (8.3)$$

Average Angular Acceleration - the change in angular velocity divided by the elapsed time. That is

$$\overline{\alpha} = \frac{\omega - \omega_0}{t - t_0} = \frac{\Delta\omega}{\Delta t} \qquad (8.4)$$

Equations of Rotational Kinematics

If the **angular acceleration is constant** (uniform), the following equations are valid for rotational motion:

$$\omega = \omega_0 + \alpha t \qquad (8.4) \qquad\qquad \theta = \omega_0 t + \frac{1}{2}\alpha t^2 \qquad (8.7)$$

$$\theta = \frac{1}{2}(\omega_0 + \omega)t \qquad (8.6) \qquad\qquad \omega^2 = \omega_0^2 + 2\alpha\theta \qquad (8.8)$$

Note the similarity of these equations to the equations of kinematics for linear motion found in Chapter 2.

Angular and Tangential Variables and Centripetal Acceleration

Tangential velocity is given by: $v_T = r\omega$ (8.9)

Tangential acceleration is given by: $a_T = r\alpha$ (8.10)

Centripetal acceleration is given by:

$$a_c = \frac{v_T^2}{r} = \frac{(r\omega)^2}{r} = r\omega^2 \qquad (8.11)$$

In **nonuniform circular motion**, the centripetal and tangential accelerations are perpendicular to one another. The magnitude of the total acceleration **a**, and the angle between **a** and $\mathbf{a_c}$, ϕ, can then be expressed as

$$a = \sqrt{a_c^2 + a_T^2} \quad \text{while} \quad \tan\phi = \frac{a_T}{a_c}$$

Rolling Motion

A point on an object rolling without slipping has a speed

$$v = v_T = r\omega \tag{8.12}$$

and an acceleration

$$a = a_T = r\alpha \tag{8.13}$$

DISCUSSION OF SELECTED SECTIONS

8.1 Rotational Motion and Angular Displacement

Example 1
Find the angle subtended by the moon as seen from earth. The moon's diameter is 3.48×10^6 m and its distance from earth is 3.85×10^8 m. Express your answer in radians and degrees.

Using equation (8.1) with $s = 3.48 \times 10^6$ m and $r = 3.85 \times 10^8$ m, we have

$$\theta = \frac{s}{r} = \frac{3.48 \times 10^6\, m}{3.85 \times 10^8\, m} = 9.04 \times 10^{-3}\, rad$$

$$9.04 \times 10^{-3}\, rad \left(\frac{360°}{2\pi\, rad}\right) = 0.52°.$$

8.2 Angular Velocity and Angular Acceleration

The following examples make use of the definitions of angular velocity and angular acceleration.

Example 2
Find the average angular velocity of the earth as it rotates on its axis. Express your answer in rad/s.

We know that the earth makes one revolution per day. We must convert "revolutions" to the unit of angular displacement, "radians", and convert "days" to "seconds". Using equation (8.2) we can write:

$$\overline{\omega} = \frac{\Delta\theta}{\Delta t} = \left(\frac{1\, rev}{day}\right)\left(\frac{2\pi\, rad}{rev}\right)\left(\frac{1\, day}{24\, h}\right)\left(\frac{1\, h}{3600\, s}\right) = 7.27 \times 10^{-5}\, rad/s.$$

Example 3
A phonograph turntable changes from 33.3 rev/min to 78.0 rev/min in 3.00 s. What is the average angular acceleration? Express your answer in rad/s^2.

Using equation (8.4):

$$\omega_0 = \left(33.3 \frac{\text{rev}}{\text{min}}\right)\left(\frac{2\pi \text{ rad}}{1 \text{ rev}}\right)\left(\frac{1 \text{ min}}{60 \text{ s}}\right) = 3.49 \text{ rad/s}$$

$$\omega = \left(78.0 \frac{\text{rev}}{\text{min}}\right)\left(\frac{2\pi \text{ rad}}{1 \text{ rev}}\right)\left(\frac{1 \text{ min}}{60 \text{ s}}\right) = 8.17 \text{ rad/s}$$

$$\alpha = \frac{\Delta\omega}{\Delta t} = \frac{\omega - \omega_0}{\Delta t} = \frac{8.17 \text{ rad/s} - 3.49 \text{ rad/s}}{3.00 \text{ s}} = 1.56 \text{ rad/s}^2.$$

8.3   The Equations of Rotational Kinematics

Equations (8.4), (8.6), (8.7), and (8.8) are the equations for rotational kinematics for constant angular acceleration. Note their similarity to the equations of kinematics for linear motion. In fact, the equations are identical if the following linear quantities are replaced by their rotational analogs:

Quantity	Linear Motion	Rotational Motion
Displacement	x	θ
Initial velocity	v_0	ω_0
Final velocity	v	ω
Acceleration	a	α
Time	t	t

Example 4

If it takes 20.0 s for the shaft of a motor to start from rest and attain its full speed of 1800 rev/min, what is the magnitude of the average angular acceleration of the shaft? What is the angular displacement of the shaft?

$$\omega = \omega_0 + \alpha t \qquad\qquad\qquad\qquad (8.4)$$

$$\omega_0 = 0$$

$$\omega = \left(1800 \frac{\text{rev}}{\text{min}}\right)\left(\frac{1 \text{ min}}{60 \text{ s}}\right)\left(\frac{2\pi \text{ rad}}{1 \text{ rev}}\right) = 188 \text{ rad/s}$$

$$t = 20.0 \text{ s}$$

$$\alpha = \frac{188 \text{ rad/s}}{20.0 \text{ s}} = 9.40 \text{ rad/s}^2$$

$$\theta = \omega_0 t + \frac{1}{2}\alpha t^2 = \frac{1}{2}(9.40 \text{ rad/s}^2)(20.0 \text{ s})^2 = 1880 \text{ rad}$$

Example 5

A grinding wheel has an initial angular velocity of 215 rad/s when the driving motor is shut off. After 10.0 s, its angular velocity is 125 rad/s. What is the angular acceleration of the wheel? What is the angular displacement during the ten second interval? Assuming the angular acceleration remains constant, find the total number of revolutions the wheel turns through in coming to rest.

$$\omega = \omega_0 + \alpha t \implies \alpha = \frac{\omega - \omega_0}{t} = \frac{125 \text{ rad/s} - 215 \text{ rad/s}}{10.0 \text{ s}} = -9.0 \text{ rad/s}^2$$

$$\theta = \omega_0 t + \frac{1}{2}\alpha t^2 = (215 \text{ rad/s})(10.0 \text{ s}) + \frac{1}{2}(-9.0 \text{ rad/s}^2)(10.0 \text{ s})^2 = 1.70 \times 10^3 \text{ rad}$$

$$\omega^2 = \omega_0^2 + 2\alpha\theta' \implies \theta' = \frac{\omega^2 - \omega_0^2}{2\alpha} = \frac{-(215 \text{ rad/s})^2}{2(-9.0 \text{ rad/s}^2)} = 2600 \text{ rad}\left(\frac{1 \text{ rev}}{2\pi \text{ rad}}\right) = 410 \text{ rev.}$$

8.4 Angular Variables and Tangential Variables

Consider two people standing on a moving merry-go-round; one person near the center and one near the outer edge of the merry-go-round. Both people experience the same angular speed ω, since the merry-go-round is a rigid body rotating about a fixed axis. However, the person on the outer edge is moving faster, since he is sweeping out a larger arc length in a given amount of time. The arc length traveled by each person, Δs, is given by $\Delta s = r\Delta\theta$, so that the linear speed is

$$v = \frac{\Delta s}{\Delta t} = r\frac{\Delta\theta}{\Delta t} = r\omega.$$

Therefore, the greater the value of r, the greater the linear or **tangential** speed. A similar expression exists for the tangential acceleration. The relationships between v_T and ω, and a_T and α are given by

$$v_T = r\omega \qquad (8.9) \qquad\qquad\qquad a_T = r\alpha \qquad (8.10)$$

A word of caution: The expression "$s = r\theta$" is valid only when the angle θ is expressed in radians. Any expression or equation derived from "$s = r\theta$" has the same restrictions. Therefore, in using equations (8.9) and (8.10), all angular quantities (ω and α) must be expressed in radians; that is ω must be expressed in rad/s and α must be expressed in rad/s^2.

Example 6

A large disk is rotating with an angular speed of 225 rad/s. Find the tangential speed at a distance of 2.0 m from the center of the disk. If the disk uniformly speeds up to 450 rad/s in 25 s, what is the tangential acceleration of this point?

Using equation (8.9) we have:

$$v_T = r\omega = (2.0 \text{ m})(225 \text{ rad/s}) = 450 \text{ m/s.}$$

In order to determine the tangential acceleration, we need to find the angular acceleration.

Using equation (8.4) we have:

$$\alpha = (\omega - \omega_0)/t = (450 \text{ rad/s} - 225 \text{ rad/s})/(25 \text{ s}) = 9.0 \text{ rad/s}^2.$$

So the tangential acceleration is:

$$a_T = r\alpha = (2.0 \text{ m})(9.0 \text{ rad/s}^2) = 18 \text{ m/s}^2.$$

8.5   Centripetal Acceleration and Tangential Acceleration

Suppose you tie a rock to the end of a string and spin it in a circle. The string holding the rock is exerting a force inward toward your hand. This force is called a **centripetal force**. Since the rock is spinning in a circle, and thus constantly changing direction, it is being accelerated by this centripetal force. This acceleration is directed INWARD, in the same direction as the force, and is known as the **centripetal acceleration**, a_c. The magnitude of this acceleration depends on how fast you spin the rock and how long the string is. That is,

$$a_c = \frac{v_T^2}{r} = \frac{(r\omega)^2}{r} = r\omega^2 \qquad (8.11)$$

Example 7
The moon revolves around the earth, making a complete revolution in 27.3 days. Assuming that the moon's orbit is a circle of radius 3.85 x 10^8 m, what is the magnitude of the acceleration of the moon toward the earth?

The moon's angular velocity is:

$$\omega = \Delta\theta/\Delta t = (2\pi \text{ rad}/27.3 \text{ d})(1 \text{ d}/24 \text{ h})(1 \text{ h}/3600 \text{ s})$$
$$\omega = 2.66 \text{ x } 10^{-6} \text{ rad/s}$$

So that the centripetal acceleration is:

$$a_c = r\omega^2 = (3.85 \text{ x } 10^8 \text{ m})(2.66 \text{ x } 10^{-6} \text{ rad/s})^2$$
$$a_c = 2.72 \text{ x } 10^{-3} \text{ m/s}^2.$$

8.6   Rolling Motion

As an automobile moves along the street, its tires roll without slipping. When no slippage occurs, we refer to the motion as rolling motion. The following equations are valid under these conditions:

$$v = v_T = r\omega \quad (8.12) \qquad\qquad a = a_T = r\alpha \quad (8.13)$$

Example 8
A wheel rolls without slipping for a linear distance of 628 m. If the diameter of the wheel is 1.00 m, how many revolutions does the wheel make in this time?

Using equation (8.1) we have:

$$\theta = s/r = (628 \text{ m})/(0.500 \text{ m})$$
$$\theta = 1260 \text{ rad}(1 \text{ rev}/2\pi \text{ rad})$$
$$\theta = 200 \text{ rev}.$$

Example 9
A car accelerates uniformly from rest and reaches a velocity of 60.0 mi/h in 7.0 s. If the wheels have a diameter of 24 inches, (a) how many revolutions do the tires make? (b) What is the angular acceleration of the wheels?

(a) The final speed of the car is v = (60.0 mi/h)(5280 ft/1 mi)(1 h/3600 s) = 88 ft/s. This corresponds to the tangential velocity of a point on the edge of the tire. The final angular velocity of a tire is therefore, using equation (8.12),

$$\omega = v/r = (88 \text{ ft/s})/(1.0 \text{ ft}) = 88 \text{ rad/s}.$$

Using equation (8.6),

$$\theta = (1/2)(\omega_0 + \omega)t$$

$$\theta = (1/2)(88 \text{ rad/s})(7.0 \text{ s})$$

$$\theta = 310 \text{ rad}(1 \text{ rev}/2\pi \text{ rad})$$

$$\theta = 49 \text{ rev}.$$

(b) Using equation (8.4),

$$\omega = \omega_0 + \alpha t$$

which gives

$$\alpha = \omega/t = (88 \text{ rad/s})/(7.0 \text{ s}) = 13 \text{ rad/s}^2.$$

Note that this last result can also be obtained using equations (2.4) and (8.13), i.e.,

$$v = v_0 + at$$

which gives

$$a = v/t = (88 \text{ ft/s})/(7.0 \text{ s}) = 13 \text{ ft/s}^2.$$

then

$$\alpha = a/r = (13 \text{ ft/s}^2)/(1 \text{ ft}) = 13 \text{ rad/s}^2.$$

PRACTICE PROBLEMS

1. An electric fan is set on its HIGH setting. After the LOW push button is depressed, the angular speed of the fan blades decreases to a value of 8.00×10^2 rev/min in 1.75 s. The deceleration is 42.0 rad/s^2. Determine the initial angular speed of the blades in rev/min.

2. An electric grinder uses a grinding wheel of 12-cm radius. It takes the electric motor 3.0 s to reach its rated speed of 1500 rev/min starting from rest. Find (a) the angular acceleration, (b) the total angular displacement, and (c) the tangential velocity at the rim of the grinding wheel at the rated speed.

3. The high speed drill of a dentist is driven by an air turbine and achieves a rotational speed of 450 000 rev/min. The radius of the cutting burr is 1.0 mm. (a) Find the linear speed of the outer edge of the cutting burr. (b) Determine the acceleration of the outer edge and express this as a multiple of the gravitational acceleration, g.

4. A small bug sits on a turntable, 10.0 cm from the center. What is the centripetal acceleration of the bug when the turntable rotates at 45 rev/min? What is the bug's tangential velocity?

5. A space station is constructed in the shape of a doughnut whose outer radius is 480 m and whose inner radius is 410 m. At what rate should the space station rotate so that its occupants will experience an acceleration of magnitude g when they are located at the outer diameter of the station? By what fraction would a person's weight change if they stood at the inner diameter of the station?

6. An automobile traveling 30.0 m/s has wheels of 0.750 m diameter. (a) What is the angular speed of the wheels about the axle? (b) If the car is brought to a stop uniformly in 10.0 s, what is the angular acceleration? (c) How many times do the wheels turn during this braking period?

7. Wheel A has a radius $r_A = 75$ cm and is coupled to wheel B, radius $r_B = 25$ cm, by belt C. Wheel A accelerates from rest at an angular acceleration of 2.0 rad/s^2. How long does it take for wheel B to reach a rotational speed of 1200 rad/s assuming that the belt does not slip?

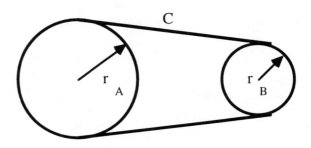

8. One end of a string is attached to the rim of a wheel 0.50 ft in radius. A small object hangs from the string as shown in the diagram below. If the wheel is given an initial angular velocity of 9.0 rad/s "into the page" and the constant angular acceleration of the wheel is 3.0 rad/s^2 "out of the page" (a) how many turns does the wheel make before coming momentarily to rest? (b) What length of string is then wrapped around the wheel? (c) At what time will the string again be entirely unwrapped from the wheel?

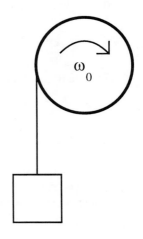

SOLUTIONS AND ANSWERS

Practice Problems

1.
$$\omega = (8.00 \times 10^2 \text{ rev/min})(1 \text{ min}/60 \text{ s})(2\pi \text{ rad}/1 \text{ rev})$$
$$\omega = 83.8 \text{ rad/s},$$

Using equation (8.4),

$$\omega_0 = \omega - \alpha t = (83.8 \text{ rad/s}) - (- 42.0 \text{ rad/s}^2)(1.75 \text{ s})$$
$$\omega_0 = 157 \text{ rad/s}(1 \text{ rev}/2\pi \text{ rad})(60 \text{ s}/1 \text{ min})$$
$$\omega_0 = \textbf{1500 rev/min.}$$

---

2. $\omega = 1500 \text{ rev/min} = 160 \text{ rad/s}, \omega_0 = 0, t = 3.0 \text{ s}, r = 0.12 \text{ s}.$

(a) Using equation (8.4) we have,

$$\alpha = (\omega - \omega_0)/t = (160 \text{ rad/s})/(3.0 \text{ s}) = \textbf{53 rad/s}^2.$$

(b) Using equation (8.7) we have,

$$\theta = \omega_0 t + (1/2)\alpha t^2 = (1/2)(53 \text{ rad/s}^2)(3.0 \text{ s})^2 = \textbf{240 rad.}$$

(c) Using equation (8.9) we have,
$$v_T = r\omega = (0.12 \text{ m})(160 \text{ rad/s}) = \textbf{19 m/s.}$$

---

3. $\omega = (4.5 \times 10^5 \text{ rev/min})(1 \text{ min}/60 \text{ s})(2\pi \text{ rad}/1 \text{ rev}) = 4.7 \times 10^4 \text{ rad/s}, r = 1.0 \times 10^{-3} \text{ m}.$

(a) Using equation (8.9) we have,

$$v_T = r\omega = (1.0 \times 10^{-3} \text{ m})(4.7 \times 10^4 \text{ rad/s}) = \textbf{47 m/s.}$$

(b) Using equation (8.11) we have

$$a_c = r\omega^2 = (1.0 \times 10^{-3} \text{ m})(4.7 \times 10^4 \text{ rad/s})^2 = 2.2 \times 10^6 \text{ m/s}^2.$$

To obtain the acceleration in terms of gravity, divide a_c by 9.8 m/s^2 to get

$$a_c = \textbf{2.3 x } \textbf{10}^5 \textbf{ g.}$$

---

4. $\omega = (45 \text{ rev/min})(2\pi \text{ rad}/1 \text{ rev})(1 \text{ min}/60 \text{ s}) = 4.7 \text{ rad/s}, r = 0.10 \text{ m}.$

Using equation (8.11) we can find the centripetal acceleration,

$$a_c = r\omega^2 = (0.10 \text{ m})(4.7 \text{ rad/s})^2 = \textbf{2.2 m/s}^2.$$

The tangential velocity can be obtained using equation (8.9),

$$v_T = r\omega = (0.10 \text{ m})(4.7 \text{ rad/s}) = \textbf{0.47 m/s.}$$

---

5. The centripetal acceleration is used to simulate gravity. Using equation (8.11): $a_c = r\omega^2 = 9.8$ m/s^2 we can solve for ω, i.e,

$$\omega = \sqrt{\frac{a_c}{r}} = \sqrt{\frac{9.8 \text{ m}/\text{s}^2}{480 \text{ m}}} = 0.14 \text{ rad}/\text{s} = \textbf{1.3 rev}/\textbf{min.}$$

At r = 410 m,

$$a_c = r\omega^2 = (410 \text{ m})(0.14 \text{ rad/s})^2 = 8.0 \text{ m/s}^2.$$

The fractional weight is then

$$(8.0 \text{ m/s}^2)/(9.8 \text{ m/s}^2) = \textbf{0.82.}$$

---

6. (a) Using equation (8.12) we have,

$$\omega = v/r = (30.0 \text{ m/s})/(0.375 \text{ m}) = \textbf{8.00 x 10}^{\textbf{1}} \textbf{ rad/s.}$$

(b) Using $\omega_0 = 80.0$ rad/s, t = 10.0 s, $\omega = 0$ and equation (8.4) we have,

$$\alpha = (\omega - \omega_0)/t = \textbf{- 8.00 } \textbf{ rad/s}^2.$$

(c) Using equation (8.6),

$$\theta = (1/2)(\omega_0 + \omega)t = (1/2)(80.0 \text{ rad/s})(10.0 \text{ s}) = 4.00 \text{ x } 10^2 \text{ rad} = \textbf{63.7 rev.}$$

---

7. In this situation, the tangential acceleration at the rim of each wheel must be the same. Using equation (8.10)

$$a_T = r_A \alpha_A = r_B \alpha_B.$$

Therefore,

$$\alpha_B = \alpha_A(r_A/r_B) = (2.0 \text{ rad/s}^2)(75 \text{ cm}/25 \text{ cm}) = 6.0 \text{ rad/s}^2.$$

We can now use equation (8.4) with $\omega_0 = 0$ and $\omega = 1200$ rad/s, to find

$$t = (\omega - \omega_0)/\alpha = (1200 \text{ rad/s})/(6.0 \text{ rad/s}^2) = \textbf{2.0 x 10}^2 \textbf{ s.}$$

---

8. (a) Use $\omega_0 = 9.0$ rad/s, $\alpha = -3.0$ rad/s^2, $\omega = 0$ and equation (8.8): $\omega^2 = \omega_0^2 + 2\alpha\theta$. Solving for θ we get,

$$\theta = -\omega_0^2/2\alpha = -(9.0 \text{ rad/s})^2/[2(-3.0 \text{ rad/s}^2) = 13.5 \text{ rad}(1 \text{ rev}/2\pi \text{ rad}) = \textbf{2.1 rev.}$$

(b) Using equation (8.1),

$$s = r\theta = (0.50 \text{ ft})(13.5 \text{ rad}) = \textbf{6.8 ft.}$$

(c) Using $\theta = -13.5$ rad, $\omega_0 = 0$, and $\alpha = -3.0$ rad/s^2, equation (8.7) gives for t,

$$t = \sqrt{\frac{2\theta}{\alpha}} = \textbf{3.0 s.}$$

CHAPTER 9
Rotational Dynamics

PREVIEW

In this chapter you will learn about the dynamics of rotational motion. The topics include Newton's second law for rotations, moment of inertia, rotational work, rotational kinetic energy, and angular momentum.

QUICK REFERENCE

Important Terms

Axis of rotation
 A line of points about which an object rotates.
Line of action
 A line drawn colinear with a force.
Lever arm
 The shortest distance from a rotation axis to a line of action of a force.
Torque
 A vector quantity whose magnitude is the magnitude of the force times the lever arm and whose direction is positive if it tends to cause a counterclockwise rotation and is negative if it tends to cause a clockwise rotation.
Translational motion
 The motion of an object in which all points of the object move in parallel paths.
Rotational motion
 The motion of a rigid object in which the points of the object move in circles about some axis.
Equilibrium
 The state of an object when it has no translational or rotational acceleration. It follows that no net torque and no net force act on an object in equilibrium.
Center of gravity (of a rigid body)
 The point at which the weight of the body can be assumed to act when calculating the torque due to the weight of the body.
Moment of inertia of an object
 A rotational quantity which plays the same role as mass in linear motion. The net torque acting on an object is proportional to the angular acceleration of the object. The constant of proportionality is the moment of inertia.
Rotational work
 The work done by a torque in rotating an object through an angle.
Rotational kinetic energy
 The total kinetic energy of all parts of a rigid object due to their rotational motion about an axis. It is equal to one-half the moment of inertia of the object times the angular speed of the object squared.
Angular momentum (of a rigid object)
 The product of the moment of inertia of the object and its angular velocity.

Equations

The magnitude of the **torque** about an axis is

$$\tau = F\ell \qquad\qquad (9.1)$$

where F is the magnitude of the force and ℓ is the lever arm.

A rigid body is in **equilibrium** when

$$\Sigma F_x = 0 \qquad \Sigma F_y = 0 \qquad \text{(4.9a and 4.9b)}$$

$$\Sigma \tau = 0 \qquad \text{(9.2)}$$

The coordinate of the **center of gravity** of a collection of objects which lie along a line is

$$x_{cg} = \frac{W_1 x_1 + W_2 x_2 + \cdots}{W_1 + W_2 + \cdots} \qquad \text{(9.3)}$$

Newton's second law for rotations
for a single mass
$$\tau = (mr^2)\alpha \qquad \text{(9.4)}$$

for a rigid collection of masses
$$\Sigma \tau = (\Sigma mr^2)\alpha \qquad \text{(9.5)}$$

$$\Sigma \tau = I\alpha \qquad \text{(9.7)}$$

Moment of inertia of a rigid body composed of many masses

$$I = \Sigma mr^2 \qquad \text{(9.6)}$$

Rotational Work

$$W_R = \tau\theta \qquad \text{(9.8)}$$

Rotational Kinetic Energy of a rigid body

$$KE_R = 1/2\, I\omega^2 \qquad \text{(9.9)}$$

Angular Momentum of a rigid body

$$L = I\omega \qquad \text{(9.10)}$$

Principles

The **principle of conservation of angular momentum** - the total angular momentum of a system remains constant if the net external torque acting on the system is zero.

DISCUSSION OF SELECTED SECTIONS

9.1 The Effects of Forces and Torques on the Motion of Rigid Objects

Torque is a vector quantity and as such can point in any direction. Many rotations caused by torques can, however, be described as either clockwise (CW) or counterclockwise (CCW) as viewed by someone looking along the axis of rotation. This is why it is convenient to classify torques as positive if they tend to produce a counterclockwise rotation or negative if they tend to produce a clockwise rotation of the object.

In order to calculate a torque due to a force using (9.1) you will need to know the magnitude of the force and the lever arm. The lever arm can usually be found from the geometry of the problem.

Example 1
A person pushes on the edge of a 0.90 m wide door with a horizontal force of 5.0 N acting at an angle of 35° to the plane of the door. What is the torque of the force about the door hinge?

The lever arm is

$$\ell = W \sin 35°$$

so the torque is

$$\tau = F\ell$$
$$\tau = FW \sin 35°$$
$$\tau = (5.0 \text{ N})(0.90 \text{ m}) \sin 35°$$
$$\tau = 2.6 \text{ N m.}$$

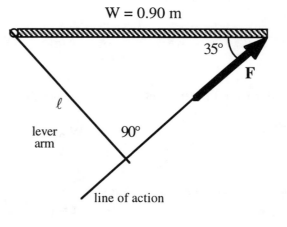

The torque is POSITIVE since the door would rotate in a counterclockwise sense if it could.

Note that the same result could have been obtained if we had found the component of the force perpendicular to the door and multiplied it by the distance from the hinge to the point where the force acts. That is,

$$\tau = (F \sin 35°)W$$

Sometimes this alternative way of calculating torques is more convenient than finding the lever arm.

9.2 Rigid Objects in Equilibrium

In solving problems involving rigid objects in equilibrium, it is usually necessary to generate a number of equations to be solved simultaneously. These equations can be derived by an application of Newton's second law for translation and rotation. That is, $\Sigma F_x = 0$, $\Sigma F_y = 0$ and $\Sigma \tau = 0$. The force equations are valid only in perpendicular directions and yield two equations at most. The torque equation, however, is valid about ANY axis and will yield at least one useful equation.

Often a judicious choice of rotation axis will result in a simple torque equation which will allow you to solve for one unknown immediately. As a general rule, try to write torque equations about axes which will eliminate torques due to any forces which are not of interest.

Example 2
A clothesline pole requires a guy wire to keep it in from being pulled over by the tension in the line. If the guy wire makes an angle of 45° to the horizontal and is attached 1.5 m above the ground, and the tension in the line is 550 N and it is attached 2.0 m above the ground, how much tension is required in the guy wire?

At first glance, this problem appears to require only a simple application of Newton's second law in the horizontal direction. However, the ground exerts forces on the pole which are completely unknown. A choice of axis through the point where the pole contacts the ground will eliminate these forces from the torque equation and will allow us to solve for the desired tension in the guy wire.

As shown by the free-body diagram, the lever arm of the tension in the line is simply 2.0 m and the lever arm of the torque due to the tension in the guy wire, T, is $\ell = (1.5 \text{ m}) \sin 45°$.

The torque equation is then

$$\Sigma\tau = 0$$

$$\Sigma\tau = + T(1.5 \text{ m}) \sin 45° - (550 \text{ N})(2.0 \text{ m}) = 0$$

$$T = 1.0 \times 10^3 \text{ N}$$

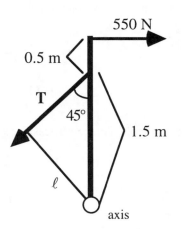

Example 3
Find the forces exerted by the ground on the pole in the previous example, if the pole has a weight of 120 N.

Newton's second law applied to the horizontal forces acting on the pole gives

$$F_h + 550 \text{ N} - (1.0 \times 10^3 \text{ N}) \sin 45° = 0$$

$$F_h = 160 \text{ N}$$

A similar application in the vertical direction gives

$$F_v - 120 \text{ N} - (1.0 \times 10^3 \text{ N}) \cos 45° = 0$$

$$F_v = 830 \text{ N}$$

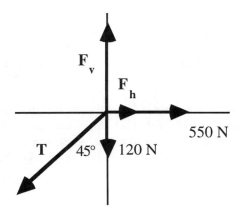

Example 4
A scaffold is constructed by placing two supports located a distance of 12 ft apart under a 16 ft board. Two painters stand on the board; one weighs 150 lb and stands 4.0 ft from one end of the board. Where must the other 180 lb painter stand so that each support carries the same weight? Neglect the weight of the board.

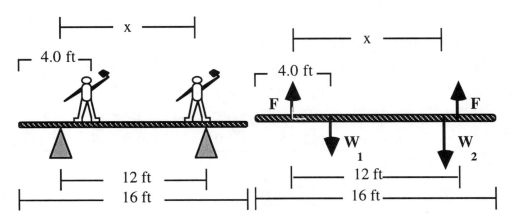

A torque equation written for an axis through the contact point between the board and the left support gives

$$(12 \text{ ft})F - (2.0 \text{ ft})W_1 - xW_2 = 0$$

$$x = (12 \text{ ft})(F/W_2) - (2.0 \text{ ft})(W_1/W_2) \tag{1}$$

Newton's second law applied in the vertical direction gives

$$2F - W_1 - W_2 = 0$$

$$F = 1/2 \, (W_1 + W_2) \tag{2}$$

Substitution of (2) into (1) yields

$$x = (4.0 \text{ ft})(W_1/W_2) + 6.0 \text{ ft}$$

$$x = (4.0 \text{ ft})(150/180) + 6.0 \text{ ft}$$

$$x = 9.3 \text{ ft}$$

The 180 lb man should stand 9.3 ft to the right of the left support.

Example 5
A 3.0 m flag pole with a mass of 2.0 kg is attached to a wall by a rigid support and makes an angle of 20.0° to the horizontal as shown. A 12 kg flag is hung so that its weight acts at a point 2.5 m from the support. In order for the flag pole to be in equilibrium, an additional force must be supplied by a slight bending of the pole. If the magnitude of this force, assumed to be perpendicular to the pole, can be at most 175 N, where is it effectively located?

The forces exerted by the wall on the flag pole, F_v and F_h, are unknown and unwanted. A good place to start is to write a torque equation about an axis through the point of contact of the pole with the wall so that these forces are eliminated.

$$Tx - W_1L_1 - W_2L_2 = 0$$

$$x = \frac{W_1L_1 + W_2L_2}{T} = \frac{(2.0 \text{ kg})(9.8 \text{ m/s}^2)(1.5 \text{ m})\cos 20.0° + (12 \text{ kg})(9.8 \text{ m/s}^2)(2.5 \text{ m})\cos 20.0°}{175 \text{ N}}$$

$$x = 1.7 \text{ m}$$

9.3 Center of Gravity

The center of gravity of an object is a point where the weight of an object can be assumed to act. The position of the center of gravity of a collection of masses arranged along a line is given by equation (9.3). The center of gravity of a group of masses lying in a plane can be found by applying (9.3) in each of the two directions, x and y.

Example 6
The center of gravity of an empty 1200 kg airplane is located 2.0 m behind the pilot's seat. When loaded, the airplane has a 75 kg pilot and a 55 kg passenger in a seat beside the pilot's seat and 95 kg of luggage located 1.0 m behind the pilot's seat. Where along the midline of the airplane is the center of gravity of the loaded airplane?

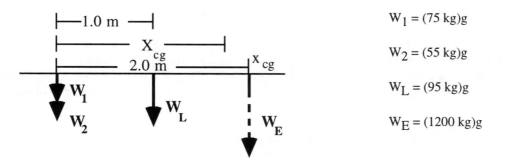

$$W_1 = (75 \text{ kg})g$$

$$W_2 = (55 \text{ kg})g$$

$$W_L = (95 \text{ kg})g$$

$$W_E = (1200 \text{ kg})g$$

Let the pilot's seat be the origin of the coordinate system. Equation (9.3) gives

$$X_{cg} = \frac{(75 \text{ kg})(0 \text{ m}) + (55 \text{ kg})(0 \text{ m}) + (95 \text{ kg})(1.0 \text{ m}) + (1200 \text{ kg})(2.0 \text{ m})}{75 \text{ kg} + 55 \text{ kg} + 95 \text{ kg} + 1200 \text{ kg}}$$

$$X_{cg} = 1.8 \text{ m}$$

Note that the common factor, g, appearing in the numerator and denominator was canceled. The center of gravity of the loaded airplane is 1.8 m behind the pilot's seat.

9.4 Newton's Second Law for Rotational Motion About a Fixed Axis

The application of Newton's second law for rotations is very similar to the procedures which you have already learned for using Newton's second law for linear motion. The main difference is that the moment of inertia of the system in question must usually be determined either from a table, such as table 9.1 in your text, or from equation (9.6).

Example 7
Three masses are rigidly located in the x-y plane as shown in the figure. What is the moment of inertia of the system about an axis through the origin and perpendicular to the x-y plane? What is the moment of inertia about the x axis?

$$I = \Sigma mr^2 = m_1r_1^2 + m_2r_2^2 + m_3r_3^2$$

For an axis through the origin the distances from the axis can be found by using the Pythagorean theorem.

$I = (2.0 \text{ kg})(13 \text{ m}^2) + (1.5 \text{ kg})(17 \text{ m}^2)$
$\quad\quad + (1.0 \text{ kg})(5.0 \text{ m}^2)$

$I = 57 \text{ kg m}^2$

The moment of inertia about the x axis is

$I = (2.0 \text{ kg})(4.0 \text{ m}^2) + (1.5 \text{ kg})(1.0 \text{ m}^2)$
$\quad\quad + (1.0 \text{ kg})(4.0 \text{ m}^2)$

$I = 14 \text{ kg m}^2$

$m_1 = 2.0 \text{ kg}$
$m_2 = 1.0 \text{ kg}$
$m_3 = 1.5 \text{ kg}$

Distances in meters

Example 8

A mechanic is "spin balancing" an automobile tire, radius R = 0.36 m, by placing it in a machine which applies a torque to accelerate the tire to a speed equivalent to traveling 95 km/h. The tire has a mass of 32 kg and has a moment of inertia equivalent to a hoop of radius 0.31 m. How much torque must the machine apply to the tire to bring it up to speed in 15 s?

Newton's second law for rotations requires that $\Sigma\tau = I\alpha$. The angular acceleration, α, can be found from rotational kinematics.

$$\alpha = \frac{\omega - \omega_0}{t} = \frac{\omega}{t}$$

A point on the rim of the tire mounted on a car traveling at a speed, v, will move about the axle with a speed v. The angular velocity of the wheel is then $\omega = v/R$ so

$$\alpha = \frac{v}{Rt}$$

The moment of inertia of the equivalent hoop is $I = Mr^2$, so the torque needed to accelerate the tire is

$$\Sigma\tau = (Mr^2)\left(\frac{v}{Rt}\right) = \frac{Mr^2 v}{Rt}$$

$$\Sigma\tau = \frac{(32 \text{ kg})(0.31 \text{ m})^2(95 \text{ km/h})(10^3 \text{ m/1 km})(1 \text{ h/3600 s})}{(0.36 \text{ m})(15 \text{ s})} = 15 \text{ N m}$$

9.5 Rotational Work and Energy

Example 9

It has been suggested that an automobile can be powered for short trips by removing energy from a spinning flywheel which has been "charged" overnight by a small electric motor. How much energy can be stored in a solid cylindrical flywheel of radius 0.5 m and mass 50 kg rotating 10 000 rpm? If the car normally requires an average of 50 hp, how long could it run if it can extract 50 % of the energy from the flywheel?

The kinetic energy of rotation of the flywheel is $KE_R = 1/2 \, I\omega^2$ where $I = 1/2 \, MR^2$ so

$$KE_R = 1/4 \, MR^2\omega^2$$

$$KE_R = (1/4)(50 \text{ kg})(0.5 \text{ m})^2[(1 \text{ X } 10^4 \text{ rev/min})(1 \text{ min/60 s})(2\pi \text{ rev}^{-1})]^2$$

$$KE_R = 3 \text{ X } 10^6 \text{ J}$$

The car has $0.5 \, KE_R$ energy available for doing work, so it can run a time $t = W/P = 0.5 \, KE_R/P$. Hence,

$$t = \frac{(0.5)(3 \text{ X } 10^6 \text{ J})}{(50 \text{ hp})(746 \text{ W/1 hp})} = 40 \text{ s}$$

9.6 Angular Momentum

The angular momentum of a system of objects is conserved if no net external torque acts on the system.

Example 10

A centrifugal clutch consists of an outer cylindrical shell mounted on a shaft and an independent inner mechanism mounted on a shaft colinear with the shell's. As the angular speed of the inner mechanism increases, it expands ultimately engaging the outer shell. After a short time the two pieces rotate with the same angular speed. Assume that the outer shell has a moment of inertia one-half that of the inner mechanism and is initially at rest while the inner mechanism is rotating with an angular speed of 150 rpm. What is the final angular speed of the clutch after the parts engage? What fraction of the kinetic energy of the clutch remains after it is engaged?

No external torques act on the clutch so angular momentum is conserved.

$$L_f = L_0$$

$$I_1\omega + I_2\omega = I_1\omega_1$$

$$I_1\omega + 1/2 \, I_1\omega = I_1\omega_1$$

$$\omega = 2/3 \, \omega_1 = (2/3)(150 \text{ rpm}) = 1.0 \text{ X } 10^2 \text{ rpm}$$

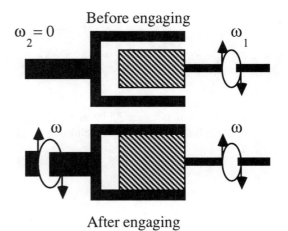

Before engaging

After engaging

The rotational kinetic energy before the parts engage is $KE_0 = 1/2 \, I_1\omega_1^2$ and after they engage it is $KE_f = 1/2 \, (I_1 + I_2)\omega^2$. The fraction remaining is

$$\frac{KE_f}{KE_0} = \frac{(I_1 + I_2)\omega^2}{I_1\omega_1^2}$$

Using $I_2 = 1/2 \, I_1$ and $\omega = 2/3 \, \omega_1$ yields

$$\frac{KE_f}{KE_0} = \frac{2}{3}$$

PRACTICE PROBLEMS

1. The crank handle on an ice cream freezer is 0.20 m from the shaft. At some point in making ice cream a boy must exert a force of 50 N to turn the crank. How much torque does the boy apply about an axis through the shaft?

2. A bolt on an automobile engine is required to be "torqued" to 95 ft lb. A mechanic wants to use a wrench to torque the bolt to this specification without having to exert more than 55 lb of force. How long should the wrench handle be?

3. A 1500 kg car is driven onto a pair of ramps so that its front wheels have been raised 0.30 m above the rear wheels. The wheel base of the car is 2.4 m and its center of gravity is located at the midpoint between the front and rear wheels. How much of the car's weight is supported by the front wheels? How much weight is supported by the rear wheels?

4. Two children whose masses are 35 kg and 55 kg are on a playground "teeter totter" which is 3.0 m long. Where must the fulcrum be placed so that the children will balance?

5. An airplane lands with a speed of 150 km/h. Before landing the airplane's wheels are not rotating. The ground exerts a constant normal force of 12 000 N on each wheel. The coefficient of kinetic friction between the runway and a tire is 0.60. A wheel has a moment of inertia of 2.9 kg m^2 and a radius of 0.15 m. How long is the skid mark that the tire leaves on the runway?

6. A 55 kg ice skater is traveling with a speed of 2.5 m/s in a circle of radius 4.0 m. The skater is held in the circle by a rope held by a friend standing at the center of the circle. If the friend slowly pulls in 1.5 m of the rope, what is the resulting speed of the skater? Neglect friction between the ice and the skater's skates.

7. Compare the initial and final kinetic energies of the skater in practice problem 6. Explain why they are not equal.

8. A bowling ball encounters a 2.50 ft vertical rise on the way back to the ball rack. Neglect frictional losses and assume that the mass of the ball is distributed uniformly. If the linear speed of the ball is 11.5 ft/s at the bottom of the rise, find the linear speed of the ball at the top.

9. A rifleman shoots a 35 g bullet at a 1.0 m high target which is a uniform 0.60 m × 0.60 m board with a mass of 9.0 kg hinged at the bottom. If the bullet enters the target at the top with a speed of 250 m/s, what is the speed of the top of the target when the target is horizontal?

SOLUTIONS AND ANSWERS

Practice Problems

1. Assuming that the boy always exerts the force perpendicular to the crank, the moment arm of the torque is just 0.2 m. Then

$$\tau = F\ell = (50.0\ \text{N})(0.20\ \text{m}) = \textbf{10.0 N m}.$$

2. The mechanic will exert the smallest force if he pushes perpendicular to and at the end of the wrench handle. The lever arm for the torque is then the length of the wrench,

$$\ell = \tau/F = 95\ \text{ft lb}/55\ \text{lb} = \textbf{1.7 ft}.$$

3. The ramp makes an angle of $\theta = \tan^{-1}(0.30/2.4) = 7.1°$ with the horizontal. The normal force exerted by the ramp on the front wheels is perpendicular to the ramp, while the weight of the car is $7.1°$ from the perpendicular to the ramp. Taking torques about an axis through the contact point between the rear wheels and the ground gives

$$F_f\ (2.4\ \text{m}) - (1500\ \text{kg})(9.80\ \text{m/s}^2)(1/2)(2.4\ \text{m})\cos 7.1° = 0$$

$$F_f = \textbf{7300 N}.$$

The rest of the weight of the car is supported by the rear wheels so

$$F_r = (1500\ \text{kg})(9.80\ \text{m/s}^2) - 7300\ \text{N} = \textbf{7400 N}.$$

4. Let x be the distance from the 35 kg child to the fulcrum. The torque equation for an axis through the fulcrum is

$$(35\ \text{kg})x - (55\ \text{kg})(3.0\ \text{m} - x) = 0.$$

Solving for x gives

$$x = \textbf{1.8 m}.$$

5. The wheel is accelerated by the torque provided by friction, $\tau = \mu F_N R = I\alpha$. The wheel will skid on the runway until its angular speed is $\omega = v/R$ where v is the speed of the airplane and R is the radius of the wheel. The wheel skids for a time, $t = \omega/\alpha$. During this time it has moved a distance, $x = vt = v\omega/\alpha$ which is the length of the skid mark. Now

$$x = (v^2/R^2)(I/\mu F_N) = \textbf{31 m}.$$

6. If friction is ignored, the angular momentum of the ice skater is conserved since no external torque acts on him. $I_f\omega_f = I_0\omega_0$. The skater's initial and final paths are circular so $\omega_f = v_f/R_f$ and $\omega_0 = v_0/R_0$. The initial and final moments of inertia are $I_f = mR_f^2$ and $I_0 = mR_0^2$. Using these in the conservation of angular momentum equation results in

$$v_f = (R_0/R_f)v_0 = (4.0/1.5)(2.5\ \text{m/s}) = \textbf{6.7 m/s}.$$

7. The initial kinetic energy of the skater is $KE_0 = 1/2\ mv_0^2$ and the final kinetic energy of the skater is

$$KE_f = 1/2\ mv_f^2 \text{ so } KE_f/KE_0 = (v_f/v_0)^2 = (6.7/2.5)^2 = \textbf{7.2}.$$

The final kinetic energy of the skater is larger than the initial kinetic energy because the skater's friend did work in pulling the skater into a smaller circle.

---

8. The total mechanical energy of the ball is conserved.

$$1/2\ mv_f^2 + 1/2\ I\omega_f^2 + mgh = 1/2\ mv_0^2 + 1/2\ I\omega_0^2$$

If the ball rolls without slipping, then $\omega_0 = v_0/R$ and $\omega_f = v_f/R$. The moment of inertia of the solid ball is $I = 2/5\ mR^2$. Using these in the conservation of energy equation and simplifying yields

$$v_f^2 = v_0^2 - (10/7)\ gh = (11.5\ ft/s)^2 - (10/7)(32.2\ ft/s^2)(2.5\ ft)$$

so

$$v_f = \textbf{4.15 ft/s.}$$

---

9. Total angular momentum is conserved during the time required for the bullet to imbed in the target. $(I + ms^2)\omega_0 = msv$ where s is the length of a side of the target, m is the mass of the bullet, v is the initial speed of the bullet, ω_0 is the resulting angular speed of the bullet and target system, and I is the moment of inertia of the target. The moment of inertia of the target is $I = 1/3\ Ms^2$ which is much larger than ms^2, so the initial angular speed is approximately $\omega_0 = 3(m/M)(v/s)$. After the bullet imbeds in the target, the total mechanical energy of the system is conserved.

$$1/2\ (I + ms^2)\omega_f^2 = 1/2\ (I + ms^2)\omega_0^2 + Mg(1/2\ s)$$

where ω_f is the angular speed of the target as it passed through the horizontal and M is the mass of the target. The mass of the bullet has been neglected since it is much less than the mass of the target. The final angular speed of the target is given by

$$\omega_f^2 = \omega_0^2 + Mgs/I = 9(m/M)^2(v/s)^2 + 3g/s$$

so that

$$\omega_f = 8.5\ rad/s$$

Now the speed of the top of the target is

$$v_f = s\omega_f = \textbf{5.1 m/s.}$$

CHAPTER 10
Simple Harmonic Motion and Elasticity

PREVIEW

In this chapter you will study the elastic properties of materials and the kinds of deformations a material can experience when subjected to outside forces. A special case, the spring, will be examined in detail. You will also study simple harmonic motion, the oscillatory motion that occurs when the restoring force of an ideal spring acts on an object.

QUICK REFERENCE

Important Terms

Ideal Spring
A spring which behaves according to the relationship $F_{Applied} = kx$ and which has no internal friction.

Restoring Force
A force which tends to restore an object to its original state. For example, the force which tends to pull a stretched spring back to its unstretched or "equilibrium" position.

Simple Harmonic Motion
The oscillatory motion that results when an elastic material is subjected to the restoring force of an ideal spring.

Period
The time required to complete one oscillatory cycle of motion.

Frequency
The number of cycles of oscillatory motion that occurs in one second.

Amplitude
The maximum distance that an oscillating object moves away from its equilibrium position.

Elasticity
The tendency for a solid object to return to its original shape after it has been stretched, compressed, or otherwise distorted by outside forces.

Young's Modulus
The proportionality constant which relates the fractional change in length of a material to the force per unit area applied to the material.

Shear Modulus
The proportionality constant which relates the shear deformation per unit length of a material to the force per unit area applied to the material.

Bulk Modulus
The proportionality constant which relates the fractional change in volume of a material to the pressure applied to the material.

Stress
The magnitude of the force per unit area required to cause an elastic deformation.

Strain
The fractional change in length or volume that results from an applied force.

Hooke's Law
The relationship which states that stress is directly proportional to strain.

Equations

Hooke's law restoring force for an ideal spring:

$$F = -kx \tag{10.2}$$

The **displacement** of an object moving with **simple harmonic motion** is:

$$x = A \cos \omega t \tag{10.3}$$

The **velocity** of an object moving with **simple harmonic motion** is:

$$v = -A\omega \sin \omega t \tag{10.7}$$

The **acceleration** of an object moving with **simple harmonic motion** is:

$$a = -A\omega^2 \cos \omega t \tag{10.9}$$

The equation for **stretching or compression** is:

$$F = Y\left(\frac{\Delta L}{L_0}\right) A \tag{10.17}$$

The equation for **shear deformation** is:

$$F = S\left(\frac{\Delta X}{L_0}\right) A \tag{10.18}$$

The **fractional volume change** is given by:

$$\Delta P = -B\left(\frac{\Delta V}{V_0}\right) \tag{10.20}$$

DISCUSSION OF SELECTED SECTIONS

10.1 The Ideal Spring and Simple Harmonic Motion

Consider the following figure in which an unstretched spring is pulled on by a force $F_{Applied}$. In the process, the spring will be stretched by an amount x from its unstrained or equilibrium position (x = 0).

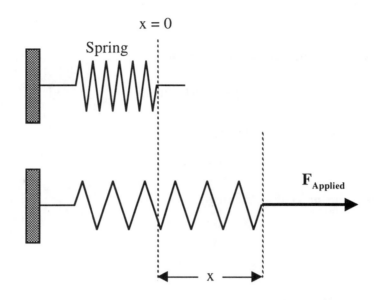

For relatively small deformations, the force required to stretch (or compress) the spring obeys the following equation:

$$F_{Applied} = kx \qquad (10.1)$$

The constant k is called the **spring constant** which is a measure of the stiffness of the spring (units for k are N/m). The larger the value of k, the harder it is to stretch the spring. Any spring that obeys equation (10.1) is said to be an **ideal spring**.

To stretch or compress a spring, a force must be applied to it. As a consequence of Newton's third law, the spring must exert an oppositely-directed force of equal magnitude. The force of the spring is a "restoring force" in the sense that it tries to pull or push the spring back to its equilibrium (x = 0) position. This restoring force can thus be written

$$F = -kx \qquad (10.2)$$

The minus sign means that the restoring force always points in a direction opposite to the direction of deformation.

Example 1
A spring has a spring constant of 155 N/m. What is the magnitude of the force needed (a) to stretch the spring by 5.00 cm from its unstretched length and (b) to compress the spring by the same amount?

(a) Using equation (10.1) we have:

$$F_{Applied} = kx = (155 \text{ N/m})(5.00 \times 10^{-2} \text{ m}) = 7.75 \text{ N.}$$

(b) The magnitude of the force required to compress the spring by the same amount (5.00 cm) is EXACTLY the same as that found in part (a), 7.75 N. That is, the same force is needed for stretching or compressing the spring by equal amounts. The difference lies in the direction of the force.

Imagine that we now attach one end of a spring to a wall, and the other end to a block which rests on a horizontal frictionless surface, as shown in the figure below. If we were to pull the block to the right, stretching the spring, and then let go, what would happen? The restoring force, given by equation (10.2), would act to pull the block back to the left. The block would begin to move to the left. However, because of the block's momentum it would not stop when it returned to its equilibrium position. It would continue to move until the spring compressed enough to stop the block. The restoring force of the spring would now push the block back to the right, and again it would begin to move (this time to the right). As it passed through its equilibrium position it would keep moving until stopped by the force of the once again stretched spring. The process would then repeat, over and over again. This is referred to as **Simple Harmonic Motion**.

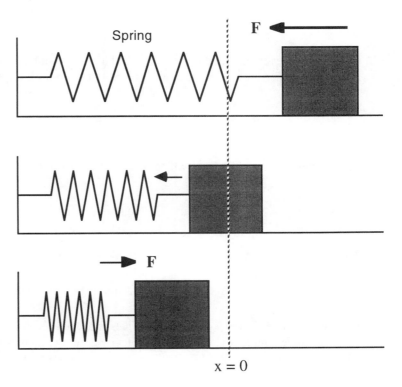

For an ideal spring, and on a frictionless surface, the block will continue to oscillate back and forth forever. We can describe this oscillatory motion by first defining a number of important quantities that are basic to any type of repetitive motion. This follows in the next section.

10.2 Simple Harmonic Motion and the Reference Circle

If we were to plot the motion depicted in the above figure, that is, the displacement of the block x as a function of time t, we would obtain the following graph. Note that a displacement of the block to the right of the equilibrium position is positive (+) and a displacement to the left of x = 0 is negative (-).

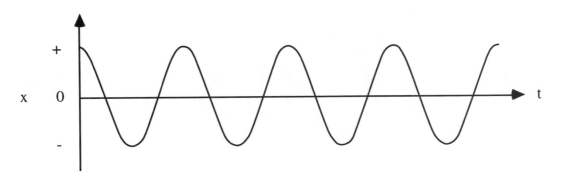

The graph above shows that the displacement is represented by a **cosine** function which can be expressed as

$$x = A \cos \omega t \qquad (10.3)$$

In equation (10.3), A is called the **amplitude** of the motion, that is, the magnitude of the maximum x-displacement of the block from its equilibrium position. The quantity, ω is the **angular frequency** where

$$\omega = \frac{2\pi}{T} \qquad (10.4)$$

and T is the **period**, the time required for the block to complete one cycle of its motion. A related quantity is the **frequency**, which represents the number of cycles of motion completed each second. We have

$$f = \frac{1}{T} \qquad (10.5)$$

The frequency has a unit of s^{-1} which is referred to as one **hertz** (Hz). Finally, the angular frequency is written

$$\omega = \frac{2\pi}{T} = 2\pi f \qquad (10.6)$$

The velocity of the block can also be represented by a sinusoidally varying function. We can write

$$v = - A\omega \sin \theta = - A\omega \sin \omega t \qquad (10.7)$$

Note the negative sign in equation (10.7). This is needed because the block begins to move to the left initially (i.e., in the negative x direction). The block's velocity is zero at each end point of its motion and it reaches its maximum value (+ or -) at the equilibrium position, x = 0. The maximum of equation (10.7) occurs when $\sin \omega t = \pm 1$. So,

$$v_{max} = A\omega \quad (\omega \text{ in rad/s}) \qquad (10.8)$$

In simple harmonic motion, the velocity is NOT constant, therefore, there IS acceleration. Furthermore, since the force varies as a function of x (i.e., the more you stretch the spring, the harder it becomes to keep it stretched), the acceleration is not constant either. The acceleration of the block can be written as

$$a = - A\omega^2 \cos \theta = - A\omega^2 \cos \omega t \qquad (10.9)$$

The acceleration is zero at x = 0 (since F = - kx = ma = 0 because x = 0) and reaches its maximum value at the endpoints of its motion (i.e., when x = A). The expression for a_{max} is obtained by setting $\cos \omega t = \pm 1$ in the above equation (10.9). We therefore have,

$$a_{max} = A\omega^2 \quad (\omega \text{ in rad/s}) \qquad (10.10)$$

Finally, it can be shown that the angular frequency is related to the mass of the block, m, and the stiffness of the spring, k, by the equation

$$\omega = \sqrt{\frac{k}{m}} \qquad (10.11)$$

Example 2

A spring whose spring constant is 196 N/m is attached to a block of mass 5.00 kg which sits on a horizontal, frictionless surface. The block is pulled 20.0 cm to the side and released. Find (a) the frequency, (b) the period, (c) the amplitude, (d) the maximum velocity, and (e) the maximum acceleration of the block; (f) write expressions for the displacement, velocity, and acceleration of the block as a function of time.

(a) Using equation (10.11),

$$\omega = \sqrt{\frac{k}{m}} = \sqrt{\frac{196 \text{ N/m}}{5.00 \text{ kg}}} = 6.26 \text{ rad/s}$$

Using equation (10.6),

$$f = \frac{\omega}{2\pi} = \frac{6.26 \text{ rad/s}}{2\pi} = 0.996 \text{ Hz}$$

(b) Using equation (10.5),

$$T = \frac{1}{f} = \frac{1}{0.996 \text{ Hz}} = 1.00 \text{ s}$$

(c) The amplitude is, $A = 20.0 \text{ cm} = 0.200 \text{ m}$

(d) Using equation (10.8),

$$v_{max} = A\omega = (0.200 \text{ m})(6.26 \text{ rad/s}) = 1.25 \text{ m/s}$$

(e) Using equation (10.10),

$$a_{max} = A\omega^2 = (0.200 \text{ m})(6.26 \text{ rad/s})^2 = 7.84 \text{ m/s}^2$$

(f) Using equations (10.3), (10.7), and (10.9), and the values of A and ω found above, we can write

$$x = A \cos \omega t = (0.200)[\cos (6.26t)] \text{ m}$$

$$v = -A\omega \sin \omega t = -(0.200 \text{ m})(6.26 \text{ rad/s})[\sin (6.26t)]$$

$$= -(1.25)[\sin (6.26t)] \text{ m/s}$$

$$a = -A\omega^2 \cos \omega t = -(0.200 \text{ m})(6.26 \text{ rad/s})^2 [\cos (6.26 \text{ t})]$$

$$= -(7.84)[\cos (6.26t)] \text{ m/s}^2.$$

10.3   Energy and Simple Harmonic Motion

When an object is attached to a spring, the spring force can do work $W_{elastic}$ on the object. If in addition, nonconservative forces do work W_{nc} on the object, the work-energy theorem discussed in Chapter 6 states that this work will result in a change in energy for the system. This can be stated mathematically as

$$W_{nc} + W_{elastic} = \Delta KE + \Delta PE_{grav}$$

The work required to stretch a spring from its initial position x_0 to its final position x_f, is given by

$$W_{elastic} = 1/2\ kx_0^2 - 1/2\ kx_f^2 \qquad (10.12)$$

The work done to stretch the spring therefore gives rise to a change in the **elastic potential energy** of the spring. In general, if a spring is stretched (or compressed) by an amount x, we can write

$$PE_{elastic} = 1/2\ kx^2 \qquad (10.13)$$

In the special case where only the forces of the spring and gravity are present, the total mechanical energy of the system, E, can be written

$$E = KE + PE_{grav} + PE_{elastic}$$

Example 3

A spring (spring constant 150 N/m) is attached to a 1.5 kg block which rests on a horizontal frictionless surface. If the block is set in motion and oscillates with an amplitude of 15 cm, what is the velocity of the block as it passes through its equilibrium position?

The conservation of energy principle states that, in the absence of friction or air resistance, the total mechanical energy of the system is conserved. That is

$$E_f = E_0$$

$$KE_f + \frac{1}{2}kx_f^2 + mgh_f = KE_0 + \frac{1}{2}kx_0^2 + mgh_0$$

Since the spring is horizontal, there is no change in PE_{grav} since $h_f = h_0$. Take the initial position to be when the spring is fully stretched, and the final position to be the equilibrium position. We can make the following substitutions: $x_0 = A = 0.15$ m, $x_f = 0$, $v_0 = 0$. We are therefore left with

$$\frac{1}{2}mv_f^2 = \frac{1}{2}kx_0^2$$

and we can solve for the final velocity,

$$v_f = x_0\sqrt{\frac{k}{m}} = (0.15 \text{ m})\sqrt{\frac{150 \text{ N/m}}{1.5 \text{ kg}}} = 1.5 \text{ m/s}.$$

10.4 The Pendulum

A **simple pendulum** consists of a particle of mass m, attached to a frictionless pivot by a thin support wire of length L, as shown in the figure.

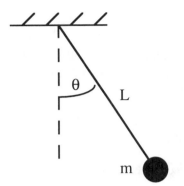

If the mass is pulled aside by a small angle, θ and released, the pendulum will swing back and forth in simple harmonic motion. It can be shown that the frequency of this motion is given by equation (10.16)

$$f = \frac{1}{2\pi} \sqrt{\frac{g}{L}}$$

Note that the frequency does NOT depend on the mass of the particle suspended from the wire! Only the length of the wire and the value for the gravitational acceleration, g, affect the pendulum's motion.

Example 4
A pendulum of length 4.0 m is pulled aside and released. Find (a) the frequency and period of the pendulum on earth, and (b) the frequency and period for the same pendulum on the moon (g' = 1.57 m/s^2).

(a) Using equation (10.16),

$$f = \frac{1}{2\pi} \sqrt{\frac{g}{L}} = \frac{1}{2\pi} \sqrt{\frac{9.80 \text{ m/s}^2}{4.0 \text{ m}}} = 0.25 \text{ Hz}$$

The period is,

$$T = 1/f = 1/(0.25 \text{ Hz}) = 4.0 \text{ s.}$$

(b) If we take this pendulum to the moon,

$$f' = \frac{1}{2\pi} \sqrt{\frac{g'}{L}} = \frac{1}{2\pi} \sqrt{\frac{1.57 \text{ m/s}^2}{4.0 \text{ m}}} = 0.10 \text{ Hz}$$

$$T' = 1/f' = 1/(0.10 \text{ Hz}) = 10 \text{ s.}$$

10.7 Elastic Deformation

Refer to the following figure in which an object of length L_0 and cross-sectional area A is subjected to a force, **F,** which pulls (or pushes) along the length of the object (i.e., perpendicular to A). The object will stretch (or compress) by an amount ΔL as shown in the figure.

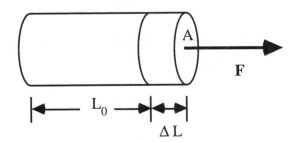

If the amount of stretching is small compared to the original length of the object, the elastic behavior of the material is maintained. We must incorporate a proportionality constant (which depends on the type of material being stretched) which relates the fractional change in length ($\Delta L/L_0$) to the force per unit area (F/A) being applied, i.e.,

$$F = Y \left(\frac{\Delta L}{L_0} \right) A \qquad (10.17)$$

Here, the proportionality constant, Y, is known as **Young's Modulus**. Solving equation (10.17) for Y shows that Young's modulus has units of force per unit area (N/m^2); the exact value depends on the nature of the material.

Example 5
A 25.0 kg mass is suspended from an aluminum wire of length 2.00 m and diameter 1.00 mm. By how much does the wire stretch?

The cross-sectional area of the wire is
$$A = \pi r^2 = \pi(5.00 \times 10^{-4} \text{ m})^2 = 7.85 \times 10^{-7} \text{ m}^2.$$

The force being applied is the weight of the 25.0 kg mass, i.e.,

$$F = mg = (25.0 \text{ kg})(9.80 \text{ m/s}^2) = 245 \text{ N}.$$

Using Table 10.1 we look up the value of Young's modulus for aluminum. We find $Y = 6.9 \times 10^{10}$ N/m^2. The amount the wire stretches is therefore,

$$\Delta L = \frac{FL_0}{YA} = \frac{(245 \text{ N})(2.00 \text{ m})}{(6.9 \times 10^{10} \text{ N/m}^2)(7.85 \times 10^{-7} \text{ m}^2)} = 9.0 \times 10^{-3} \text{ m} = 9.0 \text{ mm}.$$

It is possible to deform a solid object in a way other than stretching or compressing it. Suppose a force is applied parallel to the cross-sectional area of the object, as shown in the figure below.

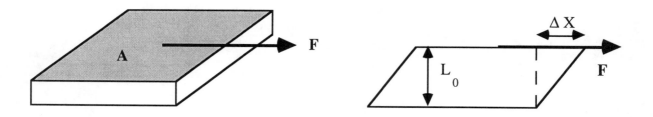

The resulting deformation in the material is called a **shear deformation**. The magnitude of the force **F** needed to produce an amount of shear ΔX for an object of cross-sectional area A and thickness L_0 is given by

$$F = S \left(\frac{\Delta X}{L_0} \right) A \qquad (10.18)$$

The constant of proportionality, S, is known as the **shear modulus**. Like Young's modulus, it has units of force per unit area (N/m^2), and depends on the nature of the material.

Example 6

A lead block of length 0.50 m, width 0.30 m, and height 0.080 m is subjected to a shearing force of 2.0×10^6 N directed perpendicular to its height. The shear modulus for lead (from Table 10.2) is $S = 5.4 \times 10^9$ N/m^2. Find the amount of shear, ΔX.

Using equation (10.18) with A = length x width = (0.50 m)(0.30 m) = 0.15 m^2 and L_0 = height = 0.080 m, we have

$$\Delta X = \frac{FL_0}{SA} = \frac{(2.0 \times 10^6 \text{ N})(0.080 \text{ m})}{(5.4 \times 10^9 \text{ N/m}^2)(0.15 \text{ m}^2)} = 2.0 \times 10^{-4} \text{ m}.$$

When a compressive force is applied to all three dimensions of a solid, the length of each dimension decreases, leading to a decrease in volume for the object. The forces acting in such situations are applied perpendicular to every available surface. Because of this, we speak of the perpendicular force per unit area, rather than the total force involved. The perpendicular force per unit area is called the **pressure.** When such a pressure is applied to a volume V, the volume changes by an amount ΔV, according to the relation

$$\Delta P = -B\left(\frac{\Delta V}{V_0}\right) \tag{10.20}$$

The proportionality constant, B, is known as the **bulk modulus of the material**. The minus sign appears in equation (10.20) because an increase in pressure (ΔP positive) always results in a decrease in volume (ΔV negative). B is always a positive quantity.

Example 7

A brass sphere of radius 1.50 m is subjected to a pressure increase of 3.0×10^6 N/m^2. What change in volume results?

The volume of the sphere is
$$V = (4/3)\pi r^3 = (4/3)\pi(1.50 \text{ m})^3 = 14.1 \text{ m}^3.$$

The bulk modulus for brass (obtained from Table 10.3) is $B = 6.7 \times 10^{10}$ N/m^2. The change in volume is therefore

$$\Delta V = \frac{-V_0 \Delta P}{B} = \frac{-(14.1 \text{ m}^3)(3.0 \times 10^6 \text{ N/m}^2)}{6.7 \times 10^{10} \text{ N/m}^2} = -6.3 \times 10^{-4} \text{ m}^3$$

10.8 Stress, Strain, and Hooke's Law

The force per unit area required to produce any elastic deformation is referred to as the **stress**. The fractional change in a quantity ($\Delta L/L_0$, $\Delta X/L_0$, or $\Delta V/V_0$) that results when a stress is applied is referred to as the **strain**. In each situation described in section 10.8, the stress was found to be directly proportional to the strain, the constants of proportionality being either Young's modulus, the shear modulus, or the bulk modulus. The relationship between stress and strain is known as **Hooke's Law**. The SI unit of stress is the **pascal** (Pa) where 1 Pa = 1 N/m^2, while the strain is a dimensionless quantity. Materials obey Hooke's law up to a certain limit. If too much stress is applied to a material, it becomes unable to "spring back" to its original size. We would say that the material has exceeded its "elastic limit". Beyond this limit the material can no longer obey Hooke's law.

Example 8

A steel cable is able to withstand a stress of 8.0×10^7 Pa up to its elastic limit. If Young's modulus for steel is $Y = 2.0 \times 10^{11}$ N/m^2, what is the strain?

Hooke's law states that the stress is proportional to the strain. In this case, since the wire is being stretched, the constant of proportionality is Young's modulus. We have, therefore,

$$Y = \frac{\text{Stress}}{\text{Strain}} \quad \Rightarrow \quad \text{Strain} = \frac{\text{Stress}}{Y} = \frac{8.0 \times 10^7 \, \text{Pa}}{2.0 \times 10^{11} \, \text{N/m}^2} = 4.0 \times 10^{-4}.$$

PRACTICE PROBLEMS

1. A 2.0 kg block hung from a vertical spring causes it to stretch by 20.0 cm. If the 2.0 kg block is replaced by a 0.50 kg mass, and the spring is stretched and released, what are the frequency and period of the oscillations?

2. A small object of mass 0.20 kg is undergoing simple harmonic motion of period 0.80s and amplitude 0.50 m. (a) What is the maximum force acting on the object? (b) What is the maximum velocity of the object? (c) If the oscillations are produced by a spring, what is the force constant of the spring?

3. A 16-lb object hangs from a spring. It undergoes simple harmonic motion in a vertical direction with an amplitude of 6.0 in and a frequency of 2.0 Hz. Find (a) the period, (b) the force constant of the spring, (c) the maximum speed of the object, (d) the maximum acceleration of the object, (e) the speed when the object is at half its maximum displacement.

4. A 1200 kg car oscillates vertically on its suspension springs with a period of 0.60 s when empty. If six persons, each of mass 80.0 kg, get into the car, how far down will the supporting springs be depressed?

5. A block of mass 0.10 kg is placed on top of a block of mass 0.20 kg. The coefficient of static friction between the blocks is 0.20. The lower block is now moved back and forth horizontally in simple harmonic oscillations with an amplitude of 6.0 cm. Keeping the amplitude constant, what is the highest frequency for which the upper block will not slip relative to the lower block?

6. A pendulum clock keeps perfect time at a location where the acceleration due to gravity is exactly 9.80 m/s^2. When the clock is moved to a higher altitude, it loses 80.0 s per day. Find the value of g at this new location.

7. A spring (spring constant of 2500 N/m) is fixed vertically to the floor. A 10.0 kg ball is placed on the spring, pushed down 0.50 m, and released. How high does the ball fly above its release position?

8. A steel wire can withstand a tensile stress of 6.0 x 10^8 N/m^2. If such a wire is used to support a mass of 750 kg, what is the minimum diameter wire that can be used?

9. A copper sphere is subjected to a pressure of 2.0 x 10^{10} Pa. What is the resulting fractional change in the volume of the sphere? What is the fractional change in the sphere's radius?

10. A rectangular block of tungsten is 5.00 cm high and has a cross-sectional area of 144 cm^2. A force of 3.6 x 10^7 N is applied to the block. (a) If the force pushes perpendicular to the block's area, by how much will the block compress? (b) If the force is applied parallel to the surface (i.e., a shearing force), what will the amount of shear deformation be?

SOLUTIONS AND ANSWERS

Practice Problems

1. Find the spring constant using equation (10.1), $F_{Applied} = kx$ which gives

$$k = F_{Applied}/x = (2.0 \text{ kg})(9.8 \text{ m/s}^2)/(0.20 \text{ m}) = 98 \text{ N/m}.$$

Now use equation (10.11),

$$\omega = \sqrt{(k/m)} = \sqrt{(98 \text{ N/m})/(0.50 \text{ kg})} = 14 \text{ rad/s}$$

to find

$$f = \omega/2\pi = \textbf{2.2 Hz}.$$

Therefore, from equation (10.5), the period is
$$T = 1/f = \textbf{0.45 s}.$$

2. (a) The maximum acceleration, equation (10.10), is

$$a_{max} = A\omega^2 = A(2\pi/T)^2$$
$$a_{max} = (0.50 \text{ m})[2\pi/(0.80 \text{ s})]^2$$
$$a_{max} = 31 \text{ m/s}^2.$$

The maximum force is obtained from

$$F_{max} = ma_{max} = (0.20 \text{ kg})(31 \text{ m/s}^2) = \textbf{6.2 N}.$$

(b) Using equation (10.8),

$$v_{max} = A\omega = A(2\pi/T) = (0.50 \text{ m})[2\pi/(0.80 \text{ s})] = \textbf{3.9 m/s}.$$

(c) Using $F_{Applied} = kx$

$$k = F_{Applied}/x = (6.2 \text{ N})/(0.50 \text{ m}) = \textbf{12 N/m}.$$

3. (a) Equation (10.5),
$$T = 1/f = 1/(2.0 \text{ Hz}) = \textbf{0.50 s},$$
(b) Equation (10.6),
$$\omega = 2\pi f = \sqrt{(k/m)},$$
solving for k,
$$k = 4\pi^2 f^2 m = 4\pi^2(2.0 \text{ Hz})^2(0.5 \text{ slugs}) = \textbf{79 lb/ft},$$
(c)
$$v_{max} = A\omega = A(2\pi f) = (0.5 \text{ ft})(2\pi)(2.0 \text{ Hz}) = \textbf{6.3 ft/s},$$
(d)
$$a_{max} = A\omega^2 = (0.5 \text{ ft})[2\pi(2.0 \text{ Hz})]^2 = \textbf{79 ft/s}^2,$$

(e) Using conservation of energy,
$$(1/2)kA^2 = (1/2)k(A/2)^2 + (1/2)mv^2.$$
Thus,
$$v^2 = (3/4)kA^2/m = (3/4)(79 \text{ lb/ft})(0.5 \text{ ft})^2/(0.5 \text{ sl}),$$
or
$$v = \textbf{5.4 ft/s}.$$

4. Find the spring constant using equation (10.11),

$$k = m\omega^2 = (1200 \text{ kg})[2\pi/(0.6 \text{ s})]^2 = 1.3 \times 10^5 \text{ N/m}.$$

Now use $F_{Applied} = kx$ to find

$$x = F_{Applied}/k = 6(80.0 \text{ kg})(9.80 \text{ m/s}^2)/(1.3 \times 10^5 \text{ N/m}) = \textbf{3.6 cm}.$$

---

5. The force of static friction for the top block (m_1) is

$$f_s = \mu m_1 g = (0.20)(0.10 \text{ kg})(9.8 \text{ m/s}^2) = 0.20 \text{ N}.$$

The maximum acceleration the top block can sustain without slipping is

$$a_{max} = f_s/m = (0.20 \text{ N})/(0.10 \text{ kg}) = 2.0 \text{ m/s}^2$$

Using equation (10.10),

$$a_{max} = A\omega^2 = A(2\pi f)^2$$

or

$$f^2 = a_{max}/4\pi^2 A = (2.0 \text{ m/s}^2)/[4\pi^2(0.06 \text{ m})],$$

then

$$f = \textbf{0.92 Hz}.$$

---

6. If the clock loses 80.0 s per day, it must be swinging slower (T greater). It loses (80.0 s/86 400 s) = 0.926 ms each second. Therefore, the new period is 1.000 926 s. If the original period is

$$T = 2\pi\sqrt{(L/g)}$$

and the new period is

$$T' = 2\pi\sqrt{(L/g')}$$

then

$$T'/T = \sqrt{(g/g')}$$

gives

$$g' = g (T/T')^2 = (9.80 \text{ m/s}^2)[(1.000 \text{ s})/(1.001 \text{ s})]^2 = \textbf{9.78 m/s}^2.$$

---

7. Using conservation of energy,

$$E_A = (1/2)kx^2 = E_B = mgh.$$

We have,

$$h = kx^2/2mg = (2500 \text{ N/m})(0.5 \text{ m})^2/[2(10.0 \text{ kg})(9.80 \text{ m/s}^2)] = \textbf{3.2 m}.$$

---

8. We know that stress = F/A where $A = (1/4)\pi d^2$ and d = diameter. Solving for d we obtain

$$d^2 = 4F/[\pi(\text{stress})] = 4(750 \text{ kg})(9.80 \text{ m/s}^2)/[\pi(6.0 \times 10^8 \text{ N/m}^2)]$$

so

$$d = \textbf{3.9} \times \textbf{10}^{-3} \textbf{ m}.$$

9. Using equation (10.20),

$$\Delta P = -B(\Delta V/V_0).$$

Solving for $(\Delta V/V_0)$ and substituting for B for copper (Table 10.3), we find

$$(\Delta V/V_0) = \Delta P/B = (2.0 \times 10^{10} \text{ Pa})/(1.3 \times 10^{11} \text{ Pa}) = \mathbf{1.5 \times 10^{-1}}.$$

The fractional change in radius is

$$(\Delta r/r_0) = (1/3)(\Delta V/V_0) = \mathbf{5.0 \times 10^{-2}}.$$

---

10. (a) Equation (10.17),

$$\Delta L = FL_0/YA = (3.6 \times 10^7 \text{ N})(5.00 \times 10^{-2} \text{ m})/[(3.6 \times 10^{11} \text{ Pa})(144 \times 10^{-4} \text{ m}^2)]$$
$$\Delta L = \mathbf{0.35 \ mm}$$

(b) Equation (10.18),
$$\Delta X = FL_0/SA = (3.6 \times 10^7 \text{ N})(5.00 \times 10^{-2} \text{ m})/[(1.5 \times 10^{11} \text{ Pa})(144 \times 10^{-4} \text{ m}^2)]$$
$$\Delta X = \mathbf{0.83 \ mm}$$

CHAPTER 11
Fluids

PREVIEW

In this chapter you will be introduced to the properties of fluids. You will learn to describe the inertial properties and the weight of a fluid in terms of its mass density and weight density. Also you will describe the forces acting on a fluid and exerted by a fluid in terms of pressure.

Pascal's principle reveals how pressures applied to fluids are transmitted to other portions of the fluid. Archimedes' principle describes how and why objects float in fluids. Using these principles, you will be able to understand such things as how hydraulic jacks work and how submarines can sometimes float on and other times dive below the surface of the ocean.

In this chapter you will also learn about flowing fluids including the equation of continuity, Bernoulli's equation, and viscous flow. Numerous applications are discussed, such as fluid flow through pipes, streamline air flow around an airplane wing, what makes a spinning baseball curve, and many others.

QUICK REFERENCE

Important Terms

Fluid
A material which has the ability to flow. Usually fluids are liquids and gases.
Mass density
The mass of a volume of a substance divided by the volume.
Specify gravity
The mass density of the material compared to the mass density of water at 4 °C.
Pressure on a surface
The perpendicular component of the force acting on the surface divided by the area of the surface.
Absolute pressure
The amount by which a pressure exceeds the pressure of a true vacuum.
Gauge pressure
The amount by which a pressure exceeds atmospheric pressure.
Buoyant force
The net force exerted by a fluid on an object which is partially or wholly immersed in the fluid. The direction of the buoyant force is up.
Steady fluid flow
Movement in which the velocity of the fluid particles at any point in the fluid is constant with time. Every particle that passes through a particular point in the fluid has the same velocity.
Incompressible fluid
A fluid in which the density of the liquid remains constant as the pressure changes.
Viscous fluid
A fluid that does not flow readily (like honey) due to the viscosity of the fluid.
Ideal fluid
An incompressible, nonviscous fluid.
Streamline
Lines representing the trajectory of the fluid particles in a steadily flowing fluid.
Mass flow rate
The mass of fluid which flows past a given point per unit time. Units are (kg/s).
Equation of continuity
The mass of fluid that enters one end of a pipe must leave at the other end. In other words, the mass flow rate must remain constant between any two positions along a tube of flowing fluid.

Equations

The **mass density** of a substance

$$\rho = \frac{m}{V} \tag{11.1}$$

The **specific gravity** of a substance

$$\text{Specific gravity} = \frac{\text{Density of substance}}{1.000 \times 10^3 \ \text{kg/m}^3} \tag{11.2}$$

The **pressure** exerted by a force acting perpendicular to a surface of area A

$$P = \frac{F}{A} \tag{11.3}$$

The **pressure within a fluid** at a depth, h, when an external pressure, P_1, is applied to the fluid

$$P_2 = P_1 + \rho g h \tag{11.4}$$

The force transmitted by a fluid from one piston to another piston

$$F_2 = F_1 \left(\frac{A_2}{A_1} \right) \tag{11.5}$$

The **equation of continuity** is:

$$\rho_1 A_1 v_1 = \rho_2 A_2 v_2 \tag{11.8}$$

For an **incompressible fluid**:

$$A_1 v_1 = A_2 v_2 \tag{11.9}$$

Bernoulli's equation is:

$$P_1 + \frac{1}{2}\rho v_1^2 + \rho g y_1 = P_2 + \frac{1}{2}\rho v_2^2 + \rho g y_2 \tag{11.11}$$

The force needed to move a layer of **viscous fluid** is:

$$F = \frac{\eta A v}{y} \tag{11.13}$$

Poiseuille's law can be written as:

$$Q = \frac{\pi R^4 (P_2 - P_1)}{8 \eta L} \tag{11.14}$$

Principles

Pascal's principle - any change in the pressure applied to a completely enclosed fluid is transmitted undiminished to all parts of the fluid and the walls of the container.

Archimedes' principle - a fluid will exert an upward buoyant force on an object wholly or partially immersed in it equal to the weight of the fluid displaced by the object.

Important Numbers

Mass density of water at 4 °C	1.00×10^3 kg/m^3 or 1.00 g/cm^3
Mass density of air at 0 °C	1.29 kg/m^3
Atmospheric pressure at sea level	1.013×10^5 Pa or 14.70 lb/in^2.

DISCUSSION OF SELECTED SECTIONS

11.1 Mass Density

Example 1
What is the volume of 1.00 kg of air?

Equation (11.1) gives the volume of air

$$V = \frac{m}{\rho} = \frac{1.00 \text{ kg}}{1.29 \text{ kg/m}^3} = 0.775 \text{ m}^3$$

This is roughly one cubic yard.

Example 2
An object weighs 450 N and has a volume of 0.083 m^3. Use the density data in Table 11.1, to determine the material from which the object is made.

Equation (11.1) gives the density of the material to be

$$\rho = \frac{m}{V} = \frac{\dfrac{450 \text{ N}}{9.8 \text{ m/s}^2}}{0.083 \text{ m}^3} = 550 \text{ kg/m}^3$$

Comparing this density with the values in Table 11.1 reveals that the object is made of PINE.

11.2 Pressure

Example 3
What is the downward force on the top of your head due to atmospheric pressure? Assume that the top of your head is a flat circle with a diameter of 7.00 inches.

$$F = P_{atm}A = (14.70 \text{ lb/in}^2)\pi(3.50 \text{ in})^2 = 566 \text{ lb.}$$

11.3 Pressure and Depth in a Static Fluid

A static fluid generates a pressure within itself due to its own weight. This pressure varies with the depth below the surface of the fluid of the point in question.

Example 4

What is the pressure due to the sea water at a depth of 500.0 m below the surface of the ocean? The density of sea water is 1025 kg/m^3.

The pressure due to the weight of the sea water is

$$P_2 - P_1 = \rho gh = (1025 \text{ kg/m}^3)(9.80 \text{ m/s}^2)(500.0 \text{ m}) = 5.02 \text{ X } 10^6 \text{ Pa}$$

This is about 730 lb/in^2!

Example 5

A small boy stands with his finger plugging a hole in a dike. The hole has a diameter of 12 mm and is located 3.4 m below the surface level of the water behind the dike. How much force must the boy exert to hold back the water?

The pressure acting on the boy's finger due to the water is ρgh. The force exerted by the water on the boy, hence the force exerted by the boy is then

$$F = \rho ghA$$
$$F = (1.00 \text{ X } 10^3 \text{ kg/m}^3)(9.80 \text{ m/s}^2)(3.4 \text{ m})\pi(6.0 \text{ X } 10^{-3} \text{ m})^2 = 3.8 \text{ N}$$

11.4 Pressure Gauges

Example 6

A gas orifice in a furnace is designed to work with a gauge pressure of 3.0 in of water as measured by an open tube manometer. What is this pressure in lb/in^2? Water has a weight density of 62.4 lb/ft^3.

The gauge pressure of the gas is the absolute gas pressure in the left arm of the tube, P_2, minus the atmospheric pressure in the right arm of the tube.

$$\text{gauge pressure} = P_2 - P_{atm} = \rho gh$$

$$\text{gauge pressure} = (62.4 \text{ lb/ft}^3)(0.25 \text{ ft})$$
$$= 15.6 \text{ lb/ft}^2$$

$$\text{gauge pressure} = (15.6 \text{ lb/ft}^2)(1 \text{ ft}^2/144 \text{ in}^2)$$
$$= 0.11 \text{ lb/in}^2$$

11.5 Pascal's Principle

Pascal's principle lies at the heart of many useful devices such as hydraulic jacks, hydraulic brake systems, etc. In any situation where a force is transmitted by a fluid, Pascal's principle is involved.

Example 7
The handle of a hydraulic jack is 15 cm long and is pivoted 2.5 cm from the input piston which has a radius of 0.60 cm. The output piston has a radius of 1.2 cm. What weight could be lifted by the jack if the person pushing on the handle is to exert no more than 110 N of force?

The weight that can be lifted is equal to the magnitude of the force exerted by the output piston. The output force is related to the input force by

$$F_2 = F_1\left(\frac{A_2}{A_1}\right) = F_1\left(\frac{R_2}{R_1}\right)^2$$

We need the force, $\mathbf{F_1}$, on the input piston. Taking torques on the handle about an axis through the pivot gives

$$F(15 \text{ cm}) - F_1'(2.5 \text{ cm}) = 0$$

$$F_1' = 6.0 \text{ F} = 660 \text{ N}.$$

The force on the input piston due to the handle is equal in magnitude to the force on the handle exerted by the input piston, F_1'.
The magnitude of the force exerted by the output piston is then

$$F_2 = (660 \text{ N})\left(\frac{1.2 \text{ cm}}{0.60 \text{ cm}}\right)^2 = 2600 \text{ N}$$

11.6 Archimedes' Principle

The buoyant force on an object in a fluid is equal to the weight of the fluid displaced by the object. The net upward force on the object is then

$$F = B - W = \rho_f g V_f - \rho_o g V_o.$$

If the object floats, then the net force on it is zero and

$$\rho_o = \rho_f\left(\frac{V_f}{V_o}\right)$$

If the object floats partially submerged, then it displaces less than its own volume of fluid. This implies that the density of the object is less than the density of the fluid. However, if the object floats submerged in the fluid, then it displaces exactly its own volume of fluid and the densities of the object and fluid must be equal.

In the case where the object sinks, it displaces its own volume of fluid and the net force is directed downward and is negative. Now the density of the object must be greater than the density of the fluid.

Example 8

Archie is playing with a small boat in the bathtub. The boat has a mass of M = 0.22 kg and a volume of V = 0.0011 m^3. Archie has several lead weights each with a mass of m = 0.028 kg. How many lead weights could Archie place in the boat before it sinks?

In order for the boat to float just submerged, its density must be equal to the density of the water. The average density of the boat with N lead weights in it is (M + Nm)/V, so

$$N = \frac{\rho_w V - M}{m} = \frac{(1.00 \times 10^3 \text{ kg/m}^3)(0.0011 \text{ m}^3) - 0.22 \text{ kg}}{0.028 \text{ kg}} = 31$$

He can place 31 lead weights in the boat before it sinks. The thirty-second weight will sink the boat.

Example 9

An empty barge 25 m long and 4.0 m wide floats with the waterline at the 1.0 m mark. That is, 1.0 m of its hull is submerged. Gravel is added to the barge until it floats with the waterline at the 3.5 m mark. What is the weight of the gravel? Assume that the barge is a rectangular solid.

Since the empty barge is floating, its weight must be equal to the buoyant force exerted by the water.

$$W_E = \rho_w g V_E = \rho_w g A h_E$$

A similar expression must hold for the loaded barge.

$$W_E + W_L = \rho_w g A h_L$$

The weight of the load of gravel is then

$$W_L = \rho_w g A (h_L - h_E)$$
$$W_L = (1.00 \times 10^3 \text{ kg/m}^3)(9.80 \text{ m/s}^2)(25 \text{ m} \times 4.0 \text{ m})(3.5 \text{ m} - 1.0 \text{ m})$$
$$W_L = 2.5 \times 10^6 \text{ N}$$

Example 10

An 80.0 kg person weighs 30.0 N when weighed in water. If body fat has a density close to that of water and the lean portion of the body has an average density of 1.060 X 10^3 kg/m^3, what is the person's percentage of body fat?

The percentage of body fat is the mass of fat divided by the total mass of the person. Also, the mass of fat, $\rho_f V_f$, and lean, $\rho_L V_L$, must add to give 80.0 kg. So

$$\rho_L V_L + \rho_f V_f = 80.0 \text{ kg}$$

$$1.060 \, V_L + V_f = 8.00 \times 10^{-2} \tag{1}$$

Archimedes' principle applied to the person weighed in water gives

$$\rho_w g V_L + \rho_w g V_f = (80.0 \text{ kg})g - 30.0 \text{ N}$$

$$V_L + V_f = 7.69 \times 10^{-2} \tag{2}$$

Solving (2) for V_L, substituting into (1) and rearranging yields $V_f = 0.025$ m^3. The mass of fat in the person is then (1.00 X 10^3 kg/m^3)(0.025 m^3) = 25 kg. This represents 25/80.0 = 0.31 or 31 % of the person's weight.

11.8   The Equation of Continuity

In the figure below, a mass of fluid flows along a tube. The fluid has a density ρ, and travels with a velocity $\mathbf{v_2}$, through a segment of the tube whose cross-sectional area is A_2. The tube then tapers off to cross-sectional area A_1, and the fluid, whose density is now ρ, now travels with velocity $\mathbf{v_1}$.

At position 1 we have: Mass Flow Rate = $\rho_1 A_1 v_1$ (11.7b)

At position 2 we have: Mass Flow Rate = $\rho_2 A_2 v_2$ (11.7a)

The **equation of continuity** states that the mass flow rate must remain constant, that is,

$$\rho_1 A_1 v_1 = \rho_2 A_2 v_2 \qquad (11.8)$$

If the fluid is incompressible, i.e., $\rho_1 = \rho_2$, equation (11.8) can be written as

$$A_1 v_1 = A_2 v_2 \qquad (11.9)$$

where $Q = Av$ is the **volume flow rate**. Note: Units are $Q = m^3/s$, $\rho = kg/m^3$, $A = m^2$, $v = m/s$.

Example 11
Water is flowing through a 3.0 mm diameter pipe at a speed of 5.0 m/s. The pipe has a constriction of 1.5 mm diameter. What is the velocity of the water through the constriction?

We assume the flow is incompressible, so equation (11.9) is valid. Noting that $A = \pi r^2$ where r is the radius of the pipe and $r = d/2$ we can write

$$A_1 v_1 = (\pi r_1^2) v_1 = A_2 v_2 = (\pi r_2^2) v_2, \text{ and solving for } v_2 \text{ yields,}$$

$$v_2 = v_1 \left(\frac{r_1}{r_2}\right)^2 = (5.0 \text{ m/s}) \left(\frac{1.5 \times 10^{-3} \text{ m}}{7.5 \times 10^{-4} \text{ m}}\right)^2 = 20 \text{ m/s}.$$

Example 12
Oil is flowing with a speed of 4.00 ft/s through a 2.00-ft diameter pipeline. How many gallons of oil flow through the pipeline in one day? (1 gal = 0.134 ft^3)

The volume flow rate is: $Q = Av = \pi r^2 v = \pi (1.00 \text{ ft})^2 (4.00 \text{ ft/s}) = 12.6 \text{ ft}^3/s.$
Converting units gives:

$$Q = \left(12.6 \frac{ft^3}{s}\right)\left(\frac{1 \text{ gal}}{0.134 \text{ ft}^3}\right)\left(\frac{3600 \text{ s}}{1 \text{ h}}\right)\left(\frac{24 \text{ h}}{1 \text{ day}}\right) = 8.12 \times 10^6 \text{ gal/day}$$

11.9   Bernoulli's Equation

For any two points (1 and 2) in the steady, irrotational flow of an incompressible, nonviscous fluid, the pressure P, the fluid speed v, and the elevation y, are related by

$$P_1 + \frac{1}{2}\rho v_1^2 + \rho g y_1 = P_2 + \frac{1}{2}\rho v_2^2 + \rho g y_2 \qquad (11.11)$$

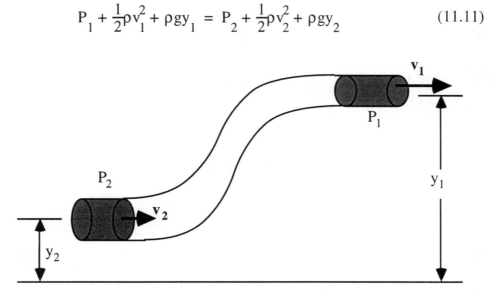

Example 13

In the diagram above, let the difference in pressure be $\Delta P = P_2 - P_1 = 5.0 \times 10^4$ Pa, take $\Delta y = y_1 - y_2 = 2.0$ m, and let $v_2 = 4.0$ m/s. What is the speed v_1 of water flowing through this pipe? (For water, $\rho = 1.0 \times 10^3$ kg/m^3)

Using equation (11.11) we have,

$$\frac{1}{2}\rho v_1^2 = (P_2 - P_1) + \rho g(y_2 - y_1) + \frac{1}{2}\rho v_2^2 , \text{ and solving for } v_1$$

$$v_1 = \sqrt{\frac{2\Delta P}{\rho} - 2g\Delta y + v_2^2} = \sqrt{\frac{2(5.0 \times 10^4 \text{ Pa})}{1.0 \times 10^3 \text{ kg/m}^3} - 2(9.8 \text{ m/s}^2)(2.0 \text{ m}) + (4.0 \text{ m/s})^2}$$

$$= 8.8 \text{ m/s}.$$

11.10   Applications of Bernoulli's Equation

For a fluid moving through a horizontal pipe ($y_1 = y_2$), equation (11.11) becomes

$$P_1 + \frac{1}{2}\rho v_1^2 = P_2 + \frac{1}{2}\rho v_2^2 \qquad (11.12)$$

Example 14

Water is flowing at a speed of 0.500 m/s through a 4.00-cm diameter hose. The hose is horizontal. (a) What is the mass flow rate of the water? (b) At what speed does the water pass through a nozzle whose effective diameter is 0.600 cm? (c) What must be the absolute pressure of the water entering the hose if the pressure at the nozzle is atmospheric pressure?

(a) Using equation (11.7) the mass flow rate is,

$$MFR = \rho A v = (1.00 \times 10^3 \text{ kg/m}^3)\pi(2.00 \times 10^{-2} \text{ m})^2(0.500 \text{ m/s}) = 0.628 \text{ kg/s}.$$

(b) Since the water is incompressible, use equation (11.9), $A_1 v_1 = A_2 v_2$, and therefore,

$$v_1 = v_2 (A_2/A_1) = v_2 (r_2/r_1)^2 = (0.500 \text{ m/s})[(2.00 \text{ cm})/(0.300 \text{ cm})]^2 = 22.2 \text{ m/s}.$$

(c) Using equation (11.12) we can write,

$$P_2 = P_1 + \frac{1}{2}\rho (v_1^2 - v_2^2)$$

$$= (1.01 \times 10^5 \text{ Pa}) + \frac{1}{2}(1.00 \times 10^3 \text{ kg/m}^3)[(22.2 \text{ m/s})^2 - (0.500 \text{ m/s})^2]$$

$$= 3.47 \times 10^5 \text{ Pa}.$$

11.11　Viscous Flow

In all real fluids, the different fluid layers do NOT all have the same velocity. The fluid at the center of a pipe moves fastest while the fluid near the walls of the pipe moves slowest. In fact, the fluid directly in contact with the wall does not move at all because it is held tightly by intermolecular forces. Such forces give rise to the **viscosity** of the fluid. Examples of fluids that are highly viscous are honey or tar. Fluids such as water, and especially oil, have low viscosities.

　　　The tangential force **F** required to move a fluid layer at a constant speed v, when the layer has an area A and is located a perpendicular distance y from an immobile surface, is given by

$$F = \frac{\eta A v}{y} \tag{11.13}$$

where η is known as the **coefficient of viscosity** or simply the viscosity. The SI unit for viscosity is the poise (P) where 1 P = 0.1 Pa s.

　　　In many cases of interest, it is necessary to know the volume flow rate in a viscous fluid. Consider two points (1 and 2) separated by a distance L in a pipe of constant radius R. If the pressure difference between these two points is $P_2 - P_1$, and η is the viscosity of the fluid in the pipe, the volume flow rate Q is given by **Poiseuille's law**, which is

$$Q = \frac{\pi R^4 (P_2 - P_1)}{8\eta L} \tag{11.14}$$

Example 15
Glycerin flows through a tube of 1.0 cm radius. If the ends of the pipe are 5.0 m apart and a pressure difference of 3.0×10^4 Pa is maintained between the ends, find the volume flow rate.

Using equation (11.14), and the value of η for glycerin found in the table we have

$$Q = \frac{\pi R^4 (P_2 - P_1)}{8\eta L} = \frac{\pi(0.010 \text{ m})^4(3.0 \times 10^4 \text{ Pa})}{8(1.5 \text{ Pa s})(5.0 \text{ m})} = 1.6 \times 10^{-5} \text{ m}^3/\text{s}.$$

PRACTICE PROBLEMS

1. The tires of a 1600 kg automobile are inflated to a pressure of 25 lb/in^2. Assuming that each tire supports 1/4 the weight of the car, calculate the area of contact between a tire and the road.

2. A research diving vessel is shaped like a sphere of radius 3.5 m. If the inside of the vessel is maintained at atmospheric pressure, what is the net force trying to crush the vessel when it is at a depth of 1500 m below the surface of the ocean?

3. What is the weight of the water in a waterbed which is 2.4 m long, 1.8 m wide and 0.23 m deep?

4. A 80.0 kg person weighs 40.0 N in fresh water and finds it difficult to stay afloat. In the Great Salt Lake the density of the water is increased due to dissolved salts and the person finds that he can float just submerged with no effort at all. What is the density of the water in the Great Salt Lake?

5. A "U" tube of volume, V, and uniform cross-section is initially filled with water. Then 1/2 V of mercury is poured into one arm of the tube and pushes the water through the tube. How much water spills out of the other arm of the tube?

6. Water is being siphoned from a fish tank. The procedure begins with filling a hose of diameter 1.0 cm with water and then placing one end of the hose in the tank and the other end 1.0 m below the surface of the water, but outside the tank. A net force then acts to push the water from the tank through the tube. Where does this force originate? What is its magnitude?

7. A garden hose having an internal diameter of 2.0 cm is connected to a lawn sprinkler that consists of an enclosure with 36 holes, each 0.10 cm in diameter. If the water in the hose has a speed of 0.75 m/s, at what speed does it leave the sprinkler holes?

8. A venturi meter has a cross section of 40.0 cm^2 at its entry and exit ports and a cross section of 25.0 cm^2 in its constricted region. Water enters at 3.5 m/s. (a) What is the speed of the water through the constricted region? (b) What is the difference in pressure between the two regions?

9. Water is moving with a speed of 8.0 m/s through a pipe of 4.0 cm diameter. The water gradually descends 15 m as the pipe diameter increases to 6.0 cm. (a) What is the speed of flow at the lower level? (b) If the pressure at the upper level is 2.0 x 10^5 Pa, what is the pressure at the lower level?

10. A stream of water (initially horizontal) flows from a small hole in the side of a tank. The stream is 2.5 m from the side of the tank after falling 0.70 m. How far below the surface of the water in the tank is the hole?

11. A straight horizontal pipe with a diameter of 1.0 cm and a length of 50.0 m carries oil with a density of $\rho = 930$ kg/m^3 and a viscosity of 0.12 Pa s. The discharge rate is 0.80 kg/s at atmospheric pressure. Find the gauge pressure at the pipe input.

12. During a heavy rain, a 3.0 m x 4.6 m family room is flooded to a depth of 15 cm. To remove the water ($\eta = 1.00$ x 10^{-3} Pa s), a pump is used that does the job in two hours. The water flows through a horizontal pipe of radius 0.64 cm and length 6.7 m. What gauge pressure does the pump produce?

SOLUTIONS AND ANSWERS

Practice Problems

1. Each tire pushes against the ground with a force $F = PA$ which is equal and opposite to the force exerted by the ground on the tire. The total force exerted by the ground on the tire is then $4F$ which must equal the weight of the car, $4F = 4PA = mg$. Solving for A gives $A = 1/4 \ mg/P$. Now

$$P = (25lb/in^2/14.7 \ lb/in^2)(1.01 \ X \ 10^5 \ Pa) = 1.7 \ X \ 10^5 \ Pa,$$

   so that,

$$A = (1/4)(1600 \ kg)(9.8 \ m/s^2)/(1.7 \ X \ 10^5 \ Pa) = \mathbf{0.023 \ m^2}.$$

2. The net pressure on the vessel is

$$P = \rho gh = (1025 \ kg/m^3)(9.8 \ m/s^2)(1500 \ m) = 1.5 \ X \ 10^7 \ Pa.$$

   The net force on the spherical surface of the vessel is then

$$F = PA = P(4\pi R^2) = (1.5 \ X \ 10^7 \ Pa)(4\pi)(3.5 \ m)^2 = \mathbf{2.3 \ X \ 10^9 \ N}.$$

3. The weight of the water is

$$W = mg = \rho gV = (1000 \ kg/m^3)(9.8 \ m/s^2)(2.4 \ m)(1.8 \ m)(0.23 \ m) = \mathbf{9700 \ N}.$$

4. In fresh water $mg - \rho_w gV = 40.0 \ N$. In the salt water $mg - \rho_s gV = 0$. The first equation gives the volume of the person to be

$$V = [\ (80.0 \ kg)(9.8 \ m/s^2) - 40.0 \ N]/[(1000 \ kg/m^3)(9.8 \ m/s^2)] = 0.076 \ m^3.$$

   The second equation then gives the density of the salt water to be

$$\rho_s = m/V = (80.0 \ kg)/(0.076 \ m^3) = \mathbf{1050 \ kg/m^3}.$$

5. Let h_w and h_m be the heights of the water column and mercury columns, respectively, measured from the mercury-water interface. If the column is not to move, the weight of the mercury above the interface must equal the weight of the water above the interface.

$$\rho_w gh_w A = \rho_m gh_m A.$$

   The length of the column of water lost is $L_w = L - h_w$ where L is the total length of the tube. Additionally, $2h_w - h_m = 1/2 \ L$. Elimination of h_w and h_m from the above equations results in

$$L_w = L - 1/2 \ L \ \{\rho_m/\rho_w \ \}/\{2\rho_m/\rho_w - 1\} = L - (1/2 \ L) \ \{13.6\}/\{27.2 - 1\} = 0.74 \ L.$$

   Hence, the volume of water lost is $V_w = \mathbf{0.74 \ V}$.

6. The force originates from the weight of the water in the portion of the siphoning tube that lies below the surface of the water in the tank. If h is the height of the water level in the tank above the end of the tube, then the net pressure at the end of the tube is

$$p = p_{atm} + \rho gh - p_{atm} = \rho gh.$$

This pressure generates a force of magnitude

$$F = pA = \rho ghA = (1000 \text{ kg/m}^3)(9.8 \text{ m/s}^2)(1.0 \text{ m})(3.14)(0.05 \times 10^{-2} \text{ m})^2 = \mathbf{7.7 \times 10^{-3} \ N}$$

to push the water from the tube.

---

7. Use $A_1v_1 = A_2v_2$, where

$$A_2 = \pi r_2^2 = \pi(1.0 \times 10^{-2} \text{ m})^2 = 3.14 \times 10^{-4} \text{ m}^2, \ v_2 = 0.75 \text{ m/s},$$
$$A_1 = 36(\pi r_1^2) = 36\pi(5.0 \times 10^{-4} \text{ m})^2 = 2.83 \times 10^{-5} \text{ m}^2.$$

Solving for v_1 we obtain,

$$v_1 = v_2(A_2/A_1) = (0.75 \text{ m/s})(3.14 \times 10^{-4} \text{ m}^2)/(2.83 \times 10^{-5} \text{ m}^2) = \mathbf{8.3 \ m/s}.$$

---

8. (a) Equation (11.9), $A_1v_1 = A_2v_2$ so that

$$v_1 = v_2(A_2/A_1) = (3.5 \text{ m/s})(40 \text{ cm}^2/25 \text{ cm}^2) = \mathbf{5.6 \ m/s}.$$

(b) Equation (11.12),

$$P_1 + (1/2)\rho v_1^2 = P_2 + (1/2)\rho v_2^2,$$

solving for the change in pressure,

$$\Delta P = (P_2 - P_1) = (1/2)\rho(v_1^2 - v_2^2) = (1/2)(1.0 \times 10^3 \text{ kg/m}^3)[(5.6 \text{ m/s})^2 - (3.5 \text{ m/s})^2] = \mathbf{9600 \ N/m^2}.$$

---

9. (a) Equation (11.9),

$$v_1 = v_2(A_2/A_1) = (8.0 \text{ m/s})(4.0 \text{ cm}/6.0 \text{ cm})^2 = \mathbf{3.6 \ m/s}.$$

(b) Solving equation (11.11) for P_1 yields, $P_1 = P_2 + (1/2)\rho(v_2^2 - v_1^2) + \rho gh$. Substituting,

$$P_1 = 2.0 \times 10^5 \text{ Pa} + (1/2)(1.0 \times 10^3 \text{ kg/m}^3)[(8.0 \text{ m/s})^2 - (3.6 \text{ m/s})^2]$$
$$+ (1.0 \times 10^3 \text{ kg/m}^3)(9.8 \text{ m/s}^2)(15 \text{ m}) = \mathbf{3.7 \times 10^5 \ Pa}.$$

---

10. Find the speed of the water as it leaves the hole. Using $v_{0y} = 0$ we can write

$$t^2 = 2y/a = 2(0.7 \text{ m})/(9.8 \text{ m/s}^2)$$

$$t = 0.38 \text{ s}.$$

Using $a_x = 0$ gives $x = v_{0x}t$ so that

$$v_{0x} = x/t = (2.5 \text{ m})/(0.38 \text{ s}) = 6.6 \text{ m/s}.$$

Using the result $v = \sqrt{2gh}$, we obtain

$$h = v^2/2g = (6.6 \text{ m/s})^2/2(9.8 \text{ m/s}^2) = \mathbf{2.2 \ m}.$$

11. The flow rate is

$$Q = (0.80 \text{ kg/s})/(930 \text{ kg/m}^3) = 8.6 \times 10^{-4} \text{ m}^3/\text{s}.$$

Now use equation (11.14) and solve for ΔP,

$$\Delta P = P_2 - P_1 = 8\eta LQ/\pi R^4 = 8(0.12 \text{ Pa s})(50.0 \text{ m})(8.6 \times 10^{-4} \text{ m}^3/\text{s})/\pi(5.0 \times 10^{-3} \text{ m})^4 = \textbf{2.1} \times \textbf{10}^{\textbf{7}} \textbf{ Pa}.$$

12. Find the volume flow rate,

$$Q = V/t = (3.0 \text{ m} \times 4.6 \text{ m} \times 0.15 \text{ m})/(2 \text{ h}) = (2.07 \text{ m}^3)/(7200 \text{ s}) = 2.9 \times 10^{-4} \text{ m}^3/\text{s}.$$

Using equation (11.14), we can solve for the gauge pressure, $P_g = P_2 - P_1$, i.e.,

$$P_g = 8\eta LQ/\pi R^4 = 8(1.00 \times 10^{-3} \text{ Pa s})(6.7 \text{ m})(2.9 \times 10^{-4} \text{ m}^3/\text{s})/\pi(0.64 \times 10^{-2} \text{ m})^4 = \textbf{2.9} \times \textbf{10}^{\textbf{3}} \textbf{ Pa}.$$

CHAPTER 12
Temperature and Heat

PREVIEW

In this chapter you will become acquainted with the concept of temperature and the three most common temperature measurement scales: Celsius, Fahrenheit and Kelvin. You will learn to convert a temperature reading in one scale to any other. You will also be introduced to various methods of measuring temperatures and will study in detail one of the most common methods: the thermal expansion of materials.

Additionally, you will study heat including the concepts of specific heat, latent heat of phase change, phase diagrams, and humidity. This chapter will serve as an introduction to later chapters involving the kinetic theory of gases and thermodynamics.

QUICK REFERENCE

Important Terms

Fahrenheit scale
The common temperature scale which assigns 32 ° to the freezing point of water and 212 ° to the boiling point of water.

Celsius scale
A scientific temperature scale which assigns 0 ° to the freezing point of water and 100 ° to the boiling point of water.

Kelvin (absolute) scale
A scientific temperature scale which has the same size degree as the Celsius scale, but assigns the lowest possible temperature as 0 °. A degree is called a Kelvin (K) and is the SI unit for temperature.

Absolute zero
The zero point on the Kelvin temperature scale. It is the lowest possible temperature that can be attained by cooling an object.

Thermal expansion
The change in physical size (length, area or volume) of a substance when its temperature changes. For most substances, the physical size increases with an increase in temperature and decreases with a decrease in temperature.

Thermal stress
A stress which develops within an object when it attempts to expand or contract in response to a temperature change, but cannot due to being held rigidly in place.

Internal energy
The energy associated with individual molecules in a gas, liquid, or solid. This energy may take the form of translational or rotational kinetic energy, vibrational energy, or potential energy.

Heat
The energy that flows from a higher temperature object to a lower temperature object because of the difference in temperatures. The SI unit of heat is the joule (J). Another unit for heat is the calorie (cal) or kilocalorie (kcal).

Specific heat capacity
The heat Q per unit mass per degree change in temperature that must be supplied or removed to change the temperature of a substance. The SI unit for specific heat is J/(kg C°).

Latent heat
The amount of heat energy per kilogram that must be added or removed when a substance changes from one phase to another (i.e., from solid to liquid - heat of fusion, or from liquid to gas - heat of vaporization). The SI unit of latent heat is J/kg.

Equilibrium vapor pressure
The pressure of the vapor phase of a substance that is in equilibrium with the liquid or solid phase. Vapor pressure depends only on temperature.

Phase diagram
A plot of pressure vs. temperature for a given substance showing the various phases possible for that particular substance.

Temperature Scale Conversion

$$T_C = \frac{5}{9}(T_F - 32.0)$$

$$T_F = \frac{9}{5}T_C + 32.0$$

$$T = T_C + 273.15 \qquad (12.1)$$

Thermal Expansion

Linear expansion

$$\Delta L = \alpha L_0 \Delta T \qquad (12.2)$$

Volume Expansion

$$\Delta V = \beta V_0 \Delta T \qquad (12.3)$$

DISCUSSION OF SELECTED SECTIONS

12.1 Common Temperature Scales

Example 1
At what temperature will the reading on the Fahrenheit scale be numerically equal to that on the Kelvin scale?

The relationship between the two scales is given by $T_F = (9/5)\,T_C + 32.0$. Setting $T_F = T_C$ and rearranging gives

$$T_F - (9/5)\,T_F = 32.0$$

$$T_F = -(5/4)(32.0) = -40.0$$

Example 2
The surface temperature of the sun is approximately 5700 °C. What is this temperature on the Fahrenheit scale?

$$T_F = (9/5)\,T_C + 32.0$$
$$T_F = (9/5)(5700) + 32.0 = 1.0 \times 10^4 \;°F$$

Example 3
Most bank thermometers alternately display the temperature in Fahrenheit and Celsius. If a bank thermometer displays 25 °F, what will it display for the Celsius temperature?

The relationship between Fahrenheit and Celsius temperatures is

$$T_C = (5/9)(T_F - 32.0)$$
$$T_C = (5/9)(25 - 32.0) = -4\ °C$$

12.2 The Kelvin Temperature Scale

The only difference between the Kelvin and Celsius scales is in the choice of the zero point. The zero point for the Celsius scale was chosen to be the freezing point of water since water is abundant and its freezing point is easily measurable. The zero of the Kelvin scale is chosen to be the lowest possible temperature, that is, absolute zero. Absolute zero corresponds to -273.15 °C. To convert between the two scales it is only necessary to adjust for this difference in zero points by adding or subtracting 273.15.

If you should need to convert from Fahrenheit to Kelvin, simply convert the Fahrenheit temperature to Celsius and then add 273.15 to get the Kelvin temperature.

Example 4
Many people keep their rooms at a temperature of 65 °F. What is this temperature on the Kelvin scale?

First convert the temperature to Celsius.

$$T_C = (5/9)(T_F - 32.0) = (5/9)(65 - 32.0) = 18\ °C$$

Now add 273.15 to get the temperature on the Kelvin scale.

$$T = 18 + 273.15 = 291\ K$$

12.4 Linear Thermal Expansion

Example 5
How much will the length of a 1.0 km section of concrete highway change if the temperature varies from -15 °C in the winter to 41 °C in the summer?

Using equation 12.2 and the value of the coefficient of linear expansion for concrete from Table 12.1, we have

$$\Delta L = \alpha LT = (12 \times 10^{-6}\ /C°)(1.0 \times 10^3\ m)[(41\ °C - (-15\ °C)]$$
$$\Delta L = 0.67\ m$$

Example 6
Two 12 ft sections of aluminum siding are placed end to end on the outside wall of a house. How large a gap should be left between the pieces to prevent buckling if the temperature can change by 55 C°?

Each piece can expand an amount

$$\Delta L = \alpha L \Delta T = (23 \times 10^{-6}\ /°C)(12\ ft)(55\ °C) = 0.015\ ft = 0.18\ in.$$

If each piece can expand in both directions, then each will expand 0.09 in. toward the other. The gap should be at least 2(0.09 in.) = 0.18 in. to keep the pieces from contacting each other and buckling.

Example 7

A glass window pane (1.100 m × 0.920 m) is fitted inside an aluminum frame with a clearance of 0.05 cm all around on a day when the temperature is 39 °C. How much must the temperature drop for the frame to begin to exert a stress on the window pane?

The changes of the largest dimensions of the pane and frame are $\Delta L_g = \alpha_g L_g \Delta T$ and $\Delta L_a = \alpha_a L_a \Delta T$. The pieces will contact when $\Delta L_a - \Delta L_g = 0.10$ cm. The necessary change in temperature is then

$$\Delta T = \frac{0.10 \times 10^{-2} \, m}{\alpha_a L_a - \alpha_g L_g} = \frac{0.10 \times 10^{-2} \, m}{(23 \times 10^{-6}/C°)(1.101 \, m) - (8.5 \times 10^{-6}/C°)(1.100 \, m)}$$

$$\Delta T = 63 \, C°$$

Similarly, for the smallest dimensions,

$$\Delta T = \frac{0.10 \times 10^{-2} \, m}{(23 \times 10^{-6}/C°)(0.921 \, m) - (8.5 \times 10^{-6}/C°)(0.920 \, m)} = 75 \, C°$$

The window pane and frame will contact along the large dimension first when the temperature drops by 63 C°.

12.5 Volume Thermal Expansion

Example 8

A diesel powered automobile has its steel tank filled to the top with 22.0 gal of diesel fuel from an underground storage tank. The fuel in the storage tank is initially at 55 °F (13 °C) and warms up to 95 °F (35 °C) as it sits in the car's tank. How much fuel spills out of the car's tank? The coefficient of volume expansion of diesel fuel is 9.5 × 10⁻⁴ /C°.

The interior volume of the car's tank expands as if it were made of steel. It will increase by $\Delta V_t = \beta_t V \Delta T$. The fuel will increase its volume by an amount $\Delta V_f = \beta_f V \Delta T$. The amount of fuel which will spill out of the tank, ΔV, is then

$$\Delta V = \Delta V_f - \Delta V_t = (\beta_f - \beta_t) V \Delta T$$
$$\Delta V = (9.5 \times 10^{-4}/C° - 36 \times 10^{-6}/C°)(22 \, gal)(22 \, C°)$$
$$\Delta V = 0.44 \, gal$$

12.7 Heat and Temperature Change: Specific Heat Capacity

In order to raise the temperature of a substance by an amount ΔT, a certain amount of heat, Q, is required. The greater the mass m of the substance, the greater the amount of heat required to raise it by ΔT. The relationship between the heat required, the mass, and the change in temperature is

$$Q = cm\Delta T \qquad (12.4)$$

where the constant of proportionality, c, is referred to as the **specific heat capacity** or simply the specific heat. The specific heat has units of J/(kg C°). Table 12.2 shows the specific heat capacities for various substances. For example, water has a specific heat of 4186 J/(kg C°). In other words, it would take 4186 J of heat to raise the temperature of 1 kilogram of water by 1 °C.

Example 9
How much heat is required to raise 5.0 kg of water from 25 °C to 55° C?

Using equation (12.4) we have:

$$Q = cm\Delta T = [4186 \text{ J/(kg C}°)](5.0 \text{ kg})(55 \text{ °C} - 25 \text{ °C}) = 6.3 \times 10^5 \text{ J.}$$

A common unit for measuring heat is the **kilocalorie** (kcal), defined as the amount of heat energy required to raise the temperature of 1 kilogram of water by 1 °C. From the above discussion it is obvious that 4186 J = 1 kcal. This conversion factor is known as the **mechanical equivalent of heat**.

Example 10
If 15 kcal of heat are added to 5.0 kg of silver, by how much will its temperature rise?

Using equation (12.4) and the specific heat of silver found in Table 12.2, we can solve for ΔT. We have

$$\Delta T = \frac{Q}{cm} = \frac{(15 \text{ kcal})(4186 \text{ J/1 kcal})}{[235 \text{ J/(kg C}°)](5.0 \text{ kg})} = 53 \text{ C}°$$

Specific heat capacity can be measured using the technique of **calorimetry**. A calorimeter is an insulated container much like a thermos for hot coffee or iced tea. An ideal calorimeter would prevent any heat from leaking in or out. However, heat can flow between the materials inside the calorimeter which have different temperatures. Cooler materials gain heat while hotter materials lose heat until a common temperature is reached. That is, until **thermal equilibrium** is achieved. In the process of reaching thermal equilibrium, the total amount of heat gained equals the total amount of heat lost.

Example 11
An aluminum cup having a mass of 250.0 g is filled with 50.0 g of water. The initial temperature of the cup and water is 25.0 °C. A 75.0-g piece of iron initially at 350.0 °C is dropped into the water. What is the final equilibrium temperature of the system assuming that no heat is lost to the outside environment?

Using equation (12.4) and the fact that the heat lost by the iron equals the heat gained by the aluminum + water,

$$\text{Heat Gained} = \text{Heat Lost}$$

$$(cm\Delta T)_{\text{aluminum}} + (cm\Delta T)_{\text{water}} = (cm\Delta T)_{\text{iron}}$$

$$[9.00 \times 10^2 \text{ J/(kg C}°)](0.250 \text{ kg})(T_f - 25.0 \text{ °C}) + [4186 \text{ J/(kg C}°)](0.0500 \text{ kg})(T_f - 25.0 \text{ °C})$$
$$= [452 \text{ J/(kg C}°)](0.0750 \text{ kg})(350.0 \text{ °C} - T_f)$$

$$T_f = 48.5 \text{ °C}$$

12.8 Heat and Phase Change: Latent Heat

There are three phases of matter that exist in nature; solid, liquid, and vapor (gas). Nearly all substances can exist in all three phases, depending on the conditions of temperature and pressure that prevail. For example, consider water at atmospheric pressure. For low temperatures (below 0 °C) water exists as a solid, in the form of ice. At intermediate temperatures (between 0 °C and 100 °C) water is a liquid, and at high temperatures (above 100 °C) it is a gas (water vapor).

We know that water can change phases, either from solid to liquid (or vice versa) or from liquid to gas. The simple act of ice melting represents a phase change.

It is important to note that phase changes always occur at a single temperature, with the addition or subtraction of heat. For example, in order to melt 1 kg of ice, the ice must be at 0 °C and we need to add 33.5×10^4 J of heat to get 1 kg of liquid water which will still be at 0 °C.

The amount of heat per kilogram required to produce a change of phase for a substance is known as the **latent heat** of that substance. The **latent heat of fusion, L_f**, refers to a phase change between the liquid and solid phases, and the **latent heat of vaporization, L_v**, refers to a change between liquid and vapor phases.

Example 12

How much heat is required to convert 250 g of ice at 0 °C to steam at 100 °C?

The problem includes three distinct parts. (1) The heat required to melt the ice, (2) the heat required to raise the temperature of the water from 0 °C to 100 °C, and (3) the heat required to convert the hot water to steam at 100 °C. Referring to Table 12.3 for the values of the latent heats of fusion and vaporization we can write

$$Q_{total} = Q_{fusion} + Q_{temp.\ change} + Q_{vaporization}$$
$$Q_{total} = mL_f + cm\Delta T + mL_v$$
$$Q_{total} = (0.25\ kg)(33.5 \times 10^4\ J/kg) + [4186\ J/(kg\ C°)](0.25\ kg)(100\ C°)$$
$$+ (0.25\ kg)(22.6 \times 10^5\ J/kg)$$
$$Q_{total} = 7.5 \times 10^5\ J.$$

12.9 Equilibrium Between Phases of Matter

Under certain conditions of temperature and pressure, a substance can exist in equilibrium in more than one phase at the same time. The pressure at which vapor coexists in equilibrium with liquid is called the **equilibrium vapor pressure**. For example, consider the following **vaporization curve** for water.

We can see that, for water, at a pressure of 1.01×10^5 Pa, the corresponding vaporization temperature is 100 °C. Thus, liquid water and vapor can exist in equilibrium at this pressure and temperature. Water, therefore, boils at 100 °C at atmospheric pressure. However, if we were to change the pressure, the corresponding temperature needed for equilibrium would change. For example, as evident from the curve, if the pressure were about 2×10^5 Pa, the corresponding temperature required for equilibrium would be about 125 °C. So water would boil at 125 °C if it were subjected to two atmospheres of pressure.

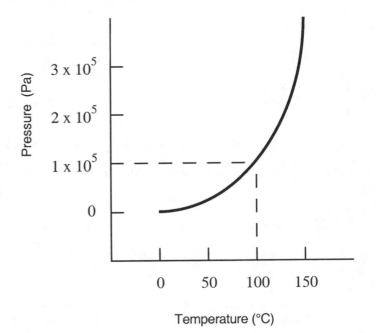

12.10   Humidity

The partial pressure of water vapor is the part of the total atmospheric pressure that is due to water vapor in the air. For example, out of the total 1.01×10^5 Pa of pressure in our atmosphere, 4.0×10^3 Pa may be due to water vapor.

The **relative humidity** is the partial pressure of water vapor at the existing temperature divided by the equilibrium vapor pressure of water at the existing temperature. That is

$$\text{Percent Relative Humidity} = \frac{\text{Partial pressure of water vapor}}{\text{Equil. Vapor Pressure at Given Temperature}} \times 100 \qquad (12.6)$$

Example 13

On a certain day the partial pressure of water vapor is 1.0×10^3 Pa. Using the vaporization curve below, determine the relative humidity if the temperature is (a) 16 °C, and (b) 26 °C.

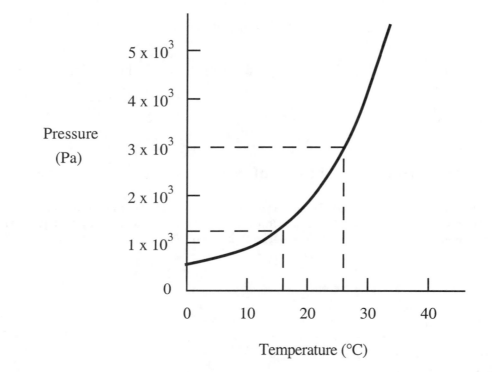

(a) At 16 °C we see that the equilibrium vapor pressure is 1.3×10^3 Pa. Therefore, using equation (12.6), we have

$$\text{Relative humidity at } 16\,°C = \frac{1.0 \times 10^3 \text{ Pa}}{1.3 \times 10^3 \text{ Pa}} \times 100 = 77\%$$

(b) At 26 °C the equilibrium vapor pressure is 3.0×10^3 Pa. Therefore,

$$\text{Relative humidity at } 26\,°C = \frac{1.0 \times 10^3 \text{ Pa}}{3.0 \times 10^3 \text{ Pa}} \times 100 = 33\%$$

When the partial pressure of water vapor equals the equilibrium vapor pressure of water at a given temperature, equation (12.6) tells us that the relative humidity is 100%. In this case, the air is said to be saturated. The temperature corresponding to this condition is known as the **dew point**. The dew point is therefore the temperature at which the water vapor in the air would condense in the form of liquid drops (dew or fog). In example 13 it can be seen that the dew point (the temperature corresponding to a vapor pressure of 1.0×10^3 Pa) is about 13 °C.

PRACTICE PROBLEMS

1. At what common temperature are the Fahrenheit and Kelvin temperatures the same?

2. A 1.5 m steel wire is heated from room temperature (22 °C) to 150 °C. How much does its length change?

3. What is the minimum number of 1/2 inch expansion joints that must be cut per mile in a concrete highway to prevent buckling if the temperature can range from - 15 °F to 105 °F?

4. A brass collar (inside radius = 2.500 cm) is to be fitted on a steel shaft (radius 2.505 cm). The collar and shaft are initially at 22.0 °C. To what temperature must the collar and shaft be heated?

5. A bolt hole in the cast iron block of an automobile engine has a diameter of 1.27 cm when the engine is at a cool temperature of 25.0 °C. What is the diameter of the hole when the engine block is at a temperature of 98.0 °C?

6. What is the temperature on the Fahrenheit temperature scale that corresponds to absolute zero?

7. An aluminum airplane has a wingspan of 9.50 m on the ground where the temperature is 25 °C. What is its wingspan when it is at an altitude where the temperature is - 35 °C?

8. A one quart glass canning jar is filled with veggies and liquid at room temperature (72 °F) and heated to 212 °F so that liquid has overflowed the jar, and cooled back to room temperature. Assume that the veggies and liquid expand like water and find the volume of the hot veggies and liquid which have spilled out of the jar.

9. Show that the change in area of a rectangular sheet of material with a linear coefficient of expansion, α, is approximately $\Delta A = 2\alpha A \Delta T$.

10. An automotive cooling system holds 15 quarts of pure water at 10.0 °C. Use Figure 12.19 in the text to determine how much the volume of the water has changed if the temperature drops to 4.0 °C.

11. A 150.0-g glass beaker contains 750 g of water initially at 25.0 °C. A 250.0-g piece of copper is taken out of a hot oven and dropped into the water. After equilibrium is reached between the copper, glass, and water, the final temperature is 32.0 °C. What was the temperature of the oven?

12. Calculate the specific heat of a metal from the following data. A container made of metal has a mass of 1.50 kg and contains 6.00 kg of water. A 0.900 kg piece of the same metal initially at a temperature of 275 °C is dropped into the water. The container and water are initially at a temperature of 17.0 °C and the final equilibrium temperature of the system is 26.0 °C.

13. A well-insulated container holds 1.5 liters of water at 35 °C. How much ice (initially at 0 °C) must be added to this water in order to produce water at 12 °C once equilibrium has been achieved? Neglect the heat capacity of the container.

14. A 50.0-g aluminum calorimeter contains water at a temperature of 26 °C. When 75-g of lead at a temperature of 85 °C is dropped into the calorimeter, the final equilibrium temperature is 28 °C. How much water was in the calorimeter can?

15. Ice at -10.0 °C and water vapor at 130 °C are brought together at atmospheric pressure in a perfectly insulated container whose heat capacity can be neglected. After thermal equilibrium is reached, the liquid phase at 50.0 °C is present. Ignoring the equilibrium vapor pressure at 50.0 °C, find the ratio of the mass of the steam to that of the ice.

16. Using the vaporization curve that accompanies this problem, find the partial pressure of water vapor on a day when the weather forecast gives the relative humidity at 60% and the temperature at 30 °C.

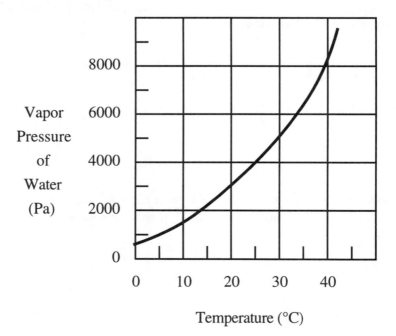

17. Using the vaporization curve which accompanies problem 16, determine the dew point when the temperature is 40.0 °C and the relative humidity is 37.5%.

18. The outdoor temperature is 15 °C and the relative humidity is 75%. The partial pressure of water vapor inside a house is the same as outdoors, but the temperature in the house is 25 °C. Using the vaporization curve for water vapor that accompanies problem 16, determine the relative humidity in the house.

SOLUTIONS AND ANSWERS

Practice Problems

1. We know

$$T = T_c + 273.15$$

and

$$T_c = 5/9 \ (T_F - 32.0)$$

so

$$T = 5/9 \ (T_F - 32.0) + 273.15. \text{ If } T_F = T,$$

then

$$(4/9)T = (5/9)(- 32.0) + 273.15$$

which gives

$$T = \mathbf{574.6} \text{ K or } °F.$$

2. Equation (12.2) applied to the steel wire gives

$$\Delta L = \alpha L \Delta T = (12 \text{ X } 10^{-6} /C°)(1.5 \text{ m})(150 \text{ °C} - 22 \text{ °C}) = \mathbf{2.3 \text{ X } 10^{-3} \text{ m}}$$

3. Equation (12.2) gives the expansion of one mile of concrete to be

$$\Delta L = \alpha L \Delta T = (12 \text{ X } 10^{-6} /C°)(5280 \text{ ft/ 1 mi})(12 \text{ in/1 ft})(5/9)(120 \text{ C°}) = 51 \text{ in.}$$

The number of 1/2 in expansion joints needed to just compensate for this increase in length is

$$(51 \text{ in})/(1/2 \text{ in}) = \mathbf{102}$$

4. Equation (12.2) gives for the expansion of the radius of the collar and for the expansion of the radius of the shaft

$$\Delta r_c = \alpha_b r_c \ (T - 22 \text{ °C})$$

and

$$\Delta r_s = \alpha_s r_s \ (T - 22 \text{ °C}).$$

We want

$$r_c + \Delta r_c = r_s + \Delta r_s$$

for the collar to just fit the shaft.

$$T = 22 \text{ °C} + (r_s - r_c)/[\alpha_b r_c + \alpha_s r_s]$$
$$T = 22 \text{ °C} + (2.505 \text{ cm} - 2.500 \text{ cm})/[(19 \text{ X } 10^{-6} /C°)(2.500 \text{ cm}) - (12 \text{ X } 10^{-6} /C°)(2.505 \text{ cm})]$$
$$T = \mathbf{309 \text{ °C}}$$

5. The bolt hole expands as if it were made of cast iron so its diameter changes by an amount

$$\Delta d = \alpha d \Delta T = (12 \text{ X } 10^{-6} /C°)(1.27 \text{ cm})(73 \text{ C°}) = 1.1 \text{ X } 10^{-3} \text{ cm.}$$

Now

$$d' = d + \Delta d = 1.27 \text{ cm} + 1.1 \text{ X } 10^{-3} \text{ cm} = \mathbf{1.27 \text{ cm.}}$$

6. We know

$$T_F = 9/5\, T_c + 32.0$$

and

$$T_c = T - 273.15; \text{ so for } T = 0$$

we have

$$T_F = 9/5\,(-273.15) + 32.0 = \textbf{-459.7 °F}$$

---

7. The change in the wingspan of the airplane is given by equation (12.2) to be

$$\Delta L = \alpha L \Delta T = (23 \times 10^{-6}\,/C°)(9.50\ m)(-60\ C°) = -0.013\ m$$

The new wingspan is then

$$9.50\ m - 0.013\ m = \textbf{9.49 m}.$$

---

8. The change in volume of the contents of the jar is

$$\Delta V_c = \beta_w V \Delta T$$

and the change in the interior volume of the jar is

$$\Delta V_j = = \beta_g V \Delta T.$$

The amount of contents which spill from the jar is then

$$\Delta V = \Delta V_c - \Delta V_g = (\beta_w - \beta_g)V\Delta T$$
$$\Delta V = (207 \times 10^{-6}\,/C° - 8.5 \times 10^{-6}\,/C°)(1.00\ qt)(5/9)(140)C° = \textbf{0.015 qt}.$$

---

9. Let the rectangular sheet have a length, a, and a width, b. Each dimension will expand by an amount given by equation (12.2).

$$\Delta a = \alpha a \Delta T \text{ and } \Delta b = \alpha b \Delta T.$$

The expanded area of the rectangle is

$$(a + \Delta a)(b + \Delta b) = ab + a\Delta b + b\Delta a + \Delta a\Delta b = ab + 2\alpha ab\Delta T + \alpha^2 ab(\Delta T)^2.$$

The change in the area of the rectangle is then

$$\Delta A = 2\alpha ab\Delta T + \alpha^2 ab(\Delta T)^2.$$

The last term is very small compared to the first term since it contains α^2, and can be neglected so

$$\Delta A \approx 2\alpha A \Delta T$$

where A = ab, the original area of the rectangle.

---

10. The mass of the water in the radiator does not change so the initial and final densities and volumes are related by $\rho_f V_f = \rho_i V_i$. The change in the volume of the water is

$$\Delta V = V_f - V_i = (\rho_i/\rho_f - 1)V_i = (999.7/999.973 - 1)(15\ qt) = \textbf{-4.1} \times 10^{-3}\ \textbf{qt}.$$

The negative sign indicates a decrease in the volume.
By the way, this answer corresponds to about 0.13 ounces.

11. We know that
$$Q_{wat} + Q_{glass} = Q_{Cu}$$
or
$$(cm\Delta T)_{wat} + (cm\Delta T)_{glass} = (cm\Delta T)_{Cu}.$$

Substitution yields,

$$[4186 \text{ J/(kg C}°)](0.750 \text{ kg})(7.0 \text{ C}°) + [840 \text{ J/(kg C}°)](0.150 \text{ kg})(7.0 \text{ C}°) = [387 \text{ J/(kg C}°)](0.250 \text{ kg})\Delta T.$$

Solving for ΔT we obtain, $\Delta T = 236$ C°. Therefore,

$$T_i = T_f + \Delta T = 32.0 \text{ °C} + 236 \text{ C}° = \mathbf{268 \text{ °C}}.$$

12. Thermal equilibrium implies that
$$Q_{met.\ cont.} + Q_{wat} = Q_{metal}$$
or
$$(cm\Delta T)_{cont} + (cm\Delta T)_{wat} = (cm\Delta T)_{met}.$$

$$(1.50 \text{ kg})(9.0 \text{ C}°)c_{met} + [4186 \text{ J/(kg C}°)](6.00 \text{ kg})(9.0 \text{ C}°) = (0.900 \text{ kg})(249 \text{ C}°)c_{met}.$$

Solving for the specific heat of the metal we obtain,

$$c_{met} = \mathbf{1070 \text{ J/(kg C}°)}.$$

13. The ice melts and then warms up to 12 °C. We have,
$$Q_{wat} = Q_{ice} + Q_{ice\ wat}.$$
Therefore, we can write,
$$(cm\Delta T)_{wat} = m_{ice}L_f + (cm\Delta T)_{ice\ wat}.$$

Using the fact that 1.5 liters of water equals 1.5 kg we have,

$$[4186 \text{ J/(kg C}°)](1.5 \text{ kg})(23 \text{ C}°) = (33.5 \times 10^4 \text{ J/kg})m_{ice} + [4186 \text{ J/(kg C}°)](12 \text{ C}°)m_{ice}.$$

Solving for the mass of the ice we obtain,
$$m_{ice} = \mathbf{0.37 \text{ kg}}.$$

14. We know that
$$\Delta Q_{alum} + \Delta Q_{wat} = \Delta Q_{lead}$$
$$(cm\Delta T)_{alum} + (cm\Delta T)_{wat} = (cm\Delta T)_{lead}.$$

Substitution yields,

$$[9.00 \times 10^2 \text{ J/(kg C}°)](0.050 \text{ kg})(2.0 \text{ C}°) + [4186 \text{ J/(kg C}°)](2.0 \text{ C}°)m_{wat} = [128 \text{ J/(kg C}°)](0.075 \text{ kg})(57 \text{ C}°).$$

Solving for the mass of the water we obtain,
$$m_{wat} = \mathbf{55 \text{ g}}.$$

15. The ice and water vapor exchange equal amounts of heat so $Q_{ice} = Q_{steam}$. Each experiences a phase change so

$$(cm\Delta T)_{ice} + m_{ice}L_f + (cm\Delta T)_{wat} = (cm\Delta T)_{steam} + m_{steam}L_v + (cm\Delta T)_{wat}.$$

Substitution yields,

$$[2.00 \times 10^3 \text{ J/(kg C°)}](10.0 \text{ C°})m_{ice} + (33.5 \times 10^4 \text{ J/kg})m_{ice} + [4186 \text{ J/(kg C°)}](50.0 \text{ C°})m_{ice} =$$
$$[2020 \text{ J/(kg C°)}](30.0 \text{ C°})m_{steam} + (22.6 \times 10^5 \text{ J/kg})m_{steam} + [4186 \text{ J/(kg C°)}](50.0 \text{ C°})m_{steam}.$$

Solving for the ratio of the mass of the steam to the mass of the ice,

$$m_{steam}/m_{ice} = \textbf{0.223}.$$

---

16. Using equation (12.6) we have, R.H. = (P.P./E.V.P.) x 100. From the curve we find that the equilibrium vapor pressure (E.V.P.) at 30 °C is 5000 Pa. The relative humidity (R.H.) is 60% so that the partial pressure of water vapor is

$$\text{P.P.} = \text{(R.H.)(E.V.P.)}/100 = (60)(5000 \text{ Pa})/100 = \textbf{3000 Pa}.$$

---

17. At 40.0 °C the equilibrium vapor pressure (from the graph) is 8000 Pa. For a relative humidity of 37.5%, the partial pressure of water vapor is

$$\text{P.P.} = \text{(R.H.)(E.V.P.)}/100 = (0.375)(8000 \text{ Pa}) = 3000 \text{ Pa}.$$

Therefore, the dew point (saturation) occurs at a temperature when the E.V.P. is 3000 Pa. From the graph we see that T = **20 °C**.

---

18. Find the partial pressure of water vapor outside,

$$\text{P.P.} = \text{(R.H.)(E.V.P. at 15 °C)}/100 = (75\%)(2000 \text{ Pa})/100 = 1500 \text{ Pa}.$$

Inside, the E.V.P. at 25 °C is 4000 Pa. Therefore,

$$\text{(R.H. at 25 °C)} = [\text{(P.P.)/(E.V.P. at 25 °C)}] \times 100.$$

Thus,

$$\text{R.H. at 25 °C} = [(1500 \text{ Pa})/(4000 \text{ Pa})] \times 100 = \textbf{38\%}.$$

CHAPTER 13
The Transfer of Heat

PREVIEW

In this chapter you will be introduced to the mechanisms of heat transfer - convection, conduction, and radiation. After completion of the chapter you will be able to explain how heat is transferred by convection and calculate the rate at which heat is transferred by conduction through various materials. In addition, you will be able to use the Stefan - Boltzmann law to calculate the rate at which energy is radiated from or absorbed by an object.

QUICK REFERENCE

Important Terms

Convection
A process in which heat energy is transferred by the flow of a fluid.

Convection Current
The flow of a fluid when heat is transferred by convection.

Natural convection
A convection process in which the fluid flows due to buoyant forces produced because the heated fluid is less dense than the surrounding cooler fluid.

Forced convection
A convection process in which an external device, such as a fan, is used to produce the fluid flow.

Conduction
A process by which heat is transferred through a material without a bulk movement of the material.

Thermal conductors
Materials which conduct heat well.

Thermal insulators
Materials which conduct heat poorly.

Radiation
A process by which energy is transferred by electromagnetic waves.

Blackbody
An idealized perfect absorber and perfect emitter of radiation.

Equations

The **heat conducted** through a bar is

$$Q = \frac{kA\Delta Tt}{L} \tag{13.1}$$

The Stefan - Boltzmann law gives the **radiant energy emitted or absorbed** by an object:

$$Q = e\sigma T^4 At \tag{13.2}$$

DISCUSSION OF SELECTED SECTIONS

13.2 Conduction

The ability of a substance to transfer heat by conduction is measured by its **thermal conductivity**, k, which appears in equation (13.1). A glance at the values of k given in Table 13.1 of the text, shows that metals have the highest thermal conductivities and are good conductors of heat, while gases generally have the lowest thermal conductivities and are poor conductors of heat. This is why most commercial insulating materials depend on trapped gases, such as air, for their insulating properties.

Example 1

Fiber glass insulation is composed of small glass fibers tangled together so that air is trapped in the spaces between them. Show that the insulating value of fiber glass insulation is due to the trapped air rather than the glass by comparing the heat which will flow through a 1.0 m^2 sheet of 3 1/2 inch thick glass in 1.0 minute and a similar sheet of air. Assume that the temperature difference maintained across the sheets is 25 C°.

The heat conducted through the glass sheet is

$$Q_g = \frac{k_g A \, \Delta T \, t}{L} = \frac{(0.80 \text{ J/s m C°})(1.0 \text{ m}^2)(25 \text{ C°})(60 \text{ s})}{8.9 \times 10^{-2} \text{ m}} = 1.3 \times 10^4 \text{ J}$$

and the heat conducted through the equivalent sheet of air is

$$Q_a = \frac{k_a A \, \Delta T \, t}{L} = \frac{(0.0256 \text{ J/s m C°})(1.0 \text{ m}^2)(25 \text{ C°})(60 \text{ s})}{8.9 \times 10^{-2} \text{ m}} = 430 \text{ J}$$

The glass conducts about thirty times as much heat as the air.

Example 2

Two identically shaped bars of different materials are placed end to end. A temperature difference, ΔT, is maintained across the bars. Find an expression for the temperature difference across the first bar. Assume that no heat is lost through the sides of the bars.

If no heat is lost through the sides of the bars, then the heat transferred through each bar in a time, t, is Q. Equation 13.1 gives for the first bar

$$Q = \frac{k_1 A \, \Delta T_1 \, t}{L}$$

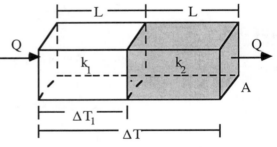

and for the second bar

$$Q = \frac{k_2 A \, (\Delta T - \Delta T_1) \, t}{L}$$

Equating and rearranging yields

$$\Delta T_1 = \left(\frac{k_2}{k_1 + k_2} \right) \Delta T$$

In the above example we can define an equivalent thermal conductivity of the two bars if we use the value that we found for ΔT_1 in the first equation.

$$Q = \frac{k_1 A\, \Delta T\, t}{L} = \left(\frac{k_1 k_2}{k_1 + k_2}\right)\left(\frac{A\, \Delta T\, t}{L}\right)$$

The two bars conduct heat energy as if they were a single bar of length, 2L, and thermal conductivity

$$k = \frac{2k_1 k_2}{k_1 + k_2}$$

Example 3
In example 2 the first bar is brass and the second bar is aluminum. The temperature of the left end is 95 °C and the temperature of the right end is 55 °C. What is the temperature at the junction between the bars? What is the equivalent conductivity of the bars?

Table 13.1 gives the thermal conductivity of brass, $k_1 = 110$ J/s m C°, and the conductivity of aluminum, $k_2 = 240$ J/s m C°. The temperature difference across the brass bar is then

$$\Delta T_1 = \left(\frac{k_2}{k_1 + k_2}\right)\Delta T = \left(\frac{240}{110 + 240}\right)(95\ ^\circ\text{C} - 55\ ^\circ\text{C})$$

$$\Delta T_1 = 27\ ^\circ\text{C}$$

The units of the conductivities have been suppressed for clarity.

The temperature at the junction between the bars is T = 95 °C - 27 °C = 68 °C.
The equivalent thermal conductivity of the bars is

$$k = \frac{2k_1 k_2}{k_1 + k_2} = \frac{2(110)(240)}{110 + 240}\ \text{J/(s m C}^\circ) = 150\ \text{J/(s m C}^\circ)$$

Example 4
Two bars of different conductivity are oriented as shown in the drawing. The upper bar transfers heat energy, Q_1, while the second bar transfers Q_2, in the same time, t. Find the total heat transferred in time, t, in terms of the temperature difference, ΔT, the length of the bars, L, and the cross-sectional area of the bars, A. Assume that no heat flows through the sides of the bars.

The heat flowing through the upper bar is

$$Q_1 = \frac{k_1 A\, \Delta T\, t}{L}$$

and the heat flowing through the lower bar is

$$Q_2 = \frac{k_2 A\, \Delta T\, t}{L}$$

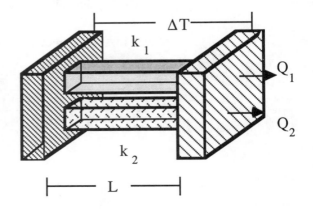

The total heat is $Q = Q_1 + Q_2$ so

$$Q = (k_1 + k_2)\left(\frac{A\,\Delta T\,t}{L}\right)$$

As before, the two bars conduct heat as if they were a single bar. This time the length of the single bar is L and the "equivalent" conductivity is $k = k_1 + k_2$.

13.3 Radiation

Both emission and absorption of radiation by an object are governed by the Stefan - Boltzmann law, equation (13.2). The temperature, T, in the equation is the Kelvin temperature of the object in the case of emission and it is the Kelvin temperature of the object's surroundings in the case of absorption.

In order to use equation (13.2), we need to know the emissivity, e, of the object in question. It ranges from e = 1 for a perfect blackbody to e = 0 for a perfect reflector of radiation. The emissivity of real objects varies with the type of radiation (visible, infrared, etc.) and is usually determined by experiment.

Example 5

The sun shines directly on one side of a flat black panel on a spacecraft which has an area of 5.0 m^2 and an emissivity of e = 0.95. The spacecraft is located 1.50×10^{11} m from the sun. How much power does the panel absorb from the sun if the sun is considered to be an ideal blackbody with a temperature of 5700 K? Assuming that the panel can only lose energy through radiation, what is the equilibrium temperature of the panel?

The total power emitted by the sun is given by equation 13.2 with T = 5700 K, the surface temperature of the sun. The radius of the sun is $R_s = 6.96 \times 10^8$ m.

$$Q_s/t = \sigma T_s^4 (4\pi R_s^2)$$

This energy spreads out over the surface of a sphere of radius, r, so that the amount reaching the panel is

$$\frac{Q}{t} = \sigma T_s^4 \left(\frac{4\pi R_s^2}{4\pi r^2}\right) A$$

$$\frac{Q}{t} = (5.67 \times 10^{-8}\ \text{J / s m}^2\ \text{K}^4)(5700\ \text{K})^4 \left(\frac{6.96 \times 10^8\ \text{m}}{1.50 \times 10^{11}\ \text{m}}\right)(5.0\ \text{m}^2) = 6400\ \text{watts}$$

The panel is in equilibrium if its temperature, T_p, remains constant. This means that the panel is losing as much power as it is gaining. It gains energy only through the one sunlit side, but it can radiate energy from both sides.

$$\text{power lost } = \text{ power gained}$$

$$e\sigma T_p^4 A = 6400\ \text{watts}$$

$$T_p = \sqrt[4]{\frac{6400\ \text{watts}}{(0.95)(5.67 \times 10^{-8}\ \text{J / s m}^2\ \text{K}^4)(10.0\ \text{m}^2)}} = 330\ \text{K}$$

PRACTICE PROBLEMS

1. How much heat per square meter is conducted in one hour through a 0.64 cm thick window pane when the inside temperature is 22 °C and the outside temperature is 0 °C?

2. How much heat per square meter is conducted in one hour through two 0.64 cm thick window panes separated by a 0.64 cm stagnant air gap when the inside temperature is 22 °C and the outside temperature is 0 °C?

3. Two cooking pots are identical in every respect except that one has a 0.64 cm thick copper bottom and the other has an aluminum bottom. How thick should the aluminum bottom be in order for it to conduct the same amount of heat from a stove burner in the same time as the copper bottom?

4. A star radiates as if it were a blackbody with a temperature of 3000 K. If it radiates the same power as the sun, what is the ratio of its radius to the radius of the sun? Assume the sun has a blackbody temperature of 6000 K.

5. An object of emissivity, 0.5, radiates a certain amount of power when it is at a temperature of 650 °C. If the object were a perfect blackbody, what temperature would be required for it to radiate the same power?

6. The top of an electric stove burner has an emissivity of 0.800 and an area of 6.00×10^{-3} m^2. The stove is in a kitchen where the ambient temperature is 27° C and is heated so that it glows cherry red (T = 727 °C). What is the net power lost by the top of the burner?

7. How much heat is lost in one hour through a 15 cm × 3.7 m × 6.1 m concrete floor if the inside temperature is 22.0 °C and the ground temperature is 13.0 °C?

8. Compare the heat lost through a brick wall with the heat lost through the same wall with 1/2 inch of styrofoam insulation. Bricks are 4.0 inches thick and have a thermal conductivity of 1.0 J/(s m C°). Assume that the inside and outside temperatures are the same for both walls.

9. A quartz radiant heater has two cylindrical tubes 40.0 cm long and 1.0 cm in radius. The heater is designed to produce 1300 watts of radiant power. What would be the temperature of the tubes if the emissivity of the tubes is 0.60?

SOLUTIONS AND ANSWERS

Practice Problems

1. Equation (13.1) applies.

    $$Q/A = kt\Delta T/L = (0.80 \text{ J/(s m C°)})(3600 \text{ s})(22 \text{ °C})/(0.64 \times 10^{-2} \text{ m}) = \mathbf{9.9 \times 10^6 \text{ J}}$$

2. Let T_i be the indoor temperature, T_1 be the temperature at the first glass-air interface, T_2 be the temperature at the second air-glass interface, and T_o be the outdoor temperature. Each layer conducts the same quantity of heat in the same time. Equation (13.1) gives for the inside pane

    $$Q/A = k_g t(T_i - T_1)/L \qquad (1)$$

    for the air gap

    $$Q/A = k_a t(T_1 - T_2)/L \qquad (2)$$

    and for the outside pane

    $$Q/A = k_g t(T_2 - T_o)/L \qquad (3)$$

    Equating (1) and (3) and rearranging gives
    $$T_i - T_1 = T_2 - T_o \qquad (4)$$
    Similarly equating (2) and (3) yields
    $$k_a T_1 - k_a T_2 = k_g T_2 \qquad (5)$$
    Solving (4) for T_1, using the result in (5) and rearranging gives

    $$T_1 = (k_g + k_a)T_i/(2k_a + k_g)$$

    $T_i = 22$ °C, $k_a = 0.0256$ J/(s m C°) and $k_g = 0.80$ J/(s m C°) so $T_1 = 21$ °C. This value used in (1) gives

    $$Q/A = (0.80 \text{ J/(s m C°)})(3600 \text{ s})(1 \text{ C°})/(0.64 \times 10^{-2} \text{ m}) = \mathbf{4.5 \times 10^5 \text{ J}}$$

3. Equation (13.1) applied to the bottom of the aluminum pot is $Q_a = k_a t A\Delta T/L_a$. A similar expression for the bottom of the copper pot is $Q_c = k_c tA\Delta T/L_c$. Equating and solving for L_a gives

    $$L_a = (k_a/k_c)L_c = (240 \text{ J/(s m C°)}/390 \text{ J/(s m C°)})(0.64 \text{ cm}) = \mathbf{0.39 \text{ cm.}}$$

4. Equation (13.2) applied to the star is $Q/t = \sigma T^4 A$ and to the sun is $Q/t = \sigma T_s^4 A_s$. Equating and solving gives

    $$A/A_s = (T_s/T)^4 \text{ or } (R/R_s) = (T_s/T)^2 = \mathbf{4}.$$

5. Equation (13.2) written for the actual object is $Q/t = (0.5)\sigma T^4 A$ and for the equivalent blackbody is $Q/t = \sigma T_b^4 A$. Equating and solving for T_b yields

    $$T_b = (0.5)^{1/4} T = (0.5)^{1/4} (650 \text{ °C} + 273) = \mathbf{776 \text{ K (or 503 °C)}}$$

6. The burner loses, due to radiation, energy at a rate of

    $$(Q/t)_{lost} = e\sigma T_b^4 A = (0.800)(5.67 \times 10^{-8} \text{ J/s m}^2 \text{ K}^4)(1.00 \times 10^3 \text{ K})^4(6.00 \times 10^{-3} \text{ m}^2) = 272 \text{ W}.$$

    The burner gains energy from the surroundings at a rate

    $$(Q/t)_{gain} = e\sigma T^4 A = (0.800)(5.67 \times 10^{-8} \text{ J/s m}^2 \text{ K}^4)(3.00 \times 10^2 \text{ K})^4(6.00 \times 10^{-3} \text{ m}^2) = 2.20 \text{ W}$$

    The net power lost by the burner is $(Q/t)_{lost} - (Q/t)_{gain} = \textbf{2.70 X 10}^2 \textbf{ W}.$

---

7. The heat lost through the floor is given by equation (13.1).

    $$Q = kAt\Delta T/L = (1.1 \text{ J/s m C}°)(3.7 \text{ m})(6.1 \text{ m})(3600 \text{ s})(9.0 \text{ C}°)/(15 \times 10^{-2} \text{ m}) = \textbf{5.4 X 10}^6 \textbf{ J}.$$

---

8. The heat lost through the brick wall in time, t, is $Q_b = k_b At\Delta T/L_b$. If ΔT_1 is the temperature difference between the inside and the brick-styrofoam interface and ΔT_2 is the temperature across the styrofoam, the heat lost through the wall is
    $$Q_s = k_b At\Delta T_1/L_b = k_s At\Delta T_2/L_s. \text{ Then } k_b\Delta T_1/L_b = k_s\Delta T_2/L_s.$$

    We also know that $\Delta T = \Delta T_1 + \Delta T_2$. Elimination of ΔT_1 from these equations yields

    $$\Delta T_2 = (k_b/L_b)\Delta T/[k_b/L_b + k_s/L_s]$$
    so that
    $$Q_s = (k_s/L_s)(k_b/L_b)At\Delta T/[k_b/L_b + k_s/L_s].$$

    Suppressing units, we find that

    $$Q_s/Q_b = k_s/L_s/[k_b/L_b + k_s/L_s] = (0.010/0.50)/[(1.0/4.0) + (0.010/0.50)] = 0.074.$$

    The heat loss through the insulated wall is **0.074 times** the loss through the brick wall.

---

9. Equation (13.2) gives

    $$T^4 = (Q/t)/(e\sigma A) = (1300 \text{ W})/[(0.60)(5.67 \times 10^{-8} \text{ J/s m}^2 \text{ K}^4)(40.0 \times 10^{-2} \text{ m})(2\pi)(1.0 \times 10^{-2} \text{ m})(2)]$$

    Thus,
    $$T = \textbf{930 K}$$

CHAPTER 14
The Ideal Gas Law and Kinetic Theory

PREVIEW

In this chapter you will gain an understanding of heat and temperature at the molecular level. The concept of an ideal gas is introduced and the kinetic theory of gases is explored. The ideal gas law is employed to study how a gas behaves under different conditions of pressure, temperature, and volume. The internal energy of a gas is defined and the process of diffusion is discussed.

QUICK REFERENCE

Important Terms

Atomic mass scale
A scale in which the mass of an atom of one element is compared to the mass of an atom of another element. The reference atom is chosen to be carbon-12, whose atomic mass is defined to be exactly twelve atomic mass units (12 u), where $1 \text{ u} = 1.6605 \times 10^{-27}$ kg.

The Mole and Avogadro's number
One mole of a substance contains as many particles (atoms or molecules) as there are atoms in 12 grams of the isotope carbon-12. There are 6.022×10^{23} atoms in 12 grams of carbon-12. The number of atoms per mole is known as Avogadro's number, N_A, where $N_A = 6.022 \times 10^{23}$ mol^{-1}.

Ideal gas law
Relates the pressure, volume, and temperature of an ideal gas. It states that $PV = nRT$ where n is the number of moles present and R is the universal gas constant.

Kinetic theory of gases
The theory which states that the temperature of an ideal gas is defined by the average kinetic energy of the particles making up the gas.

Maxwell speed distribution
The distribution of particle speeds in an ideal gas at a given temperature.

Internal energy
The sum of the various kinds of energy that the atoms or molecules of a substance possess. This includes translational and rotational kinetic energy, vibrational energy, and potential energy.

Diffusion
The process whereby solute molecules move from a region of higher concentration to a region of lower concentration.

Equations

The **ideal gas law** can be written:

$$PV = nRT \qquad (14.1)$$

An **alternative way** to write the ideal gas law is:

$$PV = NkT \qquad (14.2)$$

Boyle's law can be stated as:

$$P_i V_i = P_f V_f \qquad (14.3)$$

Charles' law is:

$$\frac{V_i}{T_i} = \frac{V_f}{T_f} \qquad (14.4)$$

The **average translational kinetic energy** of the particles in a gas is given by:

$$\overline{KE} = \frac{1}{2} m v_{rms}^2 = \frac{3}{2} kT \qquad (14.6)$$

The **internal energy** of an **ideal monatomic gas** is:

$$U = \frac{3}{2} nRT \qquad (14.7)$$

Fick's law of diffusion is:

$$m = \frac{DA \, \Delta Ct}{L} \qquad (14.8)$$

DISCUSSION OF SELECTED SECTIONS

14.1 The Mole, Avogadro's Number, and Molecular Mass

The relative masses of the atoms of different elements can be expressed in terms of their **atomic masses**. To set up the scale of atomic masses, carbon-12 is chosen as the reference element. An **atomic mass unit** is then defined as one-twelfth the mass of the carbon-12 atom. For example, the atomic mass of say, oxygen, is 15.9994 atomic mass units (u). It is approximately 16/12 the mass of the carbon-12 atom.

The **molecular mass** of a molecule is the sum of the atomic masses of its constituent atoms. So, for example, a water molecule (H_2O) has a molecular mass which is the sum of the masses of two hydrogen atoms and one oxygen atom, i.e., $2(1.00794 \text{ u}) + 15.9994 \text{ u} = 18.0153 \text{ u}$.

A **gram-mole** (or simply a mole) of a substance contains as many atoms or molecules as there are atoms in 12 grams of carbon-12. The number of atoms per mole is known as **Avogadro's number, N_A**, where $N_A = 6.022 \times 10^{23} \text{ mol}^{-1}$. One mole of a substance has a mass in grams that is equal to the atomic or molecular mass of its constituent particles. Thus, one mole of neon (atomic mass = 20.179 u) has a mass of 20.179 g.

Example 1
The chemical formula of ethyl alcohol is C_2H_5OH. (a) Determine the mass of one mole of ethyl alcohol. (b) How many molecules are contained in 175 g of ethyl alcohol?

(a) Ethyl alcohol has 2 carbon atoms, 6 hydrogen atoms, and one oxygen atom. Using the periodic table, we see that the atomic mass of ethyl alcohol is,

$$2(12.011 \text{ u}) + 6(1.00794 \text{ u}) + 15.9994 \text{ u} = 46.069 \text{ u}.$$

Therefore, one mole of ethyl alcohol has a mass of 46.069 grams.
(b) In 175 g of ethyl alcohol, there are $(175 \text{ g})/(46.069 \text{ g/mol}) = 3.80 \text{ mol}$. The number of molecules is equal to the number of moles times the number of molecules per mole (Avogadro's number, N_A):

$$(3.80 \text{ mol})(6.022 \times 10^{23} \text{ molecules/mol}) = 2.29 \times 10^{24} \text{ molecules}.$$

14.2 The Ideal Gas Law

The ideal gas law expresses the relationship between the Kelvin temperature, the pressure, the volume, and the number of moles of an ideal gas. An ideal gas is an idealized model of real gases and is a useful concept in studying the properties of gases in general. The ideal gas law can be stated as

$$PV = nRT \qquad (14.1)$$

where P is the absolute pressure (Pa), V the volume (m^3), T the temperature (K), n the number of moles, and R is the **universal gas constant** which has the value R = 8.31 J/(mol K).

An alternative way to write equation (14.1) is in terms of the total number of particles in the gas, N, rather than in terms of the number of moles, n. We have

$$PV = NkT \qquad (14.2)$$

where k is known as **Boltzmann's constant**, k = 1.38 x 10^{-23} J/K.

Example 2
A cylinder containing carbon dioxide gas (CO_2) has a volume of 0.050 m^3. At 25.0 °C the gas has an absolute pressure of 2.00 x 10^6 Pa. (a) What is the mass of carbon dioxide in the cylinder? (b) How many molecules of CO_2 are in the cylinder?

(a) First find the number of moles of CO_2 present using equation (14.1).

$$n = \frac{PV}{RT} = \frac{(2.00 \times 10^6 \text{ Pa})(0.050 \text{ m}^3)}{[8.31 \text{ J/(mol K)}](298 \text{ K})} = 40.4 \text{ mol}$$

The atomic mass of CO_2 is 2(15.9994 u) + 12.011 u = 44.010 u. The mass of CO_2 in the cylinder is therefore,

$$(44.010 \text{ g/mol})(40.4 \text{ mol}) = 1.78 \times 10^3 \text{ g} = 1.78 \text{ kg}.$$

(b) The number of CO_2 molecules in the cylinder can be obtained using equation (14.2).

$$N = \frac{PV}{kT} = \frac{(2.00 \times 10^6 \text{ Pa})(0.050 \text{ m}^3)}{(1.38 \times 10^{-23} \text{ J/K})(298 \text{ K})} = 2.43 \times 10^{25} \text{ molecules}.$$

Note that this value of N can also be obtained using Avogadro's number, that is,

$$N = nN_A = (40.4 \text{ mol})(6.022 \times 10^{23} \text{ mol}^{-1}) = 2.43 \times 10^{25} \text{ molecules}.$$

Historically, the ideal gas law developed as a result of the work of several investigators. One important law relates the conditions of pressure and volume when the temperature and mass remain fixed. **Boyle's law** states

$$P_i V_i = P_f V_f \qquad (14.3)$$

where the subscripts i and f refer to the initial and final states of the gas, respectively.

Example 3
The pressure in a tank of nitrogen is 1.75 x 10^7 Pa at 20.0 °C. If the nitrogen would occupy 30.0 m^3 at 1 atm and 20.0 °C, find the volume of the tank.

If we take the initial pressure to be 1 atm = 1.01 x 10^5 Pa, the initial volume to be 30.0 m^3, and the final pressure to be 1.75 x 10^7 Pa, equation (14.3) yields for the final volume

$$V_f = \frac{P_i V_i}{P_f} = \frac{(1.01 \times 10^5 \text{ Pa})(30.0 \text{ m}^3)}{1.75 \times 10^7 \text{ Pa}} = 0.173 \text{ m}^3.$$

If a situation arises in which the gas pressure and mass remain constant, then **Charles' law** is valid, that is,

$$\frac{V_i}{T_i} = \frac{V_f}{T_f} \qquad\qquad (14.4)$$

Example 4
A tire is inflated to a gauge pressure of 2.0 atm. (Gauge pressure is the absolute pressure minus 1 atm). As the car is driven, the temperature of the tire increases from 20.0 °C to 45.0 °C, and simultaneously, the volume of the tire increases by 5.0%. What is the gauge pressure of the tire at the higher temperature?

Since the amount of gas in the tire is constant, equation (14.1) gives

$$\frac{P_i V_i}{T_i} = \frac{P_f V_f}{T_f}$$

Using P_i = 3.0 atm, T_i = 273 + 20.0 = 293 K, T_f = 273 + 45 = 318 K, and (V_f/V_i) = 1.05 we can solve for P_f.

$$P_f = \frac{P_i V_i T_f}{T_i V_f} = \frac{(3.0 \text{ atm})(318 \text{ K})}{(293 \text{ K})(1.05)} = 3.1 \text{ atm}$$

Therefore, the final gauge pressure is $P_{gauge} = P_f - 1.0$ atm = 2.1 atm.

14.3 Kinetic Theory of Gases

A container filled with a gas contains a large number of particles which are in constant, random motion, colliding with each other and with the walls of the container. At a given temperature, the atoms and molecules are all moving at different speeds. However, it is possible to determine the most probable particle speed for the gas. As the temperature of the gas rises, the most probable particle speed increases, as shown in the following figure.

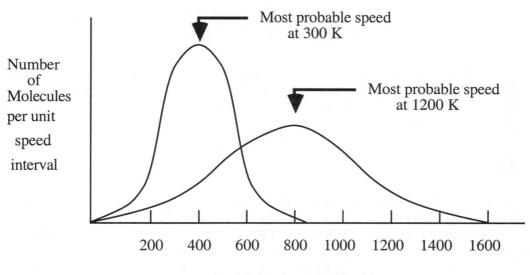

The above plots are called Maxwell distribution curves for particle speeds in a gas. Note that there is actually a wide range of velocities for each temperature, but that particles are more likely to travel at certain speeds (the most probable speed) than at others. It can be shown that the average translational kinetic energy of the particles in a gas is proportional to the temperature of the gas. That is

$$\overline{KE} = \frac{1}{2}mv_{rms}^2 = \frac{3}{2}kT \qquad (14.6)$$

where m is the mass of an individual molecule, k is Boltzmann's constant, T is the Kelvin temperature, and v_{rms} is the **root-mean-square speed** of the particles (which represents a type of average speed). If the gas can be treated as an ideal monatomic gas, the internal energy U is defined as the total kinetic energy of the N particles that make up the gas. That is

$$U = N\left(\frac{1}{2}mv_{rms}^2\right) = N\left(\frac{3}{2}kT\right) = \frac{3}{2}nRT \qquad (14.7)$$

Example 5 *Escape of hydrogen gas from the atmosphere*
Hydrogen molecules in the upper atmosphere (where the temperature is about 1500 K) have an appreciable chance of escaping the earth's gravitational attraction. The escape speed for hydrogen gas is about 10.7 km/s. Calculate the rms speed of the H_2 molecules at this altitude and comment on the presence of hydrogen gas in the atmosphere.

Use equation (14.6) to find the rms speed at 1500 K. Note that the mass of a hydrogen molecule can be found from its molecular mass and Avogadro's number, i.e.,

$$m = \frac{\text{molecular mass}}{N_A} = \frac{2(1.00794 \text{ g/mol})}{6.022 \times 10^{23} \text{ mol}^{-1}} = 3.35 \times 10^{-24} \text{ g} = 3.35 \times 10^{-27} \text{ kg}.$$

Solving equation (14.6) for the rms speed therefore yields,

$$v_{ms} = \sqrt{\frac{3kT}{m}} = \sqrt{\frac{3(1.38 \times 10^{-23} \text{ J/K})(1500 \text{ K})}{(3.35 \times 10^{-27} \text{ kg})}} = 4300 \text{ m/s.}$$

Although this rms speed is only about 0.40 of the escape speed (10 700 m/s), some molecules will have speeds many times the rms value and therefore will have sufficient energy to escape. In the 4.6 billion years since the formation of the earth, there has been ample time for essentially all of the hydrogen and helium to escape the earth's gravitational attraction and become lost in space.

14.4 Diffusion

Consider a container of water in which we drop a small amount of ink. The ink drop is initially very concentrated, but slowly disperses into the surrounding water. The process by which the ink molecules move from regions of high concentration to regions of low concentration is called **diffusion**. The ink is called the **solute** and diffuses through the water (the solvent).

The mass, m, of solute that diffuses in a time, t, through a solvent contained in a channel of length, L, and cross-sectional area, A, is

$$m = \frac{DA \, \Delta Ct}{L} \qquad\qquad (14.8)$$

where ΔC is the concentration difference between the ends of the channel and D is the diffusion constant. The SI unit for the diffusion constant is m^2/s. Equation (14.8) is known as **Fick's law of diffusion**.

Example 6
The diffusion constant of ethanol in water is 12.4×10^{-10} m^2/s. A cylinder has a cross-sectional area of 4.00 cm^2 and a length of 2.00 cm. A difference in ethanol concentration of 1.50 kg/m^3 is maintained between the ends of the cylinder. In one hour, what is the mass of the ethanol that diffuses through the cylinder?

Using A = 4.00 cm^2 = 4.00×10^{-4} m^2, and L = 2.00 cm = 2.00×10^{-2} m in equation (14.8) we obtain,

$$m = \frac{DA \, \Delta Ct}{L} = \frac{(12.4 \times 10^{-10} \text{ } m^2/s)(4.00 \times 10^{-4} \text{ } m^2)(1.50 \text{ kg/m}^3)(3600 \text{ s})}{2.00 \times 10^{-2} \text{ m}} = 1.34 \times 10^{-7} \text{ kg.}$$

PRACTICE PROBLEMS

1. How many molecules are there in 30.0 g of $C_9H_8O_4$?

2. Sodium bicarbonate, $NaHCO_3$, is one of the ingredients in baking powder. How many moles are present in a 21.0 g sample of sodium bicarbonate?

3. A 20.0 liter container of oxygen gas has a gauge pressure of 8.00 atm at 25 °C. (a) How many moles of O_2 are present in the container? (b) How many O_2 molecules are present?

4. A bubble of air, 0.010 m^3 in volume, is formed at the bottom of a lake which is 30.0 m deep and where the temperature is 8.0 °C. The bubble rises to the surface, where the water temperature is 26 °C and where the pressure is atmospheric pressure. What is the volume of the bubble just as it reaches the surface?

5. Consider an ideal gas of density ρ and pressure P. Show that the root-mean-square speed of the gas molecules is equal to $\sqrt{3P/\rho}$.

6. At what temperature is the root-mean-square speed of oxygen molecules (O_2) equal to the escape velocity from the earth? Take the escape velocity to be 11.2 km/s at the earth's surface.

7. A tank contains 40.0 mol of N_2 gas at 14.0 atm of pressure. If the volume of the tank is 75.0 liters, what is the rms speed of the molecules?

8. Assume the pressure in a room remains constant at 1.01×10^5 Pa and the air is composed only of nitrogen (N_2). The volume of the room is 60.0 m^3. When the temperature increases from 16 °C to 29 °C, what mass of air escapes from the room?

9. Neon (Ne) is a monatomic gas. At a temperature of 50.0 °C, what is the internal energy of two grams of neon?

10. A cylindrical tube of 4.00 cm diameter and 2.50 m length contains 5.0×10^{-14} kg of a solute diffusing through water. A concentration difference of 5.0×10^{-3} kg/m^3 is maintained between the ends of the tube and the diffusion constant of the solute is 7.5×10^{-10} m^2/s. How much time is required for all the solute to diffuse through the water?

SOLUTIONS AND ANSWERS

Practice Problems

---

1. The molecular mass of $C_9H_8O_4$ is: $9(12.011 \text{ u}) + 8(1.00794 \text{ u}) + 4(15.9994 \text{ u}) = 180.16$ u.

    Therefore, 1 mole has a mass of 180.16 g. In 30.0 g there are

    $$(30.0 \text{ g})/(180.16 \text{ g mol}^{-1}) = 0.167 \text{ mol}.$$
    Thus, the number of molecules is

    $$(6.022 \times 10^{23} \text{ mol}^{-1})(0.167 \text{ mol}) = \mathbf{1.01 \times 10^{23}}.$$

---

2. The molecular mass of $NaHCO_3$ is

    $$22.9898 \text{ u} + 1.00794 \text{ u} + 12.011 \text{ u} + 3(15.9994 \text{ u}) = 84.007 \text{ u}.$$

    Therefore, in 21.0 g of sodium bicarbonate there are

    $$(21.0 \text{ g})/(84.007 \text{ g mol}^{-1}) = \mathbf{0.250 \text{ mol}}.$$

---

3. Using the ideal gas law, PV = nRT, and noting that $P = P_{gauge} + 1.00 \text{ atm} = 9.00 \text{ atm} = 9.09 \times 10^5$ Pa, we have

    (a) $n = PV/RT = (9.09 \times 10^5 \text{ Pa})(20.0 \times 10^{-3} \text{ m}^3)/\{[8.31 \text{ J/(mol K)}](298 \text{ K})\}$
            $n = \mathbf{7.34 \text{ mol}}$.

    (b) Since 1 mol contains 6.022×10^{23} molecules, in 7.3 mol there are

    $$(7.3 \text{ mol})(6.022 \times 10^{23} \text{ mol}^{-1}) = \mathbf{4.4 \times 10^{24}}.$$

---

4. First find the pressure at the bottom of the lake, P_2, from

    $$P_2 = P_1 + \rho gh.$$
    We have,
    $$P_2 = 1.01 \times 10^5 \text{ Pa} + (1.00 \times 10^3 \text{ kg/m}^3)(9.80 \text{ m/s}^2)(30.0 \text{ m})$$
    $$P_2 = 3.95 \times 10^5 \text{ Pa} = 3.91 \text{ atm}.$$

    Since the amount of gas in the bubble is constant,

    $$P_1V_1/T_1 = P_2V_2/T_2.$$

    The volume of the bubble at the surface is therefore,

    $$V_1 = P_2V_2T_1/P_1T_2 = (3.91 \text{ atm})(0.010 \text{ m}^3)(299 \text{ K})/[(1.00 \text{ atm})(281 \text{ K})]$$
    $$V_1 = \mathbf{0.042 \text{ m}^3}.$$

---

5.   We know from equation (14.6) that

$$(1/2)mv_{rms}^2 = (3/2)kT.$$

From equation (14.1)

$$PV = nRT$$

which gives

$$T = PV/nR.$$

Therefore,

$$v_{rms}^2 = 3k(PV/nR)/m = 3kPV/nmR.$$

Since

$$k = R/N_A$$

and

$$n = N/N_A$$

we have

$$k/n = (R/N_A)/(N/N_A) = R/N.$$

Now

$$v_{rms}^2 = (3PV/mR)(R/N) = 3PV/mN.$$

But mN = total mass and ρ = mass/V, so

$$v_{rms}^2 = 3P/\rho,$$

and

$$v_{rms} = \sqrt{3P/\rho}.$$

---

6.   Using equation (14.6) and solving for T we get

$$T = mv_{rms}^2/3k.$$

Oxygen gas (O_2) has a molecular mass of 32.0 g/mol. Therefore,

$$m = (32.0 \text{ g/mol})/(6.022 \times 10^{23} \text{ mol}^{-1}) = 5.31 \times 10^{-23} \text{ g}$$
$$m = 5.31 \times 10^{-26} \text{ kg}.$$

The temperature is,

$$T = (5.31 \times 10^{-26} \text{ kg})(11.2 \times 10^3 \text{ m/s})^2/[3(1.38 \times 10^{-23} \text{ J/K})]$$
$$T = \mathbf{1.61 \times 10^5 \text{ K}}.$$

---

7.   Find the temperature,

$$T = PV/nR = (14.0 \text{ atm})(1.01 \times 10^5 \text{ Pa/1 atm})(75.0 \times 10^{-3} \text{ m}^3)/R(40.0 \text{ mol}).$$
$$T = 319 \text{ K}.$$

From equation (14.6),

$$v_{rms}^2 = 3kT/m,$$

where the mass of a nitrogen molecule, N_2, is given by

$$m = (28.0 \times 10^{-3} \text{ kg/mol})/(6.022 \times 10^{23} \text{ mol}^{-1}) = 4.65 \times 10^{-26} \text{ kg}.$$

Thus, the rms speed of the molecules is

$$v_{rms}^2 = 3(1.38 \times 10^{-23} \text{ J/K})(319 \text{ K})/(4.65 \times 10^{-26} \text{ kg}).$$

This gives

$$v_{rms} = \mathbf{533 \ m/s}.$$

---

8. Find the number of moles at each temperature:

$$n_1 = PV/RT_1 = (1.01 \times 10^5 \text{ Pa})(60.0 \text{ m}^3)/[R(289 \text{ K})] = 2520 \text{ mol}.$$

and

$$n_2 = PV/RT_2 = 2410 \text{ mol}.$$

The number of moles lost is 2520 - 2410 = 110 mol. The mass lost is therefore,

$$m(\text{lost}) = (110 \text{ mol})(28.01 \text{ g/mol}) = 3.0 \times 10^3 \text{ g} = \mathbf{3.0 \ kg}.$$

---

9. The amount of neon present is
$$(2.00 \text{ g})/(20.179 \text{ g/mol}) = 0.0991 \text{ mol}.$$

Equation (14.7) gives the internal energy

$$U = (3/2)nRT = (3/2)(0.0991 \text{ mol})[8.31 \text{ J/(mol K)}](323 \text{ K}) = \mathbf{3.99 \times 10^2 \ J}.$$

---

10. Using equation (14.8) and solving for t, we obtain,

$$t = mL/(DA \ \Delta C) = (5.0 \times 10^{-14} \text{ kg})(2.50 \text{ m})/[(7.5 \times 10^{-10} \text{ m}^2/\text{s})\pi(2.00 \times 10^{-2} \text{ m})^2(5.0 \times 10^{-3} \text{ kg/m}^3)]$$
$$t = \mathbf{27 \ s}.$$

CHAPTER 15
Thermodynamics

PREVIEW

In this chapter you will learn about the laws of **thermodynamics**. The zeroth law is concerned with the concept of thermal equilibrium. It requires that two systems be in thermal equilibrium if they are both in equilibrium with a third system. The first law of thermodynamics is a statement of the conservation of energy for a thermodynamic system.

You will also study the second and third laws of thermodynamics, including discussions of heat engines, Carnot's principle, and entropy. You will learn how to find the efficiency of engines, and learn about refrigerators, air conditioners, and heat pumps. Entropy, as it relates to the second law of thermodynamics, will be studied, and discussed in the context of reversible and irreversible processes.

QUICK REFERENCE

Important Terms

Thermodynamics
A branch of physics built upon the laws which govern the behavior of heat and work.
Diathermal walls
Walls which separate a system and the environment and allow heat to flow through them.
Adiabatic walls
Walls which separate the system and environment and do not allow heat to flow through them.
State of the system
The physical condition of the system which is usually described by specifying the pressure, volume and temperature.
Thermal equilibrium
The state of two or more systems in thermal contact when no heat flows between them. Systems are in thermal equilibrium when they are at the same temperature.
Isobaric process
A thermodynamic process which occurs at constant pressure.
Isochoric process
A thermodynamic process which occurs at constant volume.
Isothermal process
A thermodynamic process which occurs at constant temperature.
Adiabatic process
A thermodynamic process in which no heat flows into or out of the system.
Molar specific heat capacity
The constant, C, in the equation $Q = Cn\Delta T$ which determines the amount of heat lost or gained by n moles of a substance when its temperature changes by ΔT.
Heat engine
Any device that uses heat to perform work.
Reversible process
A process in which both the system and its environment can be returned to exactly the states they had before the process occurred.
Carnot's principle
No irreversible engine operating between two reservoirs at constant temperatures can have a greater efficiency than a reversible engine operating between the same temperatures.
Entropy
A function of state that is associated with disorder in the system and environment.

Equations

The **first law of thermodynamics** is

$$\Delta U = U_f - U_i = Q - W \qquad (15.1)$$

The **work done** during an **isobaric process** is

$$W = P \, \Delta V \qquad (15.2)$$

The **work done** during an **isothermal expansion or compression of an ideal gas** is

$$W = nRT \ln\left(\frac{V_f}{V_i}\right) \qquad (15.3)$$

The **work done** during the **adiabatic expansion** or **compression** of a monatomic **ideal gas** is

$$W = \frac{3}{2} nR(T_i - T_f) \qquad (15.4)$$

The following equation applies to an **adiabatic expansion** or **compression** of an ideal gas:

$$P_i V_i^{\gamma} = P_f V_f^{\gamma} \qquad (15.5)$$

The heat of a system Q, the molar heat capacity C, and the change in temperature ΔT are related by

$$Q = Cn\Delta T \qquad (15.6)$$

The **efficiency of a heat engine** is given by:

$$\text{Efficiency} = \frac{\text{Work Done}}{\text{Input Heat}} = \frac{W}{Q_H} \qquad (15.11)$$

We also know that:

$$Q_H = W + Q_C \qquad (15.12)$$

This leads to:

$$\text{Efficiency} = \frac{Q_H - Q_C}{Q_H} = 1 - \frac{Q_C}{Q_H} \qquad (15.13)$$

For a **Carnot engine**,

$$\text{Efficiency} = 1 - \frac{T_C}{T_H} \qquad (15.15)$$

The entropy of a system is given by:

$$\Delta S = \left(\frac{Q}{T}\right)_R \qquad (15.18)$$

Laws

Zeroth law of thermodynamics - two systems individually in thermal equilibrium with a third system are in thermal equilibrium with each other.

First law of thermodynamics - When a system absorbs or loses an amount of heat Q, and performs work W, the internal energy of the system changes by an amount ΔU, from an initial value U_i to a final value U_f:

$$\Delta U = U_f - U_i = Q - W$$

Second Law of thermodynamics - Heat flows spontaneously from a substance at a higher temperature to a substance at a lower temperature and does not flow spontaneously in the reverse direction. In terms of entropy: the total entropy of the universe does not change when a reversible process occurs ($\Delta S_{universe} = 0$) and increases when an irreversible process occurs ($\Delta S_{universe} > 0$).

DISCUSSION OF SELECTED SECTIONS

15.3 The First Law of Thermodynamics

The first law of thermodynamics is simply a statement of the conservation of energy for a thermodynamic system which can lose or gain heat and perform negative or positive work. Any net energy gained by the system increases its store of internal energy. Also, any net energy lost by the system comes from its store of internal energy.

Keep in mind that all changes in internal energy of an ideal gas are accompanied by a change in the TEMPERATURE of the system and vice versa. In the specific case where the system consists of n moles of a monatomic ideal gas, the change in internal energy accompanying a temperature change ΔT is $\Delta U = 3/2\, nR\Delta T$.

Example 1
An ideal gas performs 550 J of work while losing 150 cal of heat energy. What are the values of W, Q and ΔU? What happens to the temperature of the system during the process?

The work done, W = + 550 J, since the system does the work. The heat transferred is

$$Q = - 150 \text{ cal} = - (150 \text{ cal})(4.186 \text{ J}/1 \text{ cal}) = - 630 \text{ J.}$$

The negative sign is used to indicate that the system is losing the heat.

The first law then gives the change in internal energy of the system to be

$$\Delta U = Q - W = - 630 \text{ J} - 550 \text{ J} = - 1180 \text{ J.}$$

The internal energy of the system decreases during the process, so the system's temperature decreases.

Example 2
One mole of a monatomic ideal gas at a temperature of 57 °C absorbs 1.4 kcal of heat and does 2800 J of work on its surroundings. What is the resulting temperature of the gas?

The change in temperature of the ideal gas is

$$\Delta T = \frac{\Delta U}{\frac{3}{2} nR}$$

The first law gives the change in the internal energy to be

$$\Delta U = Q - W = (+ 1.4 \text{ kcal})(4186 \text{ J/1 kcal}) - 2800 \text{ J} = 3100 \text{ J}$$

so

$$\Delta T = \frac{3100 \text{ J}}{\frac{3}{2}(1.0 \text{ mole})(8.31 \text{ J/mole K})} = 250 \text{ K}$$

The final temperature of the gas is then

$$T_f = T_i + \Delta T = 330 \text{ K} + 250 \text{ K} = 580 \text{ K}.$$

15.4 Thermal Processes

A thermodynamic system may undergo a variety of processes. There are four basic processes, however, which are very common. These are: isobaric, isochoric, isothermal, and adiabatic processes.

Isobaric Process
This is a process in which a system maintains a constant pressure. Other quantities such as temperature, volume, and internal energy may change during the process while heat is added or taken away and work is done. The work done during an isobaric process is $W = P (V_f - V_i)$.

Example 3
When water evaporates, it does work in pushing the air away as it expands into a vapor. If 1.0 kg of water does 1.7×10^5 J of work on the atmosphere, what is the change in volume of the water as it evaporates? What is the change in internal energy of the water?

This is a constant pressure process where the pressure is that due to the atmosphere. The volume change of the system of 1.0 kg of water is

$$V_f - V_i = \frac{W}{P} = \frac{1.7 \times 10^5 \text{ J}}{1.01 \times 10^5 \text{ Pa}} = 1.7 \text{ m}^3$$

The first law requires $\Delta U = Q - W$. In order to evaporate the water an amount of energy equal to the latent heat of vaporization must be added to the water. $Q = mL_v$ so

$$\Delta U = mL_v - W = (1.0 \text{ kg})(2.26 \times 10^6 \text{ J/kg}) - 1.7 \times 10^5 \text{ J} = 2.1 \times 10^6 \text{ J}$$

Isochoric Process
No volume changes occur during this type of process. Other quantities such as pressure, temperature and internal energy may change as heat is added or taken away from the system. However, no work can be done during an isochoric process, since no mechanical motion (expansion or contraction) of the system is possible. The first law of thermodynamics then requires that $\Delta U = Q$.

Example 4
Two moles of an ideal monatomic gas at 95 °C is enclosed in a container with rigid walls. A fixed quantity of heat, $Q = 5.0$ kJ, is added to the gas. What is the final temperature of the gas?

The process is isochoric since the gas cannot expand against the rigid walls of the container so $\Delta U = Q$. The change in internal energy of an ideal gas is $\Delta U = 3/2\, nR(T_f - T_i)$, so

$$T_f = T_i + \frac{Q}{\frac{3}{2}nR} = 368\ K + \frac{5.0\ X\ 10^3\ J}{\frac{3}{2}(2.0\ mol)(8.31\ J/mol\ K)} = 5.7\ X\ 10^2\ K$$

Isothermal Process
The temperature remains constant in this type of process. Again, other quantities such as pressure and volume may change while heat is added to or taken away from the system and the system does work. The internal energy of the system will not change if it depends solely on the temperature of the system. This is true for an ideal gas and is covered in the next section.

Adiabatic process
No heat is allowed to flow into or out of the system in an adiabatic process. These processes occur in well insulated containers, or they happen so quickly that little heat has a chance to flow between the system and its environment. An example of the latter is when a gas, such as freon, expands rapidly from a pressurized container. The expansion in this case is at least approximately adiabatic.

The first law of thermodynamics gives $\Delta U = -W$ for an adiabatic process. Any work done on the system will increase its internal energy while work done by the system will decrease its internal energy. These internal energy changes often appear as temperature changes of the system. In the case of the freon expanding from its container, the temperature of the freon decreases as its does work to push the atmosphere out of the way.

15.5 Thermal Processes That Utilize an Ideal Gas

Isothermal Expansion or Compression
Since the internal energy of an ideal gas depends only on the temperature of the gas, it remains constant during an isothermal process. The first law of thermodynamics then gives $Q = W$.

The work done during an isothermal process involving an ideal gas system is $W = nRT \ln (V_f/V_i)$. It is the property of the natural logarithm that it is positive if the argument is greater than one, $(V_f > V_i)$. This corresponds to an expansion of the gas and the work is positive indicating that work is done BY the gas. According to the first law an equal amount of heat must be ADDED to the gas to maintain the constant temperature. The natural logarithm is negative if the argument is less than one $(V_f < V_i)$. In this case the work done is negative indicating that work is being done ON the gas to compress it, and an equal amount of heat must be REMOVED from the gas to maintain the constant temperature.

Example 5
A piston and cylinder arrangement is used to compress 1.0 mole of an ideal gas from an initial volume of 0.010 m^3 to a final volume of 0.0050 m^3. The temperature of the gas is maintained at a constant temperature of 150 °C. How much heat is transferred during the process? Is this heat added to or removed from the gas?

The process is isothermal since the temperature is constant throughout the process. The heat transferred is then

$$Q = W = nRT \ln (V_f/V_i)$$
$$Q = (1.0\ mol)(8.31\ J/mol\ K)(423\ K) \ln (0.0050\ m^3/0.010\ m^3) = -2400\ J$$

The negative sign indicates that heat is lost by the gas and must be REMOVED to keep the gas temperature constant.

Adiabatic Expansion or Compression

The first law of thermodynamics requires that $\Delta U = - W$ for an adiabatic process. If the system under consideration is a monatomic ideal gas, then the work done during the adiabatic process is $W = 3/2\ nR(T_i - T_f)$. Note that if the temperature of the gas rises, then the work is negative indicating that work is done ON the gas. This corresponds to the gas being compressed by an outside agent. If the temperature of the gas falls during the process, then the work is positive and work is done BY the gas. In this case the gas has expanded.

An additional relationship applicable to the adiabatic process is $P_iV_i^\gamma = P_fV_f^\gamma$ where $\gamma = c_p/c_V$.

Example 6

A piston cylinder arrangement contains 1.5 moles of ideal air ($\gamma = 7/5$) at 27 °C. The piston is quickly pushed down compressing the gas to one-half of its original volume. What is the final temperature of the gas and how much work was done during the process?

If the piston is pushed quickly enough, the process may be considered to be adiabatic since little time is available for heat to flow. The ideal gas law can be applied to find the final temperature of the gas.

For the initial state of the air

$$P_iV_i = nRT_i$$

and for the final state

$$P_fV_f = nRT_f.$$

Dividing and solving for T_f yields

$$T_f = \left(\frac{P_f}{P_i}\right)\left(\frac{V_f}{V_i}\right)T_i$$

Equation (15.5) gives

$$\frac{P_f}{P_i} = \frac{V_i^\gamma}{V_f^\gamma}$$

so the final temperature of the air is

$$T_f = \left(\frac{V_i^{\gamma-1}}{V_f^{\gamma-1}}\right)T_i = (2)^{\frac{7}{5}-1}(273+27)\ K = 396\ K$$

That is, 123 °C.

The work done during the adiabatic process is then

$$W = 3/2\ nR(T_i - T_f) = 3/2\ (1.5\ mol)(8.31\ J/mol\ K)(- 96\ K) = - 1800\ J.$$

1800 J of work is done ON the air during the process.

15.6   Specific Heat Capacities and the First Law of Thermodynamics

When a heating process occurs at constant pressure,

$$Q_{constant\ pressure} = nC_P\Delta T$$

where the molar specific heat capacity at constant pressure for a monatomic ideal gas, C_P is

$$C_P = 5/2\ R \qquad\qquad (15.7)$$

When the heating process occurs at constant volume,

$$Q_{constant\ volume} = nC_V\Delta T$$

where the molar specific heat capacity at constant volume for a monatomic ideal gas, C_V is

$$C_V = 3/2\ R \qquad\qquad (15.8)$$

The ratio of the specific heats for a monatomic ideal gas is

$$\gamma = \frac{C_P}{C_V} = \frac{\frac{5}{2}R}{\frac{3}{2}R} = \frac{5}{3} \qquad\qquad (15.9)$$

Also,

$$C_P - C_V = R \qquad\qquad (15.10)$$

Example 7

How much heat is required to change the temperature of 3.0 moles of a monatomic ideal gas by 55 K if: (a) the pressure is held constant and (b) if the volume is held constant?

(a) At constant pressure

$$Q_{constant\ pressure} = nC_P\Delta T = (3.0\ moles)(5/2)(8.31\ J/mole\ K)(55\ K) = 3400\ J.$$

(b) At constant volume

$$Q_{constant\ volume} = nC_V\Delta T = (3.0\ moles)(3/2)(8.31\ J/mole\ K)(55\ K) = 2100\ J.$$

15.8 Heat Engines

A heat engine is any device that uses heat to perform work and is represented schematically in the following diagram.

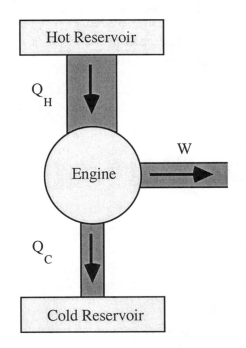

Note the main features of a heat engine:

Q_H - heat input from the hot reservoir flows into the engine.

Engine - uses available heat, Q_H, to perform work, W.

Q_C - heat rejected by the engine goes into a cold reservoir.

The efficiency of a heat engine is given as

$$\text{Efficiency} = \frac{\text{Work Done}}{\text{Input Heat}} = \frac{W}{Q_H} \qquad (15.11)$$

If there are no other losses in the engine, the principle of energy conservation requires that

$$Q_H = W + Q_C \qquad (15.12)$$

Solving for W in equation (15.12) and substituting into equation (15.11) yields,

$$\text{Efficiency} = \frac{Q_H - Q_C}{Q_H} = 1 - \frac{Q_C}{Q_H} \qquad (15.13)$$

Example 8
A heat engine does 350 J of work and rejects 750 J of heat. What is the efficiency of the engine?

In order to calculate the efficiency of the engine, we need to know what the input heat, Q_H, is. We can use equation (15.12) to find that

$$Q_H = W + Q_C = 350 \text{ J} + 750 \text{ J} = 1100 \text{ J}.$$

The efficiency of the engine can now be obtained using either equation (15.11) or (15.13). Use equation (15.11) to get

$$\text{Efficiency} = \frac{W}{Q_H} = \frac{350 \text{ J}}{1100 \text{ J}} = 0.32.$$

Therefore, the efficiency of the engine is 0.32 or 32%. This means that 32% of the heat provided to the engine is converted into work which can be used for various purposes (like turning a crankshaft and providing mechanical energy to move a car).

15.9 Carnot's Principle and the Carnot Engine

If the heat engine shown above acts "reversibly", this idealized engine is called a **Carnot engine**. A reversible process is one in which both the system and its environment (the rest of the universe) can be returned to exactly the states (of P, V, and T) they had before the process occurred.

In a Carnot engine the input heat Q_H originates from a hot reservoir at a single Kelvin temperature T_H, and all rejected heat Q_C is returned to the cold reservoir at a single Kelvin temperature T_C. The ratio of the rejected heat to the input heat can be shown to be

$$\frac{Q_C}{Q_H} = \frac{T_C}{T_H} \qquad\qquad (15.14)$$

where the temperatures MUST be expressed in **Kelvins**. Using equations (15.13) and (15.14), the efficiency of a Carnot engine can therefore be written,

$$\text{Efficiency} = 1 - \frac{Q_C}{Q_H} = 1 - \frac{T_C}{T_H} \qquad\qquad (15.15)$$

This relation gives the maximum possible efficiency for a heat engine operating between two Kelvin temperatures.

Example 9
In a steam turbine, steam heated to 655 °C enters the turbine and is exhausted at 115 °C. What is the maximum efficiency that this turbine can attain?

The maximum efficiency possible is that of a Carnot engine operating at the temperatures given. Converting to Kelvin temperature we have $T_H = 655 + 273 = 928$ K, and $T_C = 115 + 273 = 388$ K. Therefore,

$$\text{Efficiency} = 1 - \frac{T_C}{T_H} = 1 - \frac{388 \text{ K}}{928 \text{ K}} = 0.582.$$

Example 10
A Carnot engine has an efficiency of 0.700, and the temperature of its cold reservoir is 105 °C. (a) Determine the Kelvin temperature of its hot reservoir. (b) If 1.25 kcal of heat are rejected to the cold reservoir, what amount of heat is put into the engine?

(a) Solving equation (15.15) for T_H we obtain

$$T_H = \frac{T_C}{(1 - \text{Efficiency})} = \frac{378 \text{ K}}{(1 - 0.700)} = 1260 \text{ K}.$$

(b) Now solve equation (15.14) for Q_H to obtain

$$Q_H = Q_C \left(\frac{T_H}{T_C}\right) = (1.25 \text{ kcal})\left(\frac{1260 \text{ K}}{378 \text{ K}}\right) = 4.17 \text{ kcal}.$$

15.10 Refrigerators, Air Conditioners, and Heat Pumps

Refrigerators, air conditioners, and heat pumps are all similar devices which make heat flow from cold to hot. Of course, in order to do this, work must be done. The essential components of these devices are shown below.

These devices use work W to "reach into" a cold reservoir, "grab onto" an amount of heat Q_C, and deposit an amount of heat Q_H into a hot reservoir. This process is called a **refrigeration process**, but applies equally as well to air conditioners and heat pumps. Notice that the directions of the arrows symbolizing heat and work in this process are opposite to those for the heat engine. If the process is a reversible one, we have a Carnot refrigerator, air conditioner, or heat pump. Under these conditions the relation

$$Q_C/Q_H = T_C/T_H$$

applies, just as it does for the Carnot engine. In a sense, these devices work to move heat "up-hill" from a lower temperature T_C to a higher temperature T_H.

Example 11
A Carnot air conditioner uses 15 400 J of electrical energy to cool a house down to 22 °C when the outdoor temperature is 36 °C. How much heat is deposited outdoors?

First find the ratio of the heats Q_C and Q_H using equation (15.14), where $T_C = 22 + 273 = 295$ K and $T_H = 309$ K.

$$\frac{Q_C}{Q_H} = \frac{T_C}{T_H} = \frac{295K}{309K} = 0.955$$

Thus, $Q_H = Q_C/0.955$. But according to equation (15.12), $Q_C = Q_H - W$, so that $Q_H = (Q_H - W)/0.955$. As a result

$$Q_H = \frac{W}{1 - 0.955} = \frac{15\ 400\ J}{0.045} = 3.4 \times 10^5\ J$$

Example 12
Determine the coefficient of performance for the air conditioner discussed in example 11.

First we need to determine the heat removed from the cold reservoir using equation (15.12). We have

$$Q_C = Q_H - W = 3.4 \times 10^5\ J - 15\ 400\ J = 3.2 \times 10^5\ J.$$

The coefficient of performance is defined as

$$\text{Coefficient of Performance} = \frac{Q_C}{W} = \frac{3.2 \times 10^5\ J}{15\ 400\ J} = 21.$$

15.11 Entropy and the Second Law of Thermodynamics

For irreversible processes, such as those involving friction, engines operate at less than maximum efficiency. The heat dissipated by friction causes a loss in the engine's ability to perform work. This partial loss can be expressed in terms of a concept called **entropy**.

We can define the change in entropy ΔS as

$$\Delta S = \left(\frac{Q}{T}\right)_R \qquad\qquad (15.18)$$

where the subscript R means that this relation pertains to reversible processes. The total change in entropy which occurs during some process is equal to the entropy of the final state minus the entropy of the initial state.

Example 13
If 2500 J of heat flow spontaneously from a hot reservoir at 128 °C to a cold reservoir at 52 °C, determine the total change in entropy for this irreversible process.

The spontaneous flow of heat from hot to cold is irreversible. However, the relation $\Delta S = (Q/T)_R$ can be applied to a hypothetical process whereby 2500 J of heat are taken reversibly from the hot reservoir and added to the cold reservoir. The total entropy change is the algebraic sum of the entropy changes for each reservoir.

$$\Delta S_{tot} = \Delta S_C - \Delta S_H = \frac{2500\ J}{325\ K} - \frac{2500\ J}{401\ K} = +1.5\ J/K.$$

Notice that the change in entropy for the above process is greater than zero. In fact, any irreversible process always INCREASES the entropy of the universe. In other words, $\Delta S_{universe} > 0$ for an irreversible process. However, the total entropy of the universe does NOT change when a reversible process occurs. So $\Delta S_{universe} = 0$ in this case.

Example 14
Compute the change in entropy of a system consisting of 5.00 kg of ice at 0 °C which melts slowly (reversibly) to water at the same temperature. The latent heat of fusion for water is 33.5 x 10^4 J/kg.

Since the phase change occurs reversibly, we can use equation (15.18) to compute ΔS. We have

$$\Delta S = \frac{Q}{T} = \frac{(33.5\ x\ 10^4\ J/kg)(5.00\ kg)}{273\ K} = +6.14\ x\ 10^3\ J/K.$$

PRACTICE PROBLEMS

1. Three moles of an ideal gas receives 2500 J of heat energy and its temperature rises from 27 °C to 55 °C. How much work is done by the gas?

2. The P-V diagram shows three processes undergone by 1.0×10^2 mole of an ideal gas. The processes start at 1 and proceed clockwise. Identify each of the ideal gas processes and find the work done during each. (Use your best judgment to identify $1 \rightarrow 2$.) What is the total work done?

3. The internal energy of a system increases by 1350 J when the system absorbs 1150 J of heat energy at a constant pressure of 1.01×10^5 Pa. By how much does the volume of the system change?

4. Air ($\gamma = 7/5$) is compressed adiabatically to one-fifth of its original volume. By what factor did the pressure of the gas change during the compression? Assume that air behaves like an ideal gas.

5. One-half mole of an ideal gas expands isothermally at T = 250 °C to fives times its original volume. How much work was done? Was it done by or on the gas? How much heat was transferred during the process?

6. An ideal gas ($\gamma = 5/3$) expands through a nozzle of a pressurized container into the atmosphere. Assuming that the contents of the container are initially at 23°C , 10.0 atm and that the gas expands very rapidly, find the final temperature of the gas.

7. A helium filled balloon with a volume of 1.000×10^3 cm^3 and temperature of 27 °C is placed in liquid nitrogen in an open container and cools to a temperature of 77 K. Assume that helium behaves like an ideal gas at constant atmospheric pressure and find the new volume of the gas in the balloon, the change in internal energy of the gas, the work done during the process, and the heat removed from the gas.

8. A Carnot heat engine operates between 327 °C and 127 °C. It absorbs 7.50×10^4 J at the higher temperature. (a) What is the efficiency of the engine? (b) How much work per cycle is the engine capable of performing?

9. A Carnot engine operates between a hot reservoir at 455 K and a cold reservoir at 364 K. (a) If the engine absorbs 650 J of heat at the hot reservoir, how much work does it deliver? (b) If the same engine works in reverse as a refrigerator between the same two reservoirs, how much work must be supplied to remove 1200 J of heat from the cold reservoir?

10. A steam turbine takes vapor at 463 °C and exhausts it to a steam boiler at 238 °C. The turbine then receives steam from the boiler at this temperature and exhausts it to a condenser at 38.0 °C. What is the maximum efficiency of this combination?

11. In a heat pump, heat from the outdoors at -8.00 °C is transferred to a room at 22.0 °C, energy being supplied by an electric motor. How many joules of heat will be delivered to the room for each joule of electrical energy consumed, ideally?

12. A reversible engine operates in the cycle shown below, which consists of two adiabatic and two isochoric processes. The engine utilizes a monatomic ideal gas. Find the efficiency in terms of V_1 and V_2, and γ, the ratio of the heat capacities c_P and c_V.

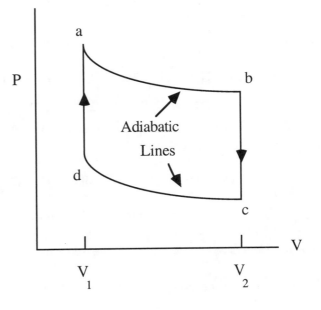

13. One end of a copper rod is in thermal contact with a reservoir at 550 K and the other end in thermal contact with a reservoir at 350 K. If 7700 J of heat are conducted from one end of the rod to the other find (a) the entropy change of each reservoir, (b) the entropy change of the rod, and (c) the entropy change of the universe.

14. If 40.0 g of ice at 0.0 °C is added to 100.0 g of water at 90.0 °C in a thermally insulated container, find the total change in entropy as the mixture comes to thermal equilibrium. Use the fact that for a change in temperature, the entropy change is $\Delta S = cm \ln(T_f/T_i)$, where c is the specific heat of the object.

SOLUTIONS AND ANSWERS

Practice Problems

---

1. The first law gives $\Delta U = Q - W$. For an ideal gas, $\Delta U = 3/2 \ nR(T_f - T_i)$ so

$$W = Q - 3/2 \ nR(T_f - T_i) = 2500 \ J - 3/2 \ (3.00 \ mol)(8.31 \ J/mol \ K)(28 \ K) = \textbf{1500 J}.$$

---

2. The process $1 \rightarrow 3$ is clearly **isochoric**. The work done is $W = \textbf{0 J}$ since the volume doesn't change. The process $2 \rightarrow 3$ is **isobaric**. The work done is

$$W = P(V_f - V_i) = (1.0 \ X \ 10^5 \ Pa)(1.0 \ m^3 - 2.0 \ m^3) = \textbf{- 1.0 X } 10^5 \textbf{ J}.$$

The process $1 \rightarrow 2$ looks isothermic. In order to decide if it is, calculate the temperatures at points 1 and 2. The ideal gas law gives $PV = nRT$ for any point. At point 1, $T_1 = 240$ K. At point 2, $T_2 = 240$ K. The process is probably **isothermal**. The work done during the isothermal process is

$$W = nRT \ ln \ (V_f/V_i) = (1.0 \ X \ 10^2 \ mol)(8.31 \ J/mol \ K)(240 \ K) \ ln \ (2.0) = \textbf{+ 1.4 X } 10^5 \textbf{ J}.$$

The total work done is then $0 \ J - 1.0 \ X \ 10^5 \ J + 1.4 \ X \ 10^5 \ J = \textbf{+ 0.4 X } 10^5 \textbf{ J}$. The positive sign indicates that the work is done by the gas.

---

3. The work done during an isobaric process is $W = P(V_f - V_i)$ so $V_f - V_i = W/P$. The first law requires that $W = Q - \Delta U$ so

$$V_f - V_i = (Q - \Delta U)/P = (1150 \ J - 1350 \ J)/(1.01 \ X \ 10^5 \ Pa) = \textbf{- 2.0 X } 10^{-3} \textbf{ m}^3.$$

The negative sign indicates that the system was compressed.

---

4. Equation (15.5) applies to an adiabatic process. Solving for the ratio of the pressures results in

$$P_f/P_i = (V_i/V_f)^\gamma = (5)^{7/5} = \textbf{9.5}$$

---

5. The work done during an isothermal process is given by equation (15.3).

$$W = nRT \ ln \ (V_f/V_i) = (1/2 \ mol)(8.31 \ J/mol \ K)(523 \ K) \ ln \ (5) = \textbf{+ 3.5 X } 10^3 \textbf{ J}.$$

The positive sign indicates that the work is done **BY the gas**.
The internal energy of an ideal gas depends ONLY on the temperature so $\Delta U = 0$ for an isothermal process. The first law then requires $Q = W = \textbf{+ 3.5 X } 10^3 \textbf{ J}$. The positive sign shows that heat is added to the gas.

---

6. Since the gas expands quickly, the process of expansion may be considered adiabatic. The ideal gas law applied to the initial and final states of the gas gives $T_f/T_i = (P_f/P_i)(V_f/V_i)$. In addition, equation (15.5) gives for the adiabatic process, $V_f/V_i = (P_i/P_f)^{1/\gamma}$. Now

$$T_f = (P_f/P_i)(P_i/P_f)^{1/\gamma} \ T_i = (1/10)(10)^{3/5} \ (296 \ K) = \textbf{120 K = - 160 °C}.$$

---

7. The pressure of the helium is essentially atmospheric throughout the process. The ideal gas law gives for the isobaric process

$$V_f = (T_f/T_i)V_i = (77 \text{ K}/3.00 \text{ X } 10^2 \text{ K})(1.00 \text{ X } 10^3 \text{ cm}^3) = \textbf{260 cm}^3 \textbf{ or } \textbf{2.60 X } 10^{-4} \textbf{ m}^3.$$

The internal energy change of the helium is $\Delta U = 3/2 \, nR(T_f - T_i)$ and the number of moles of helium is $n = PV/RT = 0.0405$ mol, so

$$\Delta U = \textbf{-113 J.}$$

The work done is

$$W = P(V_f - V_i) = (1.01 \text{ X } 10^5 \text{ Pa})(-7.40 \text{ X } 10^{-4} \text{ m}^3) = \textbf{- 74.7 J.}$$

The heat removed from the helium is found from the first law to be

$$Q = \Delta U + W = (-113 \text{ J}) + (-74.7 \text{ J}) = \textbf{- 188 J.}$$

8. (a) Equation (15.15) gives;

$$\text{Efficiency (e)} = 1 - T_C/T_H = 1 - (400 \text{ K})/(600 \text{ K}) = \textbf{0.33.}$$

(b) Equation (15.11) says $e = W/Q_H$ so

$$W = e \text{ x } Q_H = (0.33)(7.50 \text{ x } 10^4 \text{ J}) = \textbf{2.5 x } 10^4 \textbf{ J.}$$

9. Calculate the efficiency, $e = 1 - T_C/T_H = 1 - (364 \text{ K})/(455 \text{ K}) = 0.20.$

(a) $W = e \text{ x } Q_H = (0.20)(650 \text{ J}) = \textbf{130 J.}$

(b) Equation (15.13) gives $e = 1 - Q_C/Q_H$ so

$$Q_H = Q_C/(1 - e) = (1200 \text{ J})/(0.80) = 1500 \text{ J.}$$

Therefore, the work is

$$W = e \text{ x } Q_H = (0.20)(1500 \text{ J}) = \textbf{3.0 x } 10^2 \textbf{ J.}$$

10. The efficiency of the engine is simply

$$e = 1 - T_C/T_H = 1 - (311 \text{ K})/(736 \text{ K}) = \textbf{0.577.}$$

11. We know

$$Q_H = W/(1 - T_C/T_H) = (1.00 \text{ J})/(1 - 265 \text{ K}/295 \text{ K}) = \textbf{9.83 J.}$$

12. We know $e = 1 - Q_C/Q_H$. The heat enters the system during step d-a and leaves at step b-c. No heat is exchanged during segments a-b or c-d since these processes are adiabatic. Using the first law of thermodynamics we know

$$Q_H = \Delta U_{da} = (3/2)nR(T_a - T_d) \text{ and } Q_C = \Delta U_{bc} = (3/2)nR(T_b - T_c).$$

Note: We used the fact that W_{da} and W_{bc} are both zero since there is no volume change over segments d-a and b-c. The efficiency is $e = 1 - (T_b - T_c)/(T_a - T_d)$. Now using the fact that steps a-b and c-d are adiabatic, we have

$$T_a V_a{}^{\gamma-1} = T_b V_b{}^{\gamma-1} \text{ and } T_c V_c{}^{\gamma-1} = T_d V_d{}^{\gamma-1}.$$

Using $V_a = V_d = V_1$ and $V_b = V_c = V_2$, we can substitute for T_b and T_c and the volumes to obtain

$$T_b - T_c = T_a \left(\frac{V_a}{V_b} \right)^{\gamma-1} - T_d \left(\frac{V_d}{V_c} \right)^{\gamma-1} = (T_a - T_d) \left(\frac{V_1}{V_2} \right)^{\gamma-1}$$

The efficiency is therefore,

$$e = 1 - \frac{(T_b - T_c)}{(T_a - T_d)} = 1 - \left(\frac{V_1}{V_2} \right)^{\gamma-1}$$

---

13. The entropy change is given by equation (15.18), $\Delta S = Q/T$.

    (a) For the hot reservoir, $\quad \Delta S_H = Q/T_H = (-7700 \text{ J})/(550 \text{ K}) = $ **-14 J/K.**
        For the cold reservoir, $\quad \Delta S_C = Q/T_C = (+7700 \text{ J})/(350 \text{ K}) = $ **+22 J/K.**

    (b) Since the rod hasn't changed in any way, $\Delta S_{rod} = $ **0.**

    (c) The entropy change of the universe is therefore,

    $$\Delta S_{universe} = \Delta S_H + \Delta S_C + \Delta S_{rod} = \textbf{+8 J/K.}$$

---

14. First find the equilibrium temperature of the system. We have $\Delta Q_{ice} = \Delta Q_{water}$. We have $(mL_f)_{ice} + (cm\Delta T)_{ice} = (cm\Delta T)_{water}$ so that,

    $$(0.040 \text{ kg})(33.5 \times 10^4 \text{ J/kg}) + [4186 \text{ J/(kg C°)}](0.040 \text{ kg})T_f = [4186 \text{ J/(kg C°)}](0.100 \text{ kg})(90.0 \text{ °C} - T_f),$$

    from which we obtain $T_f = 41.4 \text{ °C}$. Now we can find the entropy change for each heat change.

    $\Delta S_{ice} = \Delta Q_{ice}/T = (mL_f)_{ice}/T = (0.0400 \text{ kg})(33.5 \times 10^4 \text{ J/kg})/(273 \text{ K}) = +49.1 \text{ J/K.}$

    $\Delta S_{ice \ water} = m_{ice}c_{water} \ln(T_f/T_i) = (0.0400 \text{ kg})[4186 \text{ J/(kg C°)}] \ln (314.4 \text{ K}/273 \text{ K}) = +23.6 \text{ J/K.}$

    $\Delta S_{water} = m_{water}c_{water} \ln(T_f/T_i) = (0.100 \text{ kg})[4186 \text{ J/(kg C°)}] \ln (314.4 \text{ K}/363 \text{ K}) = -60.2 \text{ J/K.}$

    Therefore, $\Delta S_{universe} = \Delta S_{ice} + \Delta S_{ice \ water} + \Delta S_{water} = +49.1 \text{ J/K} + 23.6 \text{ J/K} - 60.2 \text{ J/K} = $ **+12.5 J/K.**

CHAPTER 16
Waves and Sound

PREVIEW

This is the first of two chapters covering waves. In this chapter you will become acquainted with the concept of waves, in particular a special wave referred to as a periodic wave. You will learn to describe a periodic wave in space by its wavelength and amplitude and in time by its amplitude and period. Combining these descriptions allows us to find the speed of a wave in terms of its frequency and wavelength. The speed of waves on a stretched string will be investigated. A mathematical description of harmonic waves is given. This is very important since any complicated wave can be described as a sum of harmonic waves, as you will learn in a later chapter.

In this chapter you will also study the nature of sound. The wave properties of sound are examined, including the speed of sound in gases, liquids, and solids. You will be introduced to the concepts of sound intensity, loudness, and the decibel scale. Applications of sound are also discussed, such as sonar, ultrasound, and the Doppler effect.

QUICK REFERENCE

Important Terms

Wave
A traveling disturbance which carries energy. Mechanical waves require a medium.

Transverse wave
The wave disturbance is in a direction perpendicular to the direction of wave travel.

Longitudinal wave
The wave disturbance is parallel to the line of wave travel.

Periodic wave
The wave disturbance is repeated continuously by the source of the disturbance.

Amplitude (A)
The maximum excursion of a particle of the medium from the particle's undisturbed position.

Wavelength (λ)
The distance along the length of a wave between two equivalent points.

Period (T)
The time required for a wave to travel one wavelength.

Frequency (f)
The number of wavelengths per second which pass a given point. Note: $f = 1/T$.

Speed of a wave
The product of the wavelength and the frequency of a wave.

Condensation
A region of slightly increased air pressure produced by a vibrating source.

Rarefaction
A region of slightly reduced air pressure immediately following a condensation.

Frequency of a sound wave
The number of cycles per second (each cycle being composed of one condensation and one rarefaction) that pass a given location.

Pure tone
A sound containing a single frequency.

Pitch
A subjective quality related to the frequency of sound detected by the human ear. High pitch is high frequency and low pitch is low frequency.

Pressure amplitude

The maximum amount of change in pressure produced by a sound wave, measured relative to the undisturbed pressure.

Loudness

A subjective quality related to the pressure amplitude of a sound wave.

Sound intensity

The sound power per unit area that passes perpendicularly through a surface. Units are W/m^2.

Threshold of hearing

The smallest sound intensity that the human ear can detect.

Intensity level

A quantity which relates the intensity of a sound to some reference intensity level. The intensity level is measured in decibels.

Doppler effect

A change in the frequency of sound detected by an observer because the sound source and the observer have different velocities with respect to the medium of sound propagation.

Equations

The **speed of a periodic wave** which has a wavelength, λ, and a frequency, f, is

$$v = \lambda f \qquad (16.1)$$

The **speed of a wave traveling on a string** under a tension, F, and having a linear density m/L is

$$v = \sqrt{\frac{F}{m/L}} \qquad (16.2)$$

The displacement of a particle of the medium from its undisturbed position when a harmonic wave is moving along the + x axis is

$$y = A \sin (2\pi ft - 2\pi x/\lambda) \qquad (16.3)$$

and for motion of the wave toward - x

$$y = A \sin (2\pi ft + 2\pi x/\lambda) \qquad (16.4)$$

The **speed of sound in a gas** is:

$$v = \sqrt{\frac{\gamma kT}{m}} \qquad (16.5)$$

The **speed of sound in a liquid** is:

$$v = \sqrt{\frac{B_{ad}}{\rho}} \qquad (16.6)$$

The **speed of sound in a long, thin, solid** bar is:

$$v = \sqrt{\frac{Y}{\rho}} \qquad (16.7)$$

The **sound intensity** is defined as:

$$I = \frac{P}{A} \qquad (16.8)$$

If a sound is emitted uniformly in all directions, the **intensity at a distance r from the source** is:

$$I = \frac{P}{4\pi r^2} \qquad (16.9)$$

The **intensity level** in **decibels** is defined as:

$$\beta = (10 \text{ dB}) \log\left(\frac{I}{I_0}\right) \qquad (16.10)$$

The **Doppler shifted frequency** (general case) is:

$$f_o = f_s \left(\frac{1 \pm \dfrac{v_o}{v}}{1 \mp \dfrac{v_s}{v}}\right) \qquad (16.15)$$

DISCUSSION OF SELECTED SECTIONS

16.2 Periodic Waves

Three quantities are needed to describe periodic waves: period, wavelength, and amplitude. The wavelength and amplitude can be found from a space representation of the wave which is essentially a "snapshot" taken of the wave at an instant in time.

The period and amplitude can be found from a time representation of the wave. This is constructed by making a graph of the motion of a point in the medium. The graph is the displacement of the point from its undisturbed position versus time. This is equivalent to making a vertical position versus time graph of a small cork bobbing up and down in water as a wave passes by.

Example 1
A snapshot is taken of a periodic wave traveling on a string as shown below. What is the amplitude and wavelength of the wave?

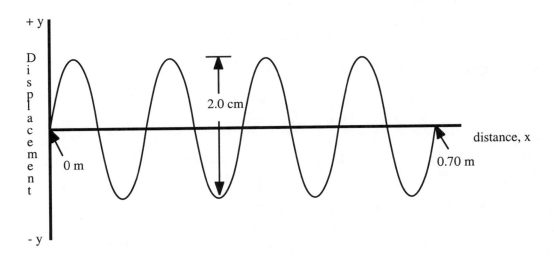

The height of the wave is seen from the "snapshot" to be 2.0 cm. The amplitude is only one-half of this, so

$$A = 1.0 \text{ cm}.$$

The wave is seen to have four wavelengths spread over a distance of 0.70 m so the wavelength

$$\lambda = 0.70 \text{ m}/4 = 0.18 \text{ m}.$$

Example 2
A point in the medium carrying the wave in example 1 is observed and a displacement versus time graph is made and shown below. Find the period, frequency, and amplitude of the wave from this graph. What is the speed of the wave?

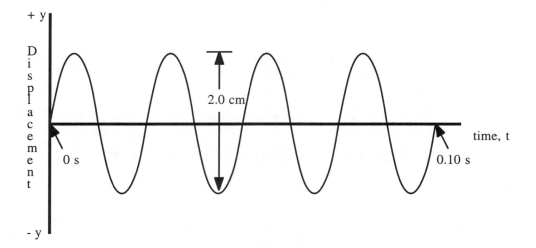

Again, the amplitude is one-half of the total height of the sine wave,

$$A = 1.0 \text{ cm}.$$

The period of the wave is found by observing that four complete oscillations occur in 0.10 s, so

$$T = (0.10 \text{ s})/4 = 0.025 \text{ s}.$$

The frequency of the wave is then

$$f = 1/T = 1/0.025 \text{ s} = 4.0 \times 10^1 \text{ Hz}.$$

The speed of the wave is

$$v = \lambda f = (0.18 \text{ m})(4.0 \times 10^1 \text{ Hz}) = 7.2 \text{ m/s}.$$

Example 3
Radio waves travel with a speed of 3.00×10^8 m/s. What is the wavelength of AM radio waves whose frequency is 640 kHz?

Equation (16.1) is true for waves of all kinds, including radio waves. The wavelength of the AM waves is

$$\lambda = v/f = (3.00 \times 10^8 \text{ m/s})/(640 \times 10^3 \text{ Hz}) = 470 \text{ m}.$$

16.3 The Speed of a Wave on a String

The speed of transverse waves on a string of uniform linear density is

$$v = \sqrt{\frac{F}{m/L}}$$

Example 4
A string of linear density, 2.0 g/m, is stretched to a tension of 4.9 N. What is the speed of the waves on the string? If the tension is increased so that the speed of the waves doubles, what is the new tension?

$$v = \sqrt{\frac{F}{m/L}} = \sqrt{\frac{4.9 \text{ N}}{(2.0 \times 10^{-3} \text{ kg/m})}} = 49 \text{ m/s}$$

Solving the above equation for the tension results in

$$F = (m/L)v^2$$

If the speed of the waves doubles to 2v, then the new tension is

$$F' = 4 \, (m/L)v^2 = 4F = 2.0 \times 10^1 \text{ N}$$

Example 5
A uniform wire carries waves whose frequency and wavelength are 450 Hz and 1.2 m, respectively. If the string is known to be under a tension of 250 N, what is the linear density of the wire?

The linear density can be found from equation (16.2): $m/L = F/v^2$.
The speed of the waves on the wire can be found from $v = \lambda f = (1.2 \text{ m})(450 \text{ Hz}) = 540$ m/s, so

$$m/L = (250 \text{ N})/(540 \text{ m/s})^2 = 8.57 \times 10^{-4} \text{ kg/m}.$$

16.4 The Mathematical Description of a Wave

If a periodic wave results from simple harmonic motion at the source, then, ideally, EACH point in the medium oscillates with simple harmonic motion. The displacement at a particular point, x =0 (time view) is
$y = A \sin (2\pi ft)$. As we saw before, the wave frozen in time looks like a sine wave, too. The displacement of each point at time zero (space view) can be written $y = A \sin (\pm 2\pi x/\lambda)$. It is customary to combine the space and time views into a single equation

$$y = A \sin (2\pi ft - 2\pi x/\lambda)$$

for a wave traveling towards +x and

$$y = A \sin (2\pi ft + 2\pi x/\lambda)$$

for a wave traveling towards -x.

Example 6
Show that equation (16.3) represents a wave traveling toward +x and that equation (16.4) represents a wave traveling toward -x.

Imagine watching a single point on a wave as it travels. This point appears to move with the speed of the wave and always has the same displacement. That is, the value of the sine function is constant. The only way for a sine function to remain constant is for its argument (the angle) to remain constant.

Equation (16.3) then gives $2\pi ft - 2\pi x/\lambda = C$, a constant. At some time, t_1, the point has a position, x_1, so $2\pi ft_1 - 2\pi x_1/\lambda = C$. At a later time, t_2, the point has a new position, x_2, so $2\pi ft_2 - 2\pi x_2/\lambda = C$. Subtracting and rearranging yields the velocity of the point on the wave

$$\frac{x_2 - x_1}{t_2 - t_1} = + \lambda f = + v$$

which is positive, indicating that the wave is traveling toward +x.

A similar treatment using equation (16.4) gives

$$\frac{x_2 - x_1}{t_2 - t_1} = - \lambda f = - v$$

which is negative. This wave is traveling toward the -x direction.

Example 7
The wave traveling on the string in example 4 has a wavelength of 0.50 m and an amplitude of 2.5 mm. Write an equation describing the wave as it travels toward the +x axis.

Equation (16.3) applies to a wave traveling toward +x. We need to find the frequency.

$$f = v/\lambda = (49 \text{ m/s})/(0.50 \text{ m}) = 98 \text{ Hz.}$$

Now

$$y = A \sin (2\pi ft - 2\pi x/\lambda) = (2.5 \text{ mm}) \sin [2\pi(98 \text{ s}^{-1})t - 2\pi x(2.0 \text{ m}^{-1})]$$

where x is in meters and t is in seconds.

Example 8
What are the wavelength, frequency and speed of the wave described by
$$y = (5.0 \text{ mm}) \sin [(450 \text{ s}^{-1})t - (25 \text{ m}^{-1})x]?$$

Comparing our wave with equation (16.3) shows that $2\pi f = 450 \text{ s}^{-1}$ and $2\pi/\lambda = 25 \text{ m}^{-1}$, so

$$f = 72 \text{ Hz and } \lambda = 0.25 \text{ m}$$

The speed of the wave is then

$$v = \lambda f = 18 \text{ m/s.}$$

16.6 The Speed of Sound

Sound travels through different materials at different speeds. For example, at room temperature sound travels at about 343 m/s in air. However, sound travels at about 1482 m/s through water and 5960 m/s through bulk steel at the same temperature. In general, sound travels slowest in gases, faster in liquids, and fastest in solids.

For a **gas**, the speed of sound depends on the temperature of the gas, and the type of molecules making up the gas. If we consider the situation for an ideal gas, we find that the speed of sound in that gas can be written

$$v = \sqrt{\frac{\gamma kT}{m}} \qquad\qquad (16.5)$$

where $\gamma = c_p/c_v$ is the ratio of specific heat capacity at constant pressure to the specific heat capacity at constant volume, m is the mass of a molecule of the gas, k is Boltzmann's constant, and T is the Kelvin temperature.

Example 9

Determine the speed of sound in air at 20.0 °C assuming the mean molecular mass of air to be 28.9×10^{-3} kg/mol and the ratio of specific heats to be $\gamma = 1.4$.

First we need to find the mass of an air molecule based on the mean molecular mass and Avogadro's number.

$$m = \frac{28.9 \times 10^{-3} \text{ kg/mol}}{6.022 \times 10^{23} \text{ mol}^{-1}} = 4.80 \times 10^{-26} \text{ kg}$$

Now use equation (16.5) to determine the speed of sound in air

$$v = \sqrt{\frac{\gamma kT}{m}} = \sqrt{\frac{(1.4)(1.38 \times 10^{-23} \text{ J/K})(293 \text{ K})}{4.80 \times 10^{-26} \text{ kg}}} = 343 \text{ m/s}$$

In a **liquid**, the speed of sound depends on the density of the material ρ and the adiabatic bulk modulus B_{ad}:

$$v = \sqrt{\frac{B_{ad}}{\rho}} \qquad\qquad (16.6)$$

In a **solid**, in the shape of a long slender bar, the speed of sound depends on the density and Young's modulus for the material:

$$v = \sqrt{\frac{Y}{\rho}} \qquad\qquad (16.7)$$

Example 10

A steel rod extends 2500 m under the ocean water. A sound is produced at one end of the rod and travels through both the rod and the water. What is the delay time between the arrival of the sound waves in the rod and the water? Take the bulk modulus for sea water to be $B_{ad} = 2.3 \times 10^9$ Pa, the density of sea water to be $\rho_w = 1025$ kg/m^3, Young's modulus for steel to be $Y = 2.0 \times 10^{11}$ Pa, and the density of steel to be $\rho_s = 7860$ kg/m^3.

$$v_w = \sqrt{\frac{B_{ad}}{\rho_w}} = \sqrt{\frac{2.3 \times 10^9 \text{ Pa}}{1025 \text{ kg/m}^3}} = 1.5 \times 10^3 \text{ m/s}; \quad v_s = \sqrt{\frac{Y}{\rho_s}} = \sqrt{\frac{2.0 \times 10^{11} \text{ Pa}}{7860 \text{ kg/m}^3}} = 5.0 \times 10^3 \text{ m/s}$$

The time it takes for each sound to travel 2500 m can now be calculated:

$$t_w = \frac{s}{v_w} = \frac{2500 \text{ m}}{1.5 \times 10^3 \text{ m/s}} = 1.7 \text{ s}; \qquad t_s = \frac{s}{v_s} = \frac{2500 \text{ m}}{5.0 \times 10^3 \text{ m/s}} = 0.50 \text{ s}$$

Therefore, the delay time between the arrival of the two waves is

$$\Delta t = t_w - t_s = 1.7 \text{ s} - 0.50 \text{ s} = 1.2 \text{ s}.$$

16.7 Sound Intensity

The **sound intensity I** is defined as the power P that passes perpendicularly through a surface divided by the area A of the surface:

$$I = \frac{P}{A} \qquad (16.8)$$

The unit of sound intensity is W/m^2. If the source emits sound uniformly in all directions, the radiated power P passes through the spherical surface area $A = 4\pi r^2$, where r is the distance to the sound source. The intensity is:

$$I = \frac{P}{4\pi r^2} \qquad (16.9)$$

Example 11
At a distance of 3.8 m from a siren, the sound intensity is 3.6×10^{-2} W/m^2. Assuming that the siren radiates sound uniformly in all directions, find the total power radiated.

Using equation (16.9) and solving for the power P gives

$$P = 4\pi r^2 I = 4\pi (3.8 \text{ m})^2 (3.6 \times 10^{-2} \text{ W/m}^2) = 6.5 \text{ W}.$$

16.8 Decibels

The **decibel (dB)** is a measurement unit encountered frequently in connection with audio equipment. The main application of the decibel concept is for comparing two sound intensities. The **intensity level β** is defined as:

$$\beta = (10 \text{ dB}) \log\left(\frac{I}{I_0}\right) \qquad (16.10)$$

where I_0 is the intensity of the reference level to which I is being compared, and is often taken as the threshold of hearing, $I_0 = 1.00 \times 10^{-12}$ W/m^2. Note that β is a dimensionless quantity and does NOT have units of W/m^2.

Example 12
The intensity level of the sound produced at a live rock concert is 120 dB above the threshold of hearing. What is the sound intensity I?

Use equation (16.10) with $I_0 = 1.00 \times 10^{-12}$ W/m^2 and solve for the intensity I. We obtain

$$I = I_0 \times 10^{\frac{\beta}{(10 \text{ dB})}} = (1.00 \times 10^{-12} \text{ W}/\text{m}^2) \times 10^{12} = 1.00 \text{ W}/\text{m}^2.$$

This represents a sound intensity that is one trillion times the threshold of hearing! However, the **loudness** will only be about 4100 times louder because each increase of 10 dB corresponds to a doubling of the loudness level. That is, a 10 dB sound will sound twice as loud as one which is 0 dB. A 20 dB sound is twice as loud as a 10 dB sound (four times louder than a 0 dB sound), and so on. Therefore, a 120 dB sound will appear about 2^{12} or approximately 4100 times as loud as a 0 dB sound.

16.9 The Doppler Effect

When a source of sound is in motion, or when a listener is moving, the frequency of the sound that is observed will change. This effect, known as the **Doppler effect**, is the change in pitch or frequency of the sound detected by an observer because the sound source and the observer have different velocities with respect to the medium of sound propagation. For example, when you hear the siren of an approaching fire truck you notice a definite change in pitch of the siren as it passes. As the truck approaches, the pitch is relatively high (that is, the frequency is high). However, as the truck passes you and moves away, the pitch drops (frequency lowers).

Let f_s represent the frequency of the sound emitted by a source. Sound travels at a speed v in the particular medium of propagation. The source may be moving with a speed v_s, and the observer may be moving with a speed v_o. The frequency observed, f_o, will be different from the emitted frequency according to the relation

$$f_o = f_s \left(\frac{1 \pm \dfrac{v_o}{v}}{1 \mp \dfrac{v_s}{v}} \right) \qquad (16.15)$$

In the numerator of the equation, the plus sign applies when the observer moves toward the source, and the minus sign applies when the observer moves away from the source. In the denominator, the minus sign is used when the source moves toward the observer, and the plus sign is used when the source moves away from the observer. The symbols v, v_o, and v_s denote numbers without algebraic sign, because the direction of travel has been taken into account by the plus and minus signs that appear directly in the equation.

Example 13

Two trucks travel at the same speed. They are far apart on adjacent lanes and thus, approach each other essentially head-on. One driver hears the horn of the other truck at a frequency that is 1.20 times the frequency he hears when the trucks are stationary. The speed of sound is 343 m/s. At what speed is each truck moving?

In this particular problem, the observer and source move with the same speed, so $v_o = v_s$. If the observed frequency is 1.20 times the emitted frequency, then $f_o / f_s = 1.20$. Equation (16.15) then yields

$$\frac{f_o}{f_s} = \left(\frac{1 + \dfrac{v_o}{v}}{1 - \dfrac{v_s}{v}} \right) = \frac{v + v_o}{v - v_s} = 1.20$$

Note that the plus sign was used in the numerator (since the observer is approaching the source) and a minus sign was used in the denominator (since the source is approaching the observer). Solving for v_o leads to

$$v + v_o = 1.20 \, (v - v_o)$$

which gives

$$2.20 \, v_o = 0.20 \, v.$$

$$v_o = \frac{0.20 \, v}{2.20} = (0.091)(343 \text{ m/s}) = 31 \text{ m/s}.$$

Thus, in order for the frequency to increase by a factor of 1.20, the observer and source are **each** moving at 31 m/s along the road.

PRACTICE PROBLEMS

1. A light wave has a wavelength of 4.55×10^{-7} m and travels with a speed of 3.00×10^8 m/s. What is the frequency of the light wave?

2. A wave on a string has a frequency of 440 Hz and a wavelength of 1.3 m. How fast does the wave travel?

3. A transverse wave on a string travels with a speed of 550 m/s and has a wavelength of 0.62 m and an amplitude of 0.50 cm. What is the maximum speed and the maximum acceleration of a point on the string?

4. A wire with a linear density of 2.5×10^{-3} kg/m is under 35 N of tension. How fast would a wave travel on the wire?

5. The waves traveling on a stretched wire travel with a speed of 180 m/s when the tension in the wire is 110 N. What is the linear density of the wire?

6. A wave has an amplitude of 0.70 mm, a wavelength of 5.5 m and travels with a speed of 150 m/s. Write an equation describing the wave.

7. A wave is described by $y = (1.0 \text{ cm}) \sin [(550 \text{ s}^{-1})t + (35 \text{ m}^{-1})x]$. What are the amplitude, wavelength, frequency and velocity of the wave?

8. A wave is described by $y = (1.5 \text{ cm}) \sin [(150 \text{ m}^{-1})x - (760 \text{ s}^{-1})t]$. How does this wave differ from one described by $y' = (1.5 \text{ cm}) \sin[(760 \text{ s}^{-1})t - (150 \text{ m}^{-1})x]$?

9. The speed of sound in a diatomic gas ($\gamma = 1.40$) is measured at 27 °C to be 280 m/s. What is the molecular mass of the gas?

10. Brass has a Young's modulus of 9.0×10^{10} Pa and a density of 8470 kg/m^3. How much time would it take for the sound to travel the length of a 5.5-m brass rod?

11. At the bottom of a lake, a sound pulse is generated which is received by an observer 15.0 m above the surface of the lake in 0.100 s. Both the lake and air are at a temperature of 20.0 °C. How deep is the lake?

12. Two sounds have intensities of 2.0×10^{-5} W/m^2 and 4.0×10^{-8} W/m^2. What is the difference in the decibel levels of the two sounds?

13. At a certain distance from a sound source, the intensity level is 75 dB. (a) What is the intensity of the sound in W/m^2? (b) How much energy falls on a 5.0 m^2 area in 25 s?

14. A listener triples his distance from a source that emits sound uniformly in all directions. By how many decibels does the sound intensity level change?

15. At 35 m from a uniformly radiating source of sound, the intensity level is 65 dB. At what distance from this source will its sound be just barely perceptible if absorption of sound by air is neglected?

16. A train whistle emits sound at a frequency of 555 Hz. A person standing next to the train track hears the whistle at a frequency of 488 Hz. What is the speed of the train? Is the train approaching or receding? Take the speed of sound to be 343 m/s.

17. A whistle of frequency 444 Hz rotates in a circle of radius 1.00 m with an angular speed of 25.0 rad/s. What is (a) the lowest frequency, and (b) the highest frequency heard by a listener a long distance away at rest with respect to the center of the circle?

18. An ambulance travels down the highway at a speed of 120 km/h. Its siren emits sound at a frequency of 660 Hz. A passenger car traveling at 88 km/h is on the other side of the highway (i.e., it's traveling in the opposite direction as the ambulance). What is the frequency of the siren heard by a passenger in the car (a) as the car and ambulance are approaching each other, and (b) as the car and ambulance are receding from one another?

SOLUTIONS AND ANSWERS

Practice Problems

---

1. $f = v/\lambda = (3.00 \times 10^8 \text{ m/s})/(4.55 \times 10^{-7} \text{ m}) = \mathbf{6.59 \times 10^{14} \text{ Hz}}$.

---

2. $v = \lambda f = (1.3 \text{ m})(440 \text{ Hz}) = \mathbf{570 \text{ m/s}}$.

---

3. The points on the string oscillate with the same frequency as the wave frequency. The maximum speed of points doing simple harmonic motion is

    $$v_{max} = 2\pi fA = 2\pi(v/\lambda)A = 2\pi(550 \text{ m/s})/(0.62 \text{ m})](0.50 \times 10^{-2} \text{ m}) = \mathbf{28 \text{ m/s}}.$$

    The maximum acceleration is
    $$a_{max} = (2\pi f)^2 A = \mathbf{1.6 \times 10^5 \text{ m/s}^2}.$$

---

4. Equation (16.2) gives $v = \sqrt{F/(m/L)} = \sqrt{35\text{N}/(2.5 \times 10^{-3} \text{kg}/\text{m})}$
    $$v = \mathbf{120 \text{ m/s}}.$$

---

5. Equation (16.2) gives $m/L = F/v^2 = (110 \text{ N})/(180 \text{ m/s})^2$
    $$m/L = \mathbf{3.4 \times 10^{-3} \text{ kg/m}}.$$

---

6. Equation (16.3) describes a wave moving toward +x. In our case,

    $$f = v/\lambda = (150 \text{ m/s})/(5.5 \text{ m}) = 27 \text{ Hz},$$
    so
    $$y = \mathbf{(0.70 \text{ mm}) \sin [(170 \text{ s}^{-1})t - (1.1 \text{ m}^{-1})x]}.$$

---

7. Compare the given equation with equation (16.4) for a wave traveling toward -x.

    $$A = \mathbf{1.0 \text{ cm}}, f = (550 \text{ s}^{-1})/2\pi = \mathbf{88 \text{ Hz}}, \lambda = 2\pi/(35 \text{ m}^{-1}) = \mathbf{0.18 \text{ m}}$$

    and the speed of the wave is
    $$v = \lambda f = 16 \text{ m/s}.$$
    The velocity is then
    $$v = \mathbf{-16 \text{ m/s}}.$$

---

8. The first of the equations can be written

    $$y = -(1.5 \text{ cm}) \sin [(760 \text{ s}^{-1})t - (150 \text{ m}^{-1})x]$$

    due to the property of the sin function that $\sin(-\theta) = -\sin\theta$. The two waves are identical and both travel toward +x. The first is the negative of the second. That is, it is "flipped" over with respect to the second wave.

---

9. Using equation (16.5),

$$v = \sqrt{(\gamma kT/m)}$$

so that

$$m = \gamma kT/v^2 = (1.4)(1.38 \times 10^{-23} \text{ J/K})(300 \text{ K})/(280 \text{ m/s})^2$$

which yields, $m = 7.4 \times 10^{-26}$ kg. The molecular mass is

$$(7.4 \times 10^{-23} \text{ g})(6.02 \times 10^{23}) = \textbf{45 g}.$$

---

10. Using equation (16.7),

$$v = \sqrt{(Y/\rho)} = \sqrt{(9.0 \times 10^{10} \text{ Pa})/(8470 \text{ kg/m}^3)} = 3.3 \times 10^3 \text{ m / s}.$$

The time it takes to travel a distance of 5.5 m is therefore,

$$t = s/v = (5.5 \text{ m})/(3.3 \times 10^3 \text{ m/s}) = \textbf{1.7 x 10}^{\textbf{-3}} \textbf{ s}.$$

---

11. The distance traveled in the water is $s_w = v_w t$. The distance in air is $s_a = v_a (0.100 \text{ s} - t) = 15.0$ m. Using $v_a = 343$ m/s we find $t = 0.056$ s. Therefore,

$$s_w = v_w t = (1482 \text{ m/s})(0.056 \text{ s}) = \textbf{83 m}.$$

---

12. The difference in intensity levels is

$$\beta = (10 \text{ dB}) \log (I_2/I_1) = (10 \text{ dB}) \log [(2.0 \times 10^{-5} \text{ W/m}^2)/(4.0 \times 10^{-8} \text{ W/m}^2)]$$
$$\beta = \textbf{27 dB}.$$

---

13. (a)

$$\beta = (10 \text{ dB}) \log (I/I_0)$$

so that

$$I = I_0 \times 10^{\beta/(10 \text{ dB})} = (1.0 \times 10^{-12} \text{ W/m}^2) \times 10^{7.5} = \textbf{3.2 x 10}^{\textbf{-5}} \textbf{ W/m}^2.$$

(b)

$$P = IA = E/t$$

gives

$$E = IAt = (3.2 \times 10^{-5} \text{ W/m}^2)(5.0 \text{ m}^2)(25 \text{ s}) = \textbf{4.0 x 10}^{\textbf{-3}} \textbf{ J}.$$

---

14. Using equation (16.9) we have

$$I_2/I_1 = (r_1/r_2)^2 = (r_1/3r_1)^2 = 1/9.$$

Therefore, the sound intensity changes by

$$\beta = (10 \text{ dB}) \log (I_2/I_1) = (10 \text{ dB}) \log (1/9) = \textbf{- 9.5 dB}.$$

15. Since
$$I_2/I_1 = (r_1/r_2)^2$$
then
$$\beta = (10 \text{ dB}) \log (I_2/I_1) = (10 \text{ dB}) \log (r_1/r_2)^2 = (20 \text{ dB}) \log (r_1/r_2) = 65 \text{ dB}.$$

Solving for r_1 yields
$$r_1 = r_2 \times 10^{\beta/(20 \text{ dB})} = (35 \text{ m}) \times 10^{(65 \text{ dB})/(20 \text{ dB})} = (35 \text{ m}) \times 10^{3.25} = \textbf{6.2 x 10}^4 \textbf{ m}.$$

16. Since the observed frequency decreases, this means the train is **receding** from the observer. Using
$$f_o = f_s [1/(1 + v_s/v)]$$
we get
$$v_s = v (f_s / f_o - 1) = (343 \text{ m/s})\{[(555 \text{ Hz})/(488 \text{ Hz})] - 1\} = \textbf{47.1 m/s}.$$

17. To find the speed of the sound source use
$$v_s = r\omega = (1.00 \text{ m})(25.0 \text{ rad/s}) = 25.0 \text{ m/s}.$$
So
$$v_s/v = 0.0729.$$
(a) For recession,
$$f_o = f_s [1/(1 + v_s/v)] = (444 \text{ Hz})[1/(1 + 0.0729)] = \textbf{414 Hz}.$$
(b) For approach,
$$f_o = f_s [1/(1 - v_s/v)] = (444 \text{ Hz})[1/(1 - 0.0729)] = \textbf{479 Hz}.$$

18. (a) Both the source and the observer are approaching, so using equation (16.15) with a + sign in the numerator and a - sign in the denominator we have,
$$f_o = f_s (1 + v_o/v)/(1 - v_s/v) = (660 \text{ Hz})(1 + 0.0713)/(1 - 0.0971)$$
$$f_o = \textbf{780 Hz}.$$

(b) Now the source and observer are receding, so use a - sign in the numerator and a + sign in the denominator. We have $f_o = f_s (1 - v_o/v)/(1 + v_s/v) = (660 \text{ Hz})(1 - 0.0713)/(1 + 0.0971) = \textbf{560 Hz}.$

CHAPTER 17
The Principle of Linear Superposition and Interference Phenomena

PREVIEW

When waves meet they combine in the region where they overlap to form a new wave. The new wave disturbance is the sum of the disturbances of the original waves. This is called the Principle of Linear Superposition. The result of the addition or superposition of the wave disturbances is called interference.

In this chapter you will learn how to superimpose waves and will study the resulting interference of two waves generated by point sources or very narrow slits (interference patterns), many waves which have diffracted through a single slit (diffraction patterns), two identical waves traveling in opposite directions (standing waves), and two waves of different frequencies (beats).

QUICK REFERENCE

Important Terms

Superposition
   The adding together of two or more wave disturbances to form a net disturbance.
Interference
   The result of the superposition of two or more waves.
Constructive interference
   The result of the superposition of two waves which meet exactly in phase. That is, crest to crest and trough to trough.
Destructive interference
   The result of the superposition of two waves which meet exactly out of phase. That is, crest to trough and trough to crest.
Diffraction
   The bending of a wave around an obstacle or edges of an opening.
Beats
   A periodic variation in the amplitude of a wave resulting from the linear superposition of waves that have slightly different frequencies.
Standing wave
   The stationary pattern of disturbance produced when identical waves traveling in opposite directions superimpose.
Node
   A point on a standing wave where the wave disturbance is zero.
Antinode
   A point on a standing wave where the wave disturbance is maximum.
Natural frequencies
   The frequencies of waves necessary to produce resonance.
Harmonics
   A set of natural frequencies of a system which are related by being integer multiples of the lowest frequency, that is, f_1, $2f_1$, $3f_1$, etc., where f_1 is the lowest frequency called the fundamental or first harmonic. The frequency $2f_1$ is called the second harmonic, the frequency $3f_1$ is called the third harmonic, etc.

Equations

The position of the **first minimum** of a **single slit diffraction pattern** can be found from

$$\sin \theta = \frac{\lambda}{D} \qquad (17.1)$$

The position of the **first minimum** of the **diffraction pattern of a circular opening** can be found from

$$\sin \theta = 1.22 \frac{\lambda}{D} \qquad (17.2)$$

The set of **natural frequencies** of a **string fixed at both ends** is given by

$$f_n = n \left(\frac{v}{2L} \right) \qquad n = 1, 2, 3, \ldots \qquad (17.3)$$

The set of **natural frequencies** of a **tube open at both ends** is given by

$$f_n = n \left(\frac{v}{2L} \right) \qquad n = 1, 2, 3, \ldots \qquad (17.4)$$

The set of **natural frequencies** of a **tube open at one end** is given by

$$f_n = n \left(\frac{v}{4L} \right) \qquad n = 1, 3, 5, \ldots \qquad (17.5)$$

Principles

The Principle of Linear Superposition - when two or more waves are present simultaneously at the same place, the resultant wave is the sum of the individual waves.

DISCUSSION OF SELECTED SECTIONS

17.2 Constructive and Destructive Interference of Sound Waves

Any time that two or more waves overlap in space an interference pattern results. The pattern may be very complicated if many waves of different amplitudes and frequencies overlap, or relatively simple if only two waves with the same amplitude and frequency are involved. Constructive interference occurs when the waves are exactly in phase and destructive interference occurs when they are exactly out of phase.

Two waves originally in phase can get out of phase if they are required to travel different distances before combining. If one wave travels λ, 2λ, 3λ, etc. farther than the other, then the two waves will be in phase and constructively interfere. If one travels $\lambda/2$, $3\lambda/2$, $5\lambda/2$, etc. farther than the other, then they will be out of phase and will destructively interfere.

Example 1
Two in-phase speakers are located 2.0 m apart and produce identical 550 Hz tones. A person stands 2.5 m from the left speaker and hears no sound. What are the possible distances of the person from the right speaker?

If the person hears no sound at all, completely destructive interference of the two sound waves arriving at his ear must be occurring. One of the sound waves must have traveled some odd multiple of $\lambda/2$, say $n\lambda/2$, farther than the other. The difference in the distance traveled is $s_R - s_L = n\lambda/2$, so

$$s_R = s_L + n\lambda/2.$$

The wavelength of the sound wave is $\lambda = v/f$. Then

$$s_R = s_L + nv/2f \; ; \; n = 1, 3, 5, \ldots$$

$$s_R = 2.5 \text{ m} + n[343/2(550) \text{ m}]$$

$$s_R = 2.5 \text{ m} + n(0.31 \text{ m}).$$

$$s_R = 2.8 \text{ m}, 3.4 \text{ m}, 4.1 \text{ m, etc.}$$

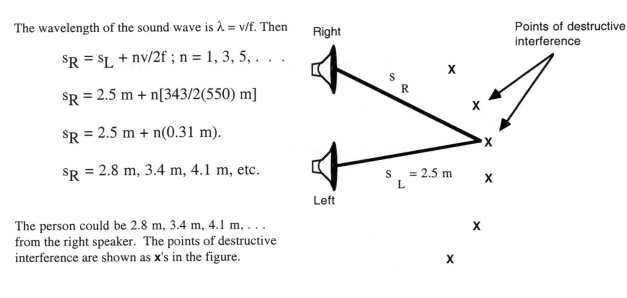

The person could be 2.8 m, 3.4 m, 4.1 m, . . . from the right speaker. The points of destructive interference are shown as **x**'s in the figure.

Example 2

Two in-phase radio towers 1.2 mi apart are broadcasting identical radio waves of frequency 550 kHz. Locate the points of constructive and destructive interference along the line that joins the two towers. Take the speed of radio waves to be 186 300 mi/s.

The wavelength is

$$\lambda = v/f = (186\ 300 \text{ mi/s})/(550\ 000 \text{ Hz})$$
$$\lambda = 0.34 \text{ mi}.$$

The maximum path difference of the two radio waves is 1.2 mi. along the line joining the towers. This corresponds to

$$1.2 \text{ mi}/\lambda = 3.5 \text{ wavelengths of path difference.}$$

There will then be 7 points of constructive interference and 6 points of destructive interference on the line between the towers, as shown in the drawing.

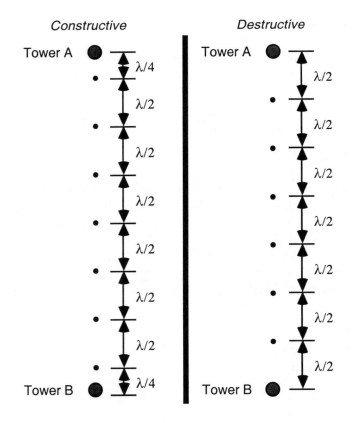

17.3 Diffraction

Whenever a wave passes close to an obstacle or edge it "bends" around the obstacle or edge. This is diffraction. It is the phenomenon which enables you to hear people on the other side of a house from you.

All waves exhibit diffraction, but sometimes the effect is so small that it escapes notice. A general rule of thumb is that diffraction of waves will be obvious when the wavelength of the wave is about the same size as the dimensions of the obstacles it encounters. This is why we hear the effect of diffracting low frequency (long wavelength) sound waves but do not see the effects of the diffraction of light (wavelength about 5×10^{-7} m) in everyday life.

Example 3

Green light whose wavelength is 5.5×10^{-7} m and sound of frequency 760 Hz pass through identical doors that are 0.92 m wide. What is the angle θ that locates the first minimum in the diffraction pattern for each wave? Take the speed of sound to be 343 m/s.

The angle θ that locates the first minimum in the diffraction pattern is given by (17.1). For the green light,

$$\theta = \sin^{-1}(\lambda/D)$$
$$\theta = \sin^{-1}[(5.5 \times 10^{-7} \text{ m})/(0.92 \text{ m})]$$
$$\theta = 3.4 \times 10^{-5} \, °.$$

Repeating the procedure for the sound which has a wavelength of $\lambda = v/f = (343 \text{ m/s})/(760 \text{ Hz}) = 0.45$ m we obtain

$$\theta = \sin^{-1}(0.45 \text{ m}/0.92 \text{ m}) = 29 \, °.$$

The angular spread is much greater for sound than for light.

17.4 Beats

Beats are heard when two sound waves of slightly different frequencies overlap. An example of this is when two slightly out of tune guitar strings are plucked simultaneously. The resulting sound intensity rises and falls at a rate which is the difference between the two sound wave frequencies. All waves exhibit the beat phenomenon. It is just much easier to detect beats in sound than in other waves such as light.

Example 4

An A-note (440.0 Hz) tuning fork is struck along with the corresponding A key on a piano. The resulting sound level rises and falls 3 times each second. The tension on the piano wire is increased by a small amount and the beat frequency increases. What was the out of tune frequency of the piano's A-note?

Since the piano and tuning fork produced 3 beats per second, the piano produced either 437 Hz or 443 Hz. Which is the correct value can be determined by considering what happens to the frequency of a wire as the tension is increased. Recall that $f = v/\lambda$ for any kind of wave and

$$v = \sqrt{\frac{F}{m/L}}$$

for the waves traveling on a wire.

As the tension, F, increases the speed of the waves on the wire increases and the frequency will increase. If the piano's note had been low, then a small increase in tension would have made it closer to 440.0 Hz and the beat frequency would have decreased. This was not observed to be the case. Hence, the piano's note was high to begin with and must have been 443 Hz.

17.5 Transverse Standing Waves

Standing wave patterns are produced when identical waves traveling in opposite directions combine. Most often one of the waves is generated by a source and the other is derived from reflection at a boundary such as a wall. In this circumstance, the relationship between the distance from the generator to the boundary and the wavelength determines whether the waves maintain the proper phase relationship to reinforce or cancel each other.

Example 5
Waves are introduced on one end of a string and meet waves which have previously been reflected from a clamped end of the string 2.0 m from the generator. Assume that the generator itself behaves approximately as a clamped end and find the possible wavelengths which can produce standing waves on the string.

Both ends of the string are fixed so any standing wave which exists on it must have stationary points (nodes) at each end. The longest wavelength which corresponds to this condition is shown in the diagram and is

$$\lambda_1 = 2L = 4.0 \text{ m}$$

The next longest wavelength is

$$\lambda_2 = L = 2.0 \text{ m}$$

and the next is

$$\lambda_3 = 2/3 \text{ L} = 1.3 \text{ m, etc.}$$

In general,

$$\lambda_n = 2L/n \quad n = 1, 2, 3, \text{ etc.}$$

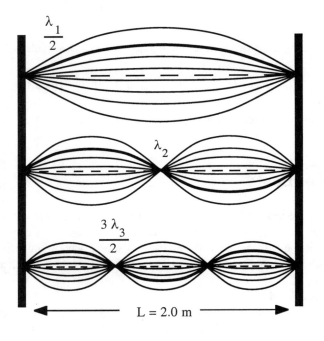

Example 6
A 30.0 m power line is observed to vibrate in its second harmonic due to excitation by the wind. The period T of the oscillation of the line is determined to be 1.5 s. What is the speed of the waves on the line?

The speed of the waves is

$$v = \lambda f = \lambda/T.$$

The wavelength of the second harmonic is just equal to the length of the line so, $\lambda = 30.0$ m. Now

$$v = (30.0 \text{ m})/(1.5 \text{ s}) = 2.0 \times 10^1 \text{ m/s}.$$

17.6 Longitudinal Standing Waves

Longitudinal waves such as sound waves can also produce standing waves by reflection from a boundary. The boundary can be any change in the environment of the sound wave such as that at the end of a tube. If the tube end is closed, then the standing wave must have a displacement node there since the air is unable to move. If the end of the tube is open then the standing wave will have a displacement antinode at the end.

Example 7

A 1.5 m tube which is closed at one end supports a standing sound wave. What are the three longest wavelengths of sound for which standing waves can exist in the tube? What are the frequencies?

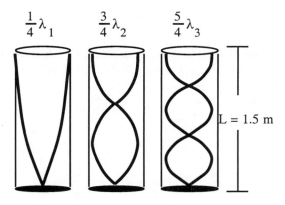

Since one end of the tube must contain a node and the other end an antinode the longest wavelength which will "fit" in the tube is such that $L = \lambda_1/4$.   This wavelength is

$$\lambda_1 = 4L = 6.0 \text{ m}.$$

The wavelength corresponds to a frequency,

$$f_1 = v/\lambda_1 = (343 \text{ m/s})/(6.0 \text{ m}) = 57 \text{ Hz}.$$

The second longest wavelength which will fit in the tube is such that $L = 3/4 \, \lambda_2$ so

$$\lambda_2 = 4/3 \, (1.5 \text{ m}) = 2.0 \text{ m}.$$

This corresponds to a frequency,

$$f_2 = (343 \text{ m/s})/(2.0 \text{ m}) = 170 \text{ Hz}.$$

The third longest wavelength is such that $L = 5/4 \, \lambda_3$, so

$$\lambda_3 = 4/5 \, L = 1.2 \text{ m}.$$

The corresponding frequency is

$$f_3 = (343 \text{ m/s})/1.2 \text{ m}) = 290 \text{ Hz}.$$

Example 8

Repeat example 7 assuming that both ends of the tube are open.
The procedure is the same as in example 7. The main difference is that the standing wave must have an antinode at EACH end of the tube. The diagram shows the situation for the three longest wavelengths.

$$\lambda_1 = 2L = 3.0 \text{ m}$$
$$f_1 = 110 \text{ Hz}$$

$$\lambda_2 = L = 1.5 \text{ m}$$
$$f_2 = 230 \text{ Hz}$$

$$\lambda_3 = 2/3 \, L = 1.0 \text{ m}$$
$$f_3 = 340 \text{ m}$$

PRACTICE PROBLEMS

In all of the following the speed of sound is assumed to be 343 m/s unless otherwise stated.

1. Two speakers generate identical 650 Hz sound waves which arrive at a person's ear after traveling 2.0 m and 2.5 m, respectively. Describe the sound heard by the person.

2. Two in-phase speakers produce identical 440 Hz sound waves. A person stands 2.0 m directly in front of one speaker and hears the loudest possible sound. What is the minimum nonzero separation of the speakers?

3. A young boy stands in a 3.5 m wide alley between two very tall buildings and shouts to a friend standing 15 m in front of one of the buildings and 5.0 m to the side of the alley. What sound frequencies of the first boy's shout will the second boy hear?

4. What must be the diameter of a speaker to produce a 15 kHz tone with a diffraction angle of 35°?

5. The same note is sounded on two slightly out of tune guitars, and a beat frequency of 7.00 Hz is heard. One of the guitars is then sounded along with a 440.0 Hz tuning fork and a beat frequency of 4.00 Hz is heard. It is judged that the tuning fork has the higher frequency. What are the possible frequencies of the note on each guitar?

6. A 1.5 m long string, fixed at both ends, has a fundamental frequency of 120 Hz. What is the speed of waves traveling on the string?

7. A 0.90 m string is stretched between rigid supports. What are the four longest wavelengths that can lead to standing waves on the string?

8. A 1.50 m long string, whose mass is 2.50 g, is stretched between fixed supports to a tension of 9.00 N. What are the natural frequencies of the string?

9. A loudspeaker is facing a perfectly reflecting wall located 3.50 m away. How many frequencies of sound in the audible range (20.0 Hz to 20.0 kHz) produced by the speaker would result in standing waves between the wall and the speaker?

10. A "pop" bottle is partially filled with pop so that a 10.0 cm air column is left at the top. A person blows across the top of the bottle producing sound waves of many frequencies in the audible range (20 Hz to 20 kHz). What is the lowest frequency of sound which will produce standing waves in the bottle, thus sounding louder?
    Note: The neck of a pop bottle is small and the shape of the bottle may have some effect. Ignore these in your calculation.

SOLUTIONS AND ANSWERS

Practice Problems

---

1. The wavelength of the sound is $\lambda = v/f = (343 \text{ m/s})/(650 \text{ Hz}) = 0.53$ m and the difference in the path lengths of the two waves is 2.5 m - 2.0 m = 0.5 m. Comparison of the wavelength and path difference shows that one wave traveled approximately one wavelength farther than the other. The waves will arrive at the person's ear in phase and will produce a **loud sound** due to constructive interference.

---

2. The minimum non-zero separation, d, of the speakers which will produce constructive interference corresponds to a one wavelength difference in the path lengths traveled by the two waves.

$$\sqrt{d^2 + (2.0 \text{ m})^2} - 2.0 \text{ m} = (343 \text{ m/s})/(440 \text{ Hz})$$

   Solving for d gives

$$d = \textbf{1.9 m}.$$

---

3. In order for the sound to reach the second boy, it must spread through at least an angle

$$\theta = \tan^{-1} (5.0 \text{ m})/(15.0 \text{ m}) = 18.4°$$

   Equation (17.1) and $\lambda f = v$ give

$$f = v/(D \sin \theta) = (343 \text{ m/s})/[(3.5 \text{ m})(\sin \theta)] = \textbf{310 Hz}.$$

   This is the **maximum frequency** which the second boy can hear.

---

4. Equation (17.2) and $\lambda = v/f$ give

$$D = 1.22(v/f)/(\sin \theta) = 1.22[(343 \text{ m/s})/(15 \times 10^3 \text{ Hz})]/(\sin 35°) = \textbf{0.049 m}.$$

---

5. The frequency produced by the guitar which is sounded with the tuning fork must be 440.0 Hz + 4 Hz = 444 Hz or 440.0 Hz - 4 Hz = 436 Hz. The tuning fork is higher so this guitar must produce **436 Hz**. The other guitar produces either 436 Hz + 7 Hz = **443 Hz** or 436 Hz - 7 Hz = **429 Hz**. It is impossible to know which frequency the guitar actually produces.

---

6. The fundamental standing wave on the string corresponds to $1/2 \lambda = L$. Then $\lambda = 2(1.5 \text{ m}) = 3.0$ m. The traveling waves on the string have this wavelength, therefore, their speed is

$$v = \lambda f = (3.0 \text{ m})(120 \text{ Hz}) = \textbf{360 m/s}.$$

---

7. The longest standing wave (fundamental) which can exist on the string must satisfy the condition

$$1/2 \lambda_1 = L \text{ or } \lambda_1 = \textbf{1.8 m}.$$

The second longest standing wave (second harmonic) which can exist on the string must satisfy the condition

$$2/2 \ \lambda_2 = L \text{ or } \lambda_2 = \textbf{0.90 m.}$$

The third longest standing wave (third harmonic) which can exist on the string must satisfy the condition

$$3/2 \ \lambda_3 = L \text{ or } \lambda_3 = \textbf{0.60 m.}$$

The fourth longest standing wave (fourth harmonic) which can exist on the string must satisfy the condition

$$4/2 \ \lambda_4 = L \text{ or } \lambda_4 = \textbf{0.45 m.}$$

---

8.   The natural frequencies of the string are given by equation (17.3). We need the speed of the waves.

$$v = \sqrt{\frac{F}{(m/L)}} = \sqrt{\frac{9.00 \text{ N}}{(2.50 \text{ X } 10^{-3} \text{ kg})/(1.50 \text{ m})}} = 73.5 \text{ m/s}$$

Now

$$f_n = \frac{nv}{2L} = n\left(\frac{73.5 \text{ m/s}}{2(1.5 \text{ m})}\right) = n(24.5 \text{ Hz})$$

The natural frequencies are **{24.5 Hz, 49.0 Hz, 73.5 Hz, etc.}**

---

9.   The situation described here is similar to a tube open at one end (speaker) and closed at the other (wall). Equation (17.5) gives the frequencies of the standing waves to be

$$f_n = \frac{nv}{4L} = \frac{n(343 \text{ m/s})}{4(3.50 \text{ m})} = n(24.5 \text{ Hz})$$

Standing waves can result from all frequencies with odd n from 1 to n = $(20.0 \text{ X } 10^3 \text{ Hz})/(24.5 \text{ Hz}) = \textbf{816.}$

---

10.   The partially filled pop bottle is a tube open at one end, so equation (17.5) applies. The lowest frequency which will produce standing waves is corresponds to n = 1.

$$f_1 = \frac{v}{4L} = \frac{343 \text{ m/s}}{4(10.0 \text{ X } 10^{-2} \text{ m})} = \textbf{858 Hz}$$

CHAPTER 18
Electric Forces and Electric Fields

PREVIEW

In this chapter you will be introduced to the concepts of electric charge, electric forces and electric fields. You will use Coulomb's law to find the forces and fields associated with point charges and charged plates. You will also learn how to "map" electric field lines and how electric fields are affected by the presence of conductors.

QUICK REFERENCE

Important Terms

Electric charge
An intrinsic property of protons (positive charge) and electrons (negative charge). It is found that like charges (same polarity) repel one another while unlike charges (opposite polarity) attract one another. The SI unit of charge is the coulomb (C).

Electrical conductors
Materials which allow electric charge to flow freely through them.

Electrical insulators
Materials which do not allow electric charge to flow freely through them.

Coulomb's law
The magnitude of the electrical force exerted by one point charge on another is directly proportional to the magnitudes of the charges and inversely proportional to the square of the distance between the charges. The direction of the force is along the line joining the charges and is repulsive for like charges and attractive for unlike charges.

Electric field
At a point in space, the electrostatic force experienced by a small test charge placed at that point divided by the charge itself. Electric field is a vector quantity. The SI unit of electric field is the newton per coulomb (N/C).

Electric field lines
A visual representation of the electric field. It can be viewed as a "map" providing information about the direction and strength of the electric field in various regions.

Gaussian surface
A hypothetical *closed* surface in the vicinity of an electric field or a charge distribution. The Gaussian surface may enclose a net charge.

Electric flux
The product of the magnitude of the normal component of the electric field at any point on a surface element and the area of the surface element. It is assumed that the magnitude of the normal component of the electric field is constant over the surface element. The SI unit of electric flux is the newton square meter per coulomb ($N \cdot m^2/C$).

Gauss' Law
The mathematical expression that relates the electric flux through a closed (Gaussian) surface to the net charge enclosed by the surface. Gauss' Law can be used to calculate the magnitude of the electric field surrounding charge distributions that possess spatial symmetry.

Equations

Coulomb's law for the magnitude of the force that one *point charge* exerts on a second *point charge* is written as

$$F = k\frac{q_1 q_2}{r^2} \qquad (18.1)$$

where q_1 and q_2 are the magnitudes of the two point charges, and r is the distance between them.

The **electric field** can be written

$$E = \frac{F}{q_0} \qquad (18.2)$$

The magnitude of the **electric field** due to a **point charge** is

$$E = \frac{kq}{r^2} \qquad (18.3)$$

The magnitude of the **electric field** between the plates of a **parallel plate capacitor** is given by

$$E = \frac{q}{\varepsilon_0 A} = \frac{\sigma}{\varepsilon_0} \qquad (18.4)$$

The **electric flux** through a closed surface can be written as

$$\Phi_E = \sum (E \cos \phi)\Delta A \qquad (18.6)$$

where E is the magnitude of the electric field at a point on the surface, and ϕ is the angle between the electric field and the normal at that point.

Gauss' Law for a closed surface is written as

$$\sum (E \cos \phi)\Delta A = \frac{Q}{\varepsilon_0} \qquad (18.7)$$

where Q is the net charge enclosed by the surface and ε_0 is the permittivity of free space.

DISCUSSION OF SELECTED SECTIONS

18.1 The Origin of Electricity
18.2 Charged Objects and the Electric Force

An intrinsic property of protons and electrons, which make up all matter, is **electric charge**. Only two types of electric charge have been discovered, positive and negative; a proton has a positive charge, and an electron has a negative charge.

The SI unit for measuring the magnitude of electric charge is the **coulomb** **(C)**. Each proton carries a charge +e, and each electron carries a charge -e, where

$$e = 1.60 \times 10^{-19} \, C$$

This charge e is the smallest amount of free charge that is possible. Charges larger than this are integer multiples of e, that is, q = Ne where N is the number of electrons (or protons) present.

Example 1

How many electrons are contained in 3.20×10^{-6} coulombs of charge?

Since the total charge is q = Ne we can write

$$N = \frac{q}{e} = \frac{3.20 \times 10^{-6} \, C}{1.60 \times 10^{-19} \, C} = 2.00 \times 10^{13} \text{ electrons}$$

Because any electric charge, q, occurs as integer multiples of the elementary charge e, electric charge is said to be **quantized**.

Since the charges on the electron and proton have identical magnitudes but opposite signs, the algebraic sum of the two charges is zero. Since most materials have equal numbers of protons and electrons, most matter is electrically neutral. However, it is possible to transfer electric charge from one object to another. Usually electrons are transferred, and the body that gains electrons acquires an excess of negative charge. The body that loses these electrons then acquires an excess of positive charge. In this sense, charge can be "separated". However, during any process, the net electric charge of an entire isolated system remains constant (is conserved). This is referred to as the **law of conservation of electric charge**.

Charge separation is important in the operation of electrical equipment. For example, batteries, alternators in cars, and electric power generators all depend on the separation of electric charges for their operation.

It is easy to demonstrate that two electrically charged objects exert a force on one another. If the two objects contain charge of the same sign, the force will be one of repulsion, i.e., the objects push one another apart. If the objects contain charge of opposite sign, the force will be one of attraction and the objects will pull each other closer together. It is a fundamental characteristic of electric charges that **like charges repel and unlike charges attract each other**. The magnitude of the force exerted by one point charge on another was experimentally determined by the French physicist Charles-Augustin Coulomb. His result, which has become known as Coulomb's law, is stated in the next section.

18.5 Coulomb's Law

Consider two "point charges", q_1 and q_2, separated by a distance r, as shown in the diagram at the right. If the charges have unlike signs, as in part (a) of the diagram, each charge is attracted to the other by a force that is directed along the line between them; +**F** is the electric force exerted on charge 1 by charge 2 and -**F** is the electric force exerted on charge 2 by charge 1. If the charges have the same sign, as in (b), each charge is repelled by the other. The forces +**F** and -**F** form an action-reaction pair as predicted by Newton's third law.

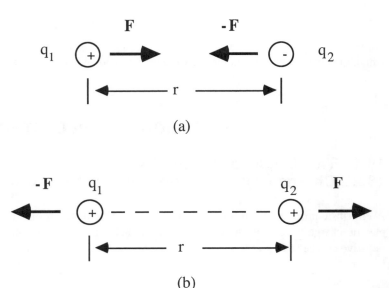

The magnitude of the force (either attractive or repulsive) is given by Coulomb's law, i.e.,

$$F = k\frac{q_1 q_2}{r^2} \qquad (18.1)$$

where k is a constant of proportionality whose value in SI units is $k = 8.99 \times 10^9$ N m^2/C^2.

Example 2
What is the force between charges $q_1 = + 1.0$ μC and $q_2 = - 2.0$ μC which are separated by a distance of 2.0 m? Equation (18.1) indicates the magnitude of the force is given by

$$F = k\frac{q_1 q_2}{r^2} = \frac{(8.99 \times 10^9 \text{ N m}^2/\text{C}^2)(1.0 \times 10^{-6} \text{ C})(2.0 \times 10^{-6} \text{ C})}{(2.0 \text{ m})^2} = 4.5 \times 10^{-3} \text{ N}.$$

Since the two charges have opposite signs, the force is one of attraction; q_1 will exert a force of 4.5×10^{-3} N which attracts q_2. According to Newton's third law, q_2 will exert a force of 4.5×10^{-3} N on q_1, which again is a force of attraction.

Example 3 *Force on a Point Charge due to Two or More Other Point Charges*

Consider four point charges arranged in the corners of a square, as shown in the diagram. Using,

$q_1 = + 3.00$ μC $\qquad q_2 = - 1.20$ μC

$q_3 = - 2.00$ μC $\qquad q_4 = - 1.20$ μC

$a = 25.0$ cm

find the net force acting on charge q_3 due to charges q_1, q_2, and q_4. Assume that all the point charges are fixed so that no motion occurs.

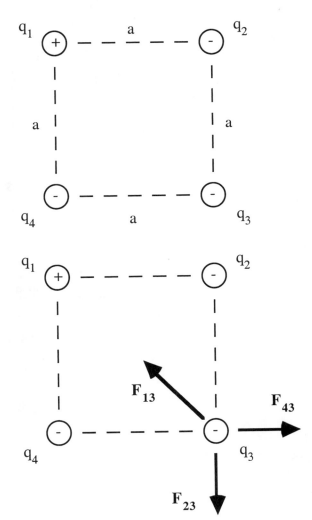

To find the net force acting on q_3, we need to find the vector sum of the forces acting on it. Note that q_3 will be repelled by q_2 and q_4, while q_3 will be attracted to q_1. If the force of charge q_1 acting on q_3 is $\mathbf{F_{13}}$, and the forces on q_3 due to charges 2 and 4 are $\mathbf{F_{23}}$ and $\mathbf{F_{43}}$, respectively, we can use the following diagram.

We begin by calculating the magnitudes of the three forces acting on q_3. We will then use the rules of vector addition to find the desired resultant.

The force exerted on the charge q_3 by the charge q_1 is, from Coulomb's law,

$$F_{13} = k \frac{q_1 q_3}{r_{13}^2} = \frac{(8.99 \times 10^9 \text{ N m}^2/\text{C}^2)(3.00 \times 10^{-6} \text{ C})(2.00 \times 10^{-6} \text{ C})}{2 (0.250 \text{ m})^2} = 0.432 \text{ N}$$

Similarly, for F_{23} and F_{43} we obtain,

$$F_{23} = k \frac{q_2 q_3}{r_{23}^2} = 0.345 \text{ N} \qquad F_{43} = k \frac{q_4 q_3}{r_{43}^2} = 0.345 \text{ N}$$

We can now add forces F_{13}, F_{23}, and F_{43} using the following vector diagram.

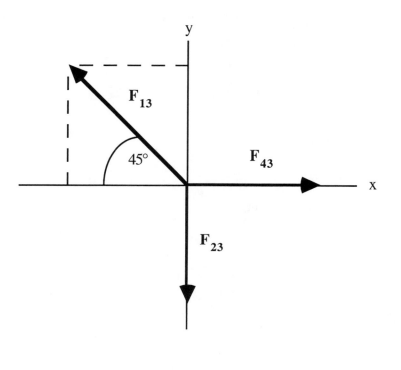

Force	x component	y component
F_{13}	$-(0.432 \text{ N}) \cos 45°$	$+(0.432 \text{ N}) \sin 45°$
	$= -0.305 \text{ N}$	$= +0.305 \text{ N}$
F_{23}	$= 0$	$= -0.345 \text{ N}$
F_{43}	$= +0.345 \text{ N}$	$= 0$
F	$= +0.040 \text{ N}$	$= -0.040 \text{ N}$

The magnitude F of the resultant is given by

$$F = \sqrt{F_x^2 + F_y^2} = \sqrt{(+0.040 \text{ N})^2 + (-0.040 \text{ N})^2} = 0.057 \text{ N}$$

The angle θ that the resultant force makes with the x axis is

$$\theta = \tan^{-1}\left(\frac{F_y}{F_x}\right) = \tan^{-1}\left(\frac{-0.040 \text{ N}}{+0.040 \text{ N}}\right) = -45°$$

where the negative angle means that the direction is clockwise with respect to the +x axis. In other words, the resultant force on q_3 is directed *below* the positive x axis.

So the general procedure for finding the force on a point charge due to two or more other point charges is to find the vector sum of the individual forces. The magnitudes of the forces are calculated from Coulomb's law. The directions of the forces are determined from the sign of each individual charge. That is, the force is either one of attraction or repulsion, depending on whether each pair of charges has unlike or like signs, respectively.

18.6 The Electric Field

A charge can experience an electrostatic force due to the presence of other charges in the environment. If a positive charge q_0 is placed at a point in space, and if other charges are present, the charge q_0 feels a force **F**, which is the vector sum of the forces exerted on it by all the other charges present. We can define the electric field **E**, that exists at this point, as the electrostatic force **F** divided by the test charge q_0. That is,

$$E = \frac{F}{q_0} \tag{18.2}$$

The electric field is a vector, and its direction is the same as the direction of the force **F** on a positive test charge. The SI unit of electric field is newtons per coulomb (N/C).

Consider the electrostatic force exerted by a point charge q on a test charge q_0. The force is given by equation (18.1),

$$F = k\frac{q\,q_0}{r^2}$$

Using equation (18.2), we can see that the magnitude of the electric field due to a point charge can be written as

$$E = \frac{kq}{r^2} \tag{18.3}$$

Example 4
Two point charges, $q_1 = -8.0\ \mu C$ and $q_2 = +4.0\ \mu C$, are separated by a distance of 0.50 m. Find the electric field (both magnitude and direction) at a point P on the line between the charges 0.20 m from q_2.

We can see from the following diagram that the net electric field at point P is due to the combined effects of both charges. Note the directions of the fields E_1 and E_2; these are based on the fact that the direction of the field is the same as the direction of the force on a POSITIVE test charge.

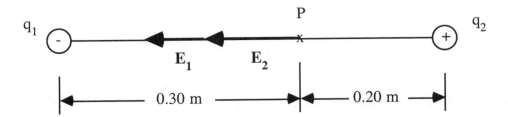

Using equation (18.3) we have,

$$E_1 = \frac{k\,q_1}{r_1^2} = \frac{(8.99 \times 10^9 \text{ N m}^2/\text{C}^2)(8.0 \times 10^{-6} \text{ C})}{(0.30 \text{ m})^2} = 8.0 \times 10^5 \text{ N/C}$$

$$E_2 = \frac{k\,q_2}{r_2^2} = \frac{(8.99 \times 10^9 \text{ N m}^2/\text{C}^2)(4.0 \times 10^{-6} \text{ C})}{(0.20 \text{ m})^2} = 9.0 \times 10^5 \text{ N/C}$$

Since E_1 and E_2 are in the same direction, the total electric field is obtained by adding them together. Thus

$$E = E_1 + E_2 = 8.0 \times 10^5 \text{ N/C} + 9.0 \times 10^5 \text{ N/C} = 1.7 \times 10^6 \text{ N/C}.$$

The field **E** is directed to the left, or towards q_1.

Now consider a **parallel plate capacitor** which consists of two parallel metal plates, each with area A. A charge +q is spread uniformly over one plate, while a charge -q is spread uniformly over the other plate. In the region between the plates and away from the edges, the electric field points from the positive plate toward the negative plate and has a magnitude of

$$E = \frac{q}{\varepsilon_0 A} = \frac{\sigma}{\varepsilon_0} \qquad (18.4)$$

where ε_0 is the **permittivity of free space** and has a value of $\varepsilon_0 = 8.85 \times 10^{-12}$ C^2/(N m^2), and $\sigma = q/A$ is the surface charge density of the plates.

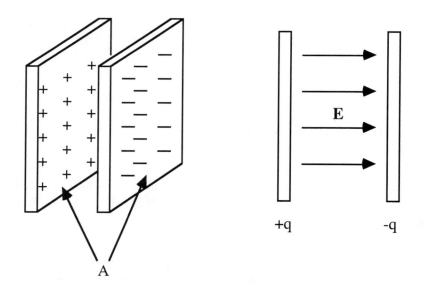

Example 5

The magnitude of the electric field between the plates of a parallel plate capacitor is 2.4×10^5 N/C. Each plate carries a charge whose magnitude is 0.15 μC. What is the area of the plates?

Using equation (18.4) and solving for A we obtain

$$A = \frac{q}{\varepsilon_0 E} = \frac{1.5 \times 10^{-7}\,C}{[8.85 \times 10^{-12}\,C^2/(N\,m^2)](2.4 \times 10^5\,N/C)} = 0.071\,m^2$$

18.9 Gauss' Law

Gauss' law relates the electric flux through a closed surface to the net charge enclosed by the surface. In order to fully understand Gauss' law, we must first discuss the concept of electric flux.

Consider a surface in the vicinity of an electric field as shown in part (a) of the figure below. Imagine dividing the surface into a number of smaller surface elements, each of area ΔA. The size of each element is chosen so that the magnitude and direction of the electric field is constant over each surface element. We now imagine constructing the line that is normal to each surface element. In general, the electric field at any point on the surface element will make an angle φ with the normal, as shown in part (b) of the figure below.

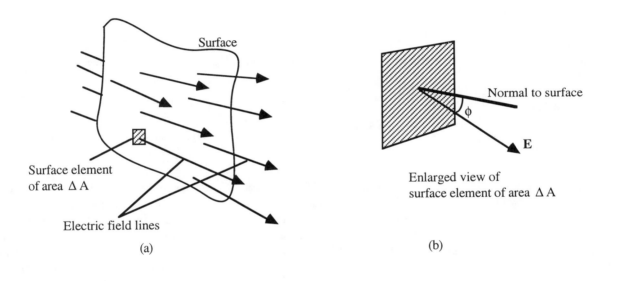

The electric flux through the entire surface is defined as

$$\Phi_E = \sum (E \cos \phi)\Delta A \qquad (18.6)$$

where E is the magnitude of the electric field at a point on the surface, and φ is the angle between the electric field and the normal at that point. The summation is carried out over the entire surface; in other words, the quantity (E cos φ)ΔA is summed for all surface elements that comprise the surface.

In general, surfaces can be *open* or *closed*. An open surface always has an edge, a rim, or some boundary. Examples include a sheet of paper, a lamp shade, and half of an egg shell. Closed surfaces, on the other hand, do not possess an edge or boundary. Examples of closed surfaces include a sphere and an unbroken egg shell. Gauss's law pertains only to *closed* surfaces, appropriately named *Gaussian surfaces*. Clearly, the surface elements on a closed surface have both an inward and an outward normal. In applying Equation (18.6) to closed surfaces, the angle φ is, by convention, the angle that the electric field makes with the outward normal. The electric flux through a closed surface will be positive if the electric field lines leave the surface and negative if the electric field lines enter the surface.

Gauss' law states that the electric flux Φ_E through a Gaussian surface is equal to the net charge Q enclosed by the surface divided by ε_0, the permittivity of free space:

$$\sum (E \cos \phi)\Delta A = \frac{Q}{\varepsilon_0}. \tag{18.7}$$

From Equation (18.7), we see that the flux through a closed surface is positive if it encloses a net positive charge, and negative if the surface encloses a net negative charge.

Gauss's law is an important calculational tool in many circumstances:

1. It can be used to determine the electric flux through a closed surface without performing the summation in Equation (18.6) if the net enclosed charge Q is known.

2. It can be used to determine the net charge within a closed surface, if the flux through the surface is known (or can be determined).

3. It can be used to determine the magnitude of the electric field for charge distributions that possess a high degree of spatial symmetry.

The next two examples illustrate the first two circumstances described above.

Example 6
A point charge of + 5.0 μC is placed at the center of a cube of side length 0.25 m. Determine the electric flux through the cube.

From Gauss' law

$$\Phi_E = \frac{Q}{\varepsilon_0} = \frac{+5.0 \times 10^{-6} \text{ C}}{8.85 \times 10^{-12} \text{ C}^2/(\text{N} \cdot \text{m}^2)} = 5.6 \times 10^5 \text{ N} \cdot \text{m}^2/\text{C} .$$

While this problem can be solved using Equation (18.6), the method is tedious and requires the use of integral calculus. Gauss' law provides a quick and convenient method that requires only simple arithmetic. Notice that the answer does not depend on the fact that the charge is at the center of the cube. In fact, the electric flux through the cube will be the same regardless of the position of the charge as long as it remains inside the cube. Certainly, if the charge is moved to a different position in the cube, the magnitude of the electric field at points on the surface of the cube will change; however, Gauss' law predicts that the electric field will change in such a way that the flux through the cube remains the same.

Example 7
One day, the electric field in the atmosphere near ground level is 110 N/C. Assume that the magnitude of this field is the same everywhere around the earth and that the direction of the field is radially inward. The radius of the earth is 6.38 x 10⁶ m. Calculate the net charge (magnitude and sign) on the earth.

For the purposes of this problem, imagine that the earth is *just* surrounded by a spherical Gaussian surface (i.e., the radius of the Gaussian surface is essentially the same as that of the earth). The electric flux through the Gaussian surface is given by

$$\Phi_E = \sum (E \cos \phi)\Delta A$$

where ϕ is the angle between the electric field and the *outward* normal to the surface element ΔA. Combining this with Gauss' law, we have

$$\frac{Q}{\varepsilon_0} = \sum (E \cos \phi)\Delta A.$$

Since the electric field is uniform in magnitude and directed radially inward, the angle between the electric field and the outward normal is 180° for each surface element ΔA. Furthermore, since the electric field is uniform in magnitude, E can be factored out of the summation. The sum $\Sigma \Delta A$ must equal the surface area of the spherical Gaussian surface. Solving for Q gives

$$Q = \varepsilon_0(E \cos \phi)\Sigma \Delta A = \varepsilon_0(E \cos \phi)(4\pi R_E^2)$$

$$Q = [8.85 \times 10^{-12}\ C^2/(N \cdot m^2)](110\ N/C)(\cos 180°)[4\pi(6.38 \times 10^6\ m)^2] = -5.0 \times 10^5\ C$$

As described above, Gauss' law can also be used to calculate the magnitude of the electric field in the vicinity of a charge distribution provided that the charge distribution (and hence the electric field) possesses sufficient symmetry. The symmetry of the charge distribution is the key factor. Imagine that we surround a point charge with an arbitrarily shaped Gaussian surface. We can deduce from Example 6 that the flux through this surface is the same regardless of its shape; however, we cannot, in general, use Gauss' law to solve for the electric field on the Gaussian surface. For an arbitrarily shaped Gaussian surface, the magnitude of the electric field will, in general, vary from point to point on the surface. Hence, we cannot solve for E in Equation (18.7). In order to solve for E, we must be able to "factor" the E out of the summation. This can only be done if the magnitude of **E** is *constant* over the Gaussian surface.

We must then use the symmetry of the charge distribution as a guide in constructing a Gaussian surface over which E is constant. For a charge distribution that possesses spherical symmetry, we can deduce that the magnitude of the electric field should be the same at all points equidistant from the distribution. Therefore, if we surround such a distribution with a spherical Gaussian surface that is concentric with the distribution, we can assume that E is constant on the surface. For a line of charge, we can deduce that the magnitude of the electric field should be the same at all points equidistant from the line of charge. In this case, such points would lie on the surface of a cylinder that is concentric with the line of charge; hence we would construct a cylindrical Gaussian surface.

Example 8
A solid conducting sphere of radius R carries a total charge -q uniformly distributed over the surface of the sphere. Determine the magnitude and direction of the electric field for any point at a distance r from the center of the sphere if (a) r > R (i.e., at any point outside the sphere), and (b) r < R (i.e., at any point inside the sphere).

Since we want to find **E**, we must perform the calculation with a Gaussian surface over which **E** is constant. We can assume that since the charge distribution possesses spherical symmetry, the magnitude of the electric field (if it exists), should be the same at all points equidistant from the center of the sphere. Thus, in both parts (a) and (b) we will use spherical Gaussian surfaces.

(a) To find the magnitude of the electric field outside the charged sphere, we construct a spherical Gaussian surface of radius r (r > R) concentric with the charged sphere. The flux through this Gaussian surface is

$$\Phi_E = \Sigma(E \cos \phi)\Delta A.$$

Since the charge is spread uniformly over the surface of the sphere, the charge distribution possesses spherical symmetry. Therefore, we expect the electric field to be directed radially. Since the charge is negative, the field is radially inward; $\phi = 180°$ for each surface element, and $(E \cos 180°) = -E$. Furthermore, since the charge distribution is uniform, we expect the electric field to be uniform in magnitude over the Gaussian surface. Hence, E can be factored out of the summation.

$$\Phi_E = -\Sigma(E\Delta A) = -E(\Sigma \Delta A),$$

where $\Sigma \Delta A$ is the sum of the area elements that make up the Gaussian surface. This sum must equal the surface area of the spherical Gaussian surface or

$$\Phi_E = -E(\Sigma \Delta A) = -E(4\pi r^2)$$

The explicit statement of Gauss' law becomes

$$-E(4\pi r^2) = \frac{Q}{\varepsilon_0},$$

where Q is the net charge enclosed by the Gaussian surface. From the statement of the problem, the net charge Q enclosed by the Gaussian surface is -q. Thus we have

$$-E(4\pi r^2) = \frac{-q}{\varepsilon_0}$$

Solving for E gives

$$E = \frac{q}{4\pi\varepsilon_0 r^2}. \qquad (r > R)$$

Since the charge is negative the direction of the field must be radially *inward*. Note that the direction of the electric field is deduced from the fact that the charge is negative. Gauss' law is used only to find the magnitude of the field.

(b) To find the magnitude of the electric field inside the charged sphere, we construct a spherical Gaussian surface of radius r (r < R) concentric with the charged sphere. We can argue, as we did in part (a), that the field (if it exists) should possess spherical symmetry. Using the arguments similar to those of part (a), we find that the explicit statement of Gauss' law for this situation is

$$E(4\pi r^2) = \frac{Q}{\varepsilon_0}.$$

Since we have a conducting sphere, however, all of the charge resides on the outer surface of the sphere. The Gaussian surface encloses zero net charge so Q = 0. Thus we obtain

$$E = 0. \qquad (r < R)$$

PRACTICE PROBLEMS

1. How many electrons must be removed from an electrically neutral silver dollar to give it a charge of +3.8 μC?

2. Three charges are located along a straight line, as shown below. What is the net electric force on charge q_2?

$q_1 = + 7.5$ μC $q_2 = - 4.0$ μC $q_3 = + 9.0$ μC

x

x = 0   x = 0.40 m   x = 1.00 m

3. Three charges are located at the corners of an equilateral triangle, as shown below. What is the net electric force (magnitude and direction) acting on the top charge, q_1? Take d = 0.50 m.

$q_1 = + 3.5$ μC

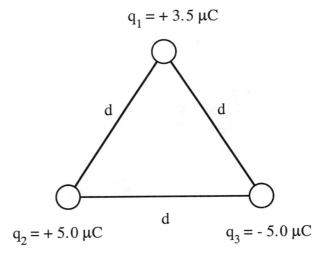

d   d

d

$q_2 = + 5.0$ μC $q_3 = - 5.0$ μC

4. Two charges, $q_1 = +5.0$ μC and $q_2 = -3.0$ μC, are separated by a distance of 1.0 m. Find the spot along the line between the charges where the net electric field is zero.

5. Four charges are arranged at the corners of a square, as shown below. Take a = 0.25 m. If all the charges have the same magnitude of 6.0 μC, what is the magnitude and direction of the electric field at P, the center of the square?

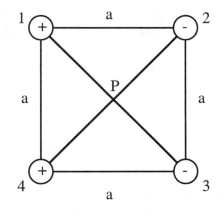

6. A negative charge - q is fixed to one corner of a rectangle, as in the drawing. What positive charge must be fixed to corner A and what positive charge must be fixed to corner B, so the total electric field at the remaining corner is zero? Express your answer in terms of q.

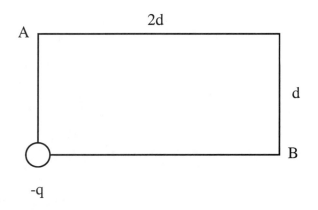

SOLUTIONS AND ANSWERS

Practice Problems

---

1. We know that $q = Ne$ so,

$$N = q/e = (3.8 \times 10^{-6} \text{ C})/(1.6 \times 10^{-19} \text{ C})$$

$$N = 2.4 \times 10^{13}.$$

---

2. The forces on q_2 are shown in the diagram below.

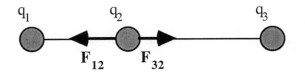

Using Coulomb's law:

$$F_{12} = kq_1q_2/r_{12}^2 = (8.99 \times 10^9 \text{ N m}^2/\text{C}^2)(7.5 \times 10^{-6} \text{ C})(4.0 \times 10^{-6} \text{ C})/(0.40 \text{ m})^2 = 1.7 \text{ N}.$$
$$F_{32} = kq_3q_2/r_{32}^2 = (8.99 \times 10^9 \text{ N m}^2/\text{C}^2)(9.0 \times 10^{-6} \text{ C})(4.0 \times 10^{-6} \text{ C})/(0.60 \text{ m})^2 = 0.90 \text{ N}.$$

The net electric force on q_2 is therefore, $F = F_{12} - F_{32} = $ **0.8 N to the left**.

---

3. The forces acting on q_1 are shown in the following diagram,
   We have:

   $$F_{21} = kq_2q_1/r_{21}^2$$
   $$F_{21} = (9.0 \times 10^9 \text{ N m}^2/\text{C}^2)(5.0 \text{ μC})(3.5 \text{ μC})/(0.50 \text{ m})^2$$
   $$F_{21} = 0.63 \text{ N}$$

   Similarly,
   $$F_{31} = kq_3q_1/r_{31}^2 = 0.63 \text{ N}.$$

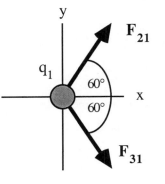

   So,
   $$F_x = (F_{21})_x + (F_{31})_x$$
   $$F_x = 2(0.63 \text{ N}) \cos 60° = 0.63 \text{ N}.$$

   and
   $$F_y = (F_{21})_y + (F_{31})_y = 0.$$

Therefore, the net force acting on q_1 is **0.63 N in the +x direction**.

---

4. Since q_1 and q_2 are of opposite sign, the electric field can not be zero in the region between the charges. Instead, $E = 0$ along the line joining the charges but away from the smaller charge, as in the figure below.

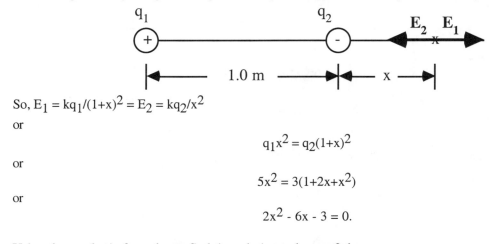

So, $E_1 = kq_1/(1+x)^2 = E_2 = kq_2/x^2$

or

$$q_1x^2 = q_2(1+x)^2$$

or

$$5x^2 = 3(1+2x+x^2)$$

or

$$2x^2 - 6x - 3 = 0.$$

Using the quadratic formula we find the solution to be x = **3.4 m**.

---

5. The total field at the center of the square is the vector sum of the fields due to each point charge.

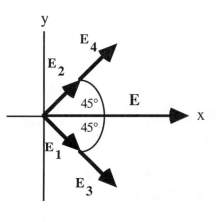

   Since each charge is equidistant from the center of the square, and since each charge has the same magnitude, the magnitudes of E_1, E_2, E_3, and E_4 are all the same. We have

$E = kq/r^2 = (8.99 \times 10^9 \text{ N m}^2/\text{C}^2)(6.0 \times 10^{-6} \text{ C})/(0.18 \text{ m})^2$
$E = 1.7 \times 10^6 \text{ N/C}.$

$\mathbf{E_{tot}} = \mathbf{E_1} + \mathbf{E_2} + \mathbf{E_3} + \mathbf{E_4}$. From the diagram we see that $(\mathbf{E_{tot}})_y = 0$ and

$(\mathbf{E_{tot}})_x = 4E \cos 45° = \mathbf{4.8 \times 10^6}$ **N/C**.

---

6. The field at the empty corner of the rectangle is the vector sum of the fields due to q_A, q_B, and q_-. We can see from the diagram that $\theta = \tan^{-1} (d/2d) = 26.57°$. Since the diagonal is of length $d\sqrt{5}$,

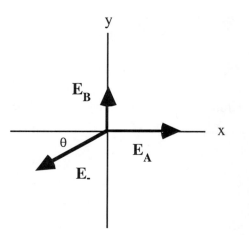

$$E_- = kq/(5d^2),$$

Also,

$$E_A = kq_A/(4d^2),$$

directed in the +x direction, and

$$E_B = kq_B/d^2,$$

directed in the +y direction. From the diagram we see that

$$(E_-)_x = E_A, \text{ and } (E_-)_y = E_B.$$

Thus,

$(E_-)_x = E_- \cos 26.57° = (kq \cos 26.57°)/(5d^2)$

$(E_-)_x = kq_A/(4d^2)$

So,

$$q_A = (4/5)q \cos 26.57° = \textbf{0.716 q}.$$

Now

$$(E_-)_y = E_- \sin 26.57° = (kq \sin 26.57°)/(5d^2)$$
$$(E_-)_y = kq_B/d^2$$

So,

$$q_B = (1/5)q \sin 26.57° = \textbf{0.0895 q}.$$

CHAPTER 19
Electric Potential Energy and the Electric Potential

PREVIEW

In the previous chapter you were introduced to the electric force and electric field. Both of these are vector quantities. Since the electric force is conservative, it is possible to define a scalar electric potential energy of a charged object in an electric field in a way similar to that of defining a gravitational potential energy.

Two other useful concepts, electric potential and electric potential difference, closely related to electric potential energy, will be developed in this chapter. A discussion of how charged objects lose or gain electric energy as they move through electric potential differences is presented.

The capacitance of a capacitor will also be discussed.

QUICK REFERENCE

Important Terms

Electric potential energy (EPE)
  The energy a charged object has due to its position in an electric field.
Electric potential
  The electric potential energy per unit charge that is available at a given point in an electric field.
Electrical potential difference
  The difference in the electric potential between two points in an electric field.
Equipotential surface
  A surface within a region of electric field on which all points are at the same electric potential. No work is done in moving a charged object at constant speed along an equipotential surface.
Capacitor
  A device that can store charge, and hence electric energy.
Capacitance
  A measure of the ability of a capacitor to store charge. It is the charge stored per unit of applied voltage.
Dielectric
  An insulating material. Dielectrics are often used between the plates of a capacitor to raise the capacitance.

Equations

The **work done** by the electric force when a charged particle moves from a point A to a point B in the electric field is

$$W_{AB} = EPE_A - EPE_B \qquad (19.1)$$

The **work done by the electric force per unit charge** when an object of charge q_0 moves from point A to point B in an electric field is

$$\frac{W_{AB}}{q_0} = \frac{EPE_A - EPE_B}{q_0} \qquad (19.2)$$

The **electric potential** at a point in an electric field is found by placing a small charge, q_0, at the point and dividing its electric potential energy by the charge.

$$V = \frac{EPE}{q_0} \qquad (19.3)$$

The **electric potential difference** between two points A and B in an electric field is

$$V_B - V_A = \frac{EPE_B - EPE_A}{q_0} = \frac{-W_{AB}}{q_0} \qquad (19.4)$$

or in "delta" notation

$$\Delta V = \frac{\Delta(EPE)}{q_0} = \frac{-W_{AB}}{q_0} \qquad (19.4)$$

The **potential difference** between two points A and B in the electric field due to a point charge is

$$V_B - V_A = \frac{kq}{r_B} - \frac{kq}{r_A} \qquad (19.5)$$

The **electric potential** at a point a distance, r, from a **point charge** is

$$V = \frac{kq}{r} \qquad (19.6)$$

The **electric field** is related to the **gradient of the electric potential** by

$$E = -\frac{\Delta V}{\Delta s} \qquad (19.7)$$

The magnitude of the **charge** on each plate of a **capacitor** is

$$q = CV \qquad (19.8)$$

The **capacitance of a parallel plate capacitor** is

$$C = \frac{\kappa \varepsilon_0 A}{d} \qquad (19.10)$$

The **energy stored in a capacitor** is

$$\text{Energy} = \frac{1}{2}CV^2 \qquad (19.11)$$

The **energy density** between the plates **of a capacitor** is

$$\text{Energy density} = \frac{1}{2}\kappa \varepsilon_0 E^2 \qquad (19.12)$$

DISCUSSION OF SELECTED SECTIONS

19.1 Potential Energy

A very close analogy can be made between the theories of gravity and electricity since the force laws have the same mathematical form. The main difference to keep in mind is that while there exists only one kind of mass and the gravitational force is always attractive, there exist two kinds of electric charge and the electric force can be either attractive or repulsive. In either theory the concept of potential energy is of central importance. Equation (19.1) can be viewed as a defining relationship for the change in electric potential energy of a charge as it moves from point A to point B.

Example 1

A proton, located at point A in an electric field, has an electric potential energy of $EPE_A = 3.20 \times 10^{-19}$ J. The proton experiences an average electric force of 0.80×10^{-9} N, directed to the right. The proton then moves to point B, which is a distance of 1.00×10^{-10} m to the right of point A. What is the electric potential energy of the proton at point B ?

The work done by the electric force is, from Equation 6.1, the product of the magnitude of the force and the magnitude of the displacement:

$$W_{AB} = Fs = (0.80 \times 10^{-9} \text{ N})(1.00 \times 10^{-10} \text{ m}) = 0.80 \times 10^{-19} \text{ J}$$

Equation (19.1) gives the EPE at the final position to be

$$EPE_B = -W_{AB} + EPE_A$$
$$EPE_B = -0.80 \times 10^{-19} \text{ J} + 3.20 \times 10^{-19} \text{ J} = 2.40 \times 10^{-19} \text{ J}$$

19.2 The Electric Potential Difference

The electric potential difference between two points in an electric field is related by equation (19.4) to the difference in the potential energy that a charge would have if moved between the points. The existence of a difference in potential, of course, implies that there is such a thing as electric potential at a point in the electric field. The electric potential is given by (19.3).

It is important to understand that to speak of a change in EPE one must actually have a charge and move it. We can, however, speak of the electric potential difference between two points in an electric field without actually moving a charge between them. In other words, electric potential energy is something that "belongs" to a charge, while electric potential is something which "belongs" to the electric field. Why this is true will become more apparent when we discuss the electric potential due to a point charge. The common name for electric potential difference is "voltage".

Example 2
What is the potential difference between the initial and final positions of the proton in example 1 above? Which point has the higher potential?

Equation (19.4) gives the difference in electric potential between points A and B to be

$$\Delta V = -W_{AB}/q_0 = -(0.80 \times 10^{-19} \text{ J})/(1.60 \times 10^{-19} \text{ C}) = -0.50 \text{ V}$$

Now $\Delta V = V_B - V_A = -0.50$ V. The potential at point B is lower by 0.50 V than the potential at point A.

A few rules regarding the relationship between the electric field, potential, potential energy and the motion of charges placed in the field are in order at this point.

> If a POSITIVE charge is placed in an electric field and released, it will
> 1. move to decrease its potential energy.
> 2. move toward a region of lower potential.
> 3. move ALONG the electric field lines.

> If a NEGATIVE charge is placed in an electric field and released, it will
> 1. move to decrease its potential energy.
> 2. move toward a region of higher potential.
> 3. move OPPOSITE the electric field lines.

Example 3

A charged object, q = + 2.0 μC and m = 1.5×10^{-15} kg, is placed at rest at point B and released. Which point will it move toward? How much kinetic energy will it have if it arrives at the point? How fast will it be traveling then?

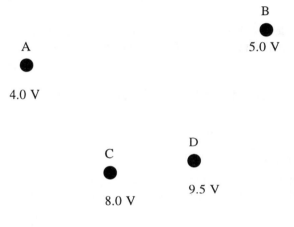

The charge is positive and must move toward a region of lower potential. Point A is the only region of lower potential. The charge will move towards point A.

Energy is conserved as the charge moves from point B to point A.

$$KE_A + EPE_A = KE_B + EPE_B$$

The kinetic energy at point B is zero and equation (19.3) applies at each point so

$$KE_A = EPE_B - EPE_A = qV_B - qV_A$$
$$KE_A = (2.0 \times 10^{-6} \text{ C})(5.0 \text{ V} - 4.0 \text{ V})$$
$$KE_A = 2.0 \times 10^{-6} \text{ J}$$

The speed of the charged object is

$$v = \sqrt{\frac{2KE}{m}} = \sqrt{\frac{2(2.0 \times 10^{-6} \text{ J})}{1.5 \times 10^{-15} \text{ kg}}} = 5.2 \times 10^4 \text{ m/s}$$

19.3 The Electric Potential Difference Created by Point Charges

The potential difference between two points in the region around a point charge is given by (19.5) and the electric potential a distance, r, from the charge is given by (19.6). Note that neither of these equations depends on there being a test charge, q_0, in the field of the point charge, q, since q_0 divided out.

The potential at a point due to any number of point charges can be found by simply finding the potential at the point due to each alone and adding the potentials.

The electric potential energy of a group of charges is found by assembling the group, one charge at a time, and determining the electric potential energy at each step. The electric potential energy of the group is the sum of the electric potential energies determined for each step.

19.4 Equipotential Surfaces and Their Relation to the Electric Field

The electric field and electrical potential are alternative ways of describing the effect a charge or charge distribution has on the space around it. If one is known at each point in space, then it is possible to find the other at each point and vice versa. In general this requires using the calculus, but for the special case where the electric field is constant or an average field is desired, equation (19.7) can be used. Keep in mind that the electric field always points in a direction opposite to the change of the potential and perpendicular to equipotential surfaces.

Example 4

The metal contacts of an electric wall socket are about 1.0 cm apart and are maintained at a potential difference of 120 V. What is the average electric field strength between the contacts? What is the direction of the electric field if the left contact is the higher potential? The lower potential? Treat the potential difference between the contacts as being constant in time.

The magnitude of the average electric field between the contacts is given by equation (19.7) without the minus sign.

$$E = \Delta V/\Delta s = 120 \text{ V}/(1.0 \text{ X } 10^{-2} \text{ m}) = 1.2 \text{ X } 10^4 \text{ V/m}.$$

The electric field is directed toward decreasing potential, so it would point right if the left contact is the higher potential and left if the left contact is the lower potential.

19.5 Capacitors and Dielectrics

A capacitor can store charge q = CV when potential difference (voltage), V, is applied across its plates.

Example 5

How much charge can be placed on the plates of a 1.0×10^3 μF capacitor by a 12 V battery? How large a voltage would be required to place the same amount of charge on a 1.0 μF capacitor?

Equation (19.8) gives the charge on the 1.0×10^3 μF capacitor to be

$$q = CV = (1.0 \text{ X } 10^{-3} \text{ F})(12 \text{ V}) = 1.2 \text{ X } 10^{-2} \text{ C}.$$

The voltage required to place the same charge on a 1.0 μF capacitor is

$$V = q/C = (1.2 \text{ X } 10^{-2} \text{ C})/(1.0 \text{ X } 10^{-6} \text{ F}) = 1.2 \text{ X } 10^4 \text{ V}.$$

The capacitance of a capacitor depends on its GEOMETRY and the material between its plates. The insertion of a material with a higher dielectric constant between the plates of a capacitor results in a higher capacitance. If the same charge is retained on the plates, then the potential difference across the capacitor plates will drop.

Example 6

An empty parallel plate capacitor ($C_0 = 25$ μF) is charged with a 12 V battery. The battery is disconnected and the region between the plates of the capacitor is filled with pure water. What is the capacitance, charge, and voltage for the water-filled capacitor?

The geometry of the capacitor has not changed, so the new capacitance is

$$C = \kappa \varepsilon_0 A/d = \kappa C_0 = (80.4)(25 \text{ μF}) = 2.0 \text{ X } 10^3 \text{ μF}.$$

The charge on the capacitor cannot change since the battery has been disconnected, so

$$q = q_0 = C_0 V_0 = (25 \ \mu F)(12 \ V) = 3.0 \times 10^2 \ \mu C.$$

The new voltage across the capacitor plates is

$$V = q/C = (3.0 \times 10^2 \ \mu C)/(2.0 \times 10^3 \ \mu F) = 0.15 \ V.$$

Energy is stored in a charged capacitor as electric potential energy due to the separation of charge on the plates of the capacitor. The amount of energy stored can be expressed several ways. Energy $= \frac{1}{2}qV$ or $\frac{q^2}{2C}$ or $\frac{1}{2}CV^2$. One form may be derived from any other by using $q = CV$.

Example 7
How much energy is stored by a 25 μF capacitor charged to a potential difference of 12 V?

The energy stored is
$$\text{Energy} = 1/2 \ CV^2 = 1/2 \ (25 \times 10^{-6} \ F)(12 \ V)^2 = 1.8 \times 10^{-3} \ J.$$

Example 8
A capacitor is to be built which will store enough energy to light a 75 W light bulb for 15 min. If the available potential difference is 150 V, what capacitance is needed?

The needed capacitance is

$$C = 2(\text{energy})/V^2 = 2(75 \ W)(15 \ \text{min})(60 \ s/ \ 1 \ \text{min})/(150 \ V)^2 = 6.0 \ F.$$

This is a very large capacitance. Such capacitors are quite large in size and are not generally available.

PRACTICE PROBLEMS

1. An electron can move from a point of 35.0 V potential to a point of 125 V potential. How much kinetic energy can the electron gain? If it started from rest, how fast would it be traveling when it arrives at the 125 V point?

2. Location A is 3.00 m from a charge of -3.00×10^{-8} C, while location B is 2.00 m from the charge. Find the potential difference $V_B - V_A$ between the two points, and state which point is at the higher potential.

3. What is the electrical potential difference between a point 1.00×10^{-10} m from an electron and another point 5.00×10^{-10} m from the electron?

4. A 0.47 μF parallel plate capacitor has 7.1×10^{-5} C of charge on its plates which are separated by 0.052 mm. What is the magnitude of the electric field between the plates?

5. An empty 15 μF parallel plate capacitor is charged by a 12 V battery. The battery remains connected, and the region between the plates of the capacitor is filled with pure water. What is the new capacitance of the capacitor and the new charge on the plates of the capacitor?

6. A positive and a negative charge have the same magnitude of charge and are located 2.0 mm apart. Where in the region around the charges is the electrical potential zero?

7. Sketch the electric field lines and equipotential surfaces in the region around two positive charges of equal magnitude separated by a small distance. Label the regions of relatively high and low potential.

8. The electron in the Bohr model of the hydrogen atom is assumed to move in a circular orbit while the proton in the nucleus is stationary. The total mechanical energy E of the electron is the sum of the kinetic energy and electric potential energy EPE. What is the ratio EPE/E?

9. An air filled parallel plate capacitor has a plate separation of 0.15 mm. What plate area is required for the capacitor to have a capacitance of 33 μF?

10. Two parallel plate capacitors, one air filled and the other filled with paper, are otherwise identical and hold the same charge, q. Which capacitor has stored more energy? How much more?

SOLUTIONS AND ANSWERS

Practice Problems

---

1. The kinetic energy gained by the electron is equal to the EPE lost by the electron.

$$K = eV_2 - eV_1 = (1.60 \text{ X } 10^{-19} \text{ C})(125 \text{ V} - 35.0 \text{ V}) = \textbf{1.44 X } 10^{-17} \textbf{ J.}$$

The speed acquired by the electron is

$$v = \sqrt{\frac{2K}{m}} = \sqrt{\frac{2(1.44 \times 10^{-17} \text{ J})}{9.11 \times 10^{-31} \text{ kg}}} = \textbf{5.62} \times \textbf{10}^6 \textbf{m/s.}$$

---

2. The potential at point A is $V_A = kq/r_A$ and the potential at point B is $V_B = kq/r_B$.
   Now
   $$V_B - V_A = kq[1/r_B - 1/r_A]$$
   $$V_B - V_A = (8.99 \text{ X } 10^9 \text{ N m}^2/\text{C}^2)(-3.00 \text{ X } 10^{-8} \text{ C})[1/(2.00 \text{ m}) - 1/(3.00 \text{ m})] = \textbf{- 45 V.}$$

   Point A is at the higher potential since the difference, $V_B - V_A$, is negative.

---

3. Equation (19.5) gives for a point charge

   $$V_B - V_A = kq[1/r_B - 1/r_A]$$
   $$V_B - V_A = (8.99 \text{ X } 10^9 \text{ N m}^2/\text{C}^2)(-1.60 \text{ X } 10^{-19} \text{ C})[1/(5.00 \text{ X } 10^{-10} \text{ m}) - 1/(1.00 \text{ X } 10^{-10} \text{ m})] = \textbf{11.5 V}$$

---

4. The electric field between the plates is constant and its magnitude is $E = V/d$. The potential across the plates is $V = q/C$ so
   $$E = q/(Cd)$$
   $$E = (7.1 \text{ X } 10^{-5} \text{C})/[(0.47 \text{ X } 10^{-6} \text{ F})(0.052 \text{ X } 10^{-3} \text{ m})] = \textbf{2.9 X } 10^6 \textbf{ V/m.}$$

---

5. The new capacitance is given by equation (19.10) $C = \kappa\varepsilon_0 A/d$, but the capacitance of the empty capacitor is

   $$C_0 = \varepsilon_0 A/d \text{ so } C = \kappa C_0 = (80.4)(15 \text{ }\mu\text{F}) = \textbf{1210 }\mu\textbf{F.}$$

   The new charge on the plates is
   $$q = CV = (1210 \text{ X } 10^{-6} \text{ F})(12 \text{ V}) = \textbf{1.45 X } 10^{-2} \textbf{ C.}$$

---

6. Let r_1 be the distance from the negative charge to a point where the electrical potential is zero and r_2 be a similar distance from the positive charge. Then

   $$kq/r_1 = kq/r_2, \text{ or } r_1 = r_2.$$

   All points equidistant from the charges have zero potential.

---

7. Place a small positive test charge in the space around the two charges, release it and trace out its path as shown in the first drawing. The dark arrows represent the repulsive forces exerted on the test charge by the positive charges and the light arrows represent the path of the test charge, hence an electric field line. Repeat this for different electric field lines to obtain the pattern shown in the second drawing.

---

8. The EPE of the electron in its circular orbit is

    $$EPE = k(e)(-e)/r = -ke^2/r \text{ where } e = 1.6 \times 10^{-19} \text{ C.}$$

    The total energy of the electron is
    $$E = KE + EPE = 1/2\ mv^2 - ke^2/r.$$

    Since the electron is in a circular orbit, the electrical force must be the centripetal force, so

    $$mv^2/r = ke^2/r^2.$$

    Hence, $KE = 1/2\ mv^2 = ke^2/(2r)$. Now

    $$EPE/E = (-ke^2/r)/[-ke^2/(2r)] = \mathbf{2.}$$

---

9. Equation (19.10) applies if we take $\kappa = 1$ for air.

    $$A = Cd/\varepsilon_0 = (33 \times 10^{-6} \text{ F})(0.15 \times 10^{-3} \text{ m}^2)/(8.85 \times 10^{-12} \text{ C}^2/\text{N m}^2) = \mathbf{560\ m^2.}$$

---

10. The energy stored in a capacitor is given by equation (19.11). Using equation (19.8) we obtain

    $$\text{energy} = 1/2\ CV^2 = q^2/2C.$$

    From this we see that a smaller capacitor will hold more energy for the same charge than a larger capacitor. The capacitance of the air-filled capacitor is $C_a = \varepsilon_0 A/d$ and the capacitance of the paper filled capacitor is $C_p = \kappa\varepsilon_0 A/d$. Hence, $C_p = \kappa C_a = (3.3)C_a$. The air capacitor has the smallest capacitance and will store more energy than the paper capacitor. The ratio of the energy stored in the air capacitor to the energy stored in the paper capacitor is $C_p/C_a = 3.3$. **The air capacitor stores 3.3 times more energy.**

CHAPTER 20
Electric Circuits

PREVIEW

In this chapter you will be introduced to the basic concepts needed to analyze electric circuits. Current is defined and Ohm's law is used in simple circuits. The concepts of resistance and resistivity are examined in detail along with electric power, and dc and ac circuits.

This chapter also deals with circuits containing more than one device. The devices used in the circuits are resistors, voltage sources and capacitors. Methods of analyzing circuits containing devices connected in series and parallel by constructing equivalent circuits are investigated. In addition a general method of circuit analysis utilizing Kirchhoff's laws is presented. Kirchhoff's laws can be applied to any circuit, but are particularly useful when the devices cannot be regarded as being connected in either series or parallel.

The construction of ammeters, voltmeters, and ohmmeters is discussed along with how they are connected to a circuit to measure current, voltage and resistance.

QUICK REFERENCE

Important Terms

Electromotive force (emf)
   The maximum electric potential difference (in volts) that exists between the terminals of an electric generator, such as a battery.
Electric current
   The rate of flow of electric charge, $I = \Delta q/(\Delta t)$. The SI unit for current is the ampere (A).
Direct current (dc)
   The electric current that results when charges flow in only one direction.
Alternating current (ac)
   The electric current that results when the direction of charge flow changes from moment to moment.
Conventional current
   The hypothetical flow of positive charges that would accomplish the same effect in the circuit as the movement of negative electrons that actually does occur in the wires.
Electrical resistance
   The ratio of the voltage applied across a piece of material to the current that flows in that material. The SI unit of resistance is the ohm (Ω).
Ohm's law
   The relation between current, voltage and a constant resistance for a material, $V = IR$.
Resistivity
   The proportionality constant ρ for a material which relates the resistance R, the length of the material L, and the material's cross-sectional area A. That is, $R = \rho L/A$.
Electric power
   The power P delivered to a circuit in which a current I results from a voltage V is $P = IV$.
Series wiring
   Circuit devices are wired together so that each device MUST carry the same current.
Parallel wiring
   Circuit devices are wired together so that each MUST have the same voltage drop.
Equivalent resistance
   A single resistance which would demand the same current from the same source as does a network of resistors.
Internal resistance
   The resistance to current flow through a voltage source such as a battery which arises from the resistance of the materials from which the source is made.

Terminal voltage

The actual voltage output of a voltage source such as battery. It is the emf of the source minus the voltage drop across the internal resistance of the source.

Equivalent capacitor

A single capacitor which would store the same charge or energy as does a network of capacitors.

Galvanometer

An electrical meter which contains a moving coil and is used as the basic component of analog ammeters and voltmeters.

Ammeter

A galvanometer in parallel with a shunt resistance. The ammeter measures current.

Voltmeter

A galvanometer in series with a large resistance. The voltmeter measures potential difference (voltage).

Time constant (τ)

The product of resistance and capacitance, RC. In a discharging series RC circuit the time constant represents the time needed for the charge on the capacitor to fall to $1/e$ (37 %) of its initial value at time zero. In a charging RC circuit the time constant represents the time required for the charge on the capacitor to rise to 63 % of its fully-charged value.

Equations

Electric current,

$$I = \frac{\Delta q}{\Delta t} \qquad (20.1)$$

Ohm's law,

$$V = IR \qquad (20.2)$$

Resistivity,

$$R = \rho \frac{L}{A} \qquad (20.3)$$

Temperature coefficient of resistivity,

$$\rho = \rho_0[1 + \alpha(T - T_0)] \qquad (20.4)$$

$$R = R_0[1 + \alpha(T - T_0)] \qquad (20.5)$$

Electric power,

$$P = IV \qquad (20.6)$$

For alternating currents,

$$V = V_0 \sin 2\pi ft \qquad (20.7)$$

$$I = I_0 \sin 2\pi ft \qquad (20.8)$$

$$P = I_0 V_0 \sin^2 2\pi ft \qquad (20.9)$$

$$\overline{P} = \frac{1}{2} I_0 V_0 \qquad (20.10)$$

$$\overline{P} = \left(\frac{I_0}{\sqrt{2}}\right)\left(\frac{V_0}{\sqrt{2}}\right) = I_{ms} V_{ms} \qquad (20.11)$$

$$I_{rms} = \frac{I_0}{\sqrt{2}} \qquad (20.12)$$

$$V_{rms} = \frac{V_0}{\sqrt{2}} \qquad (20.13)$$

$$V_{rms} = I_{rms}R \qquad (20.14)$$

$$\overline{P} = I_{rms}V_{rms} \qquad (20.15a)$$

$$\overline{P} = I_{rms}^2 R \qquad (20.15b)$$

$$\overline{P} = \frac{V_{rms}^2}{R} \qquad (20.15c)$$

Series Circuits
The **equivalent resistance** of a network of **resistors** connected in series is

$$R_S = R_1 + R_2 + R_3 + \ldots \qquad (20.16)$$

The **equivalent capacitance** of a network of **capacitors** connected in series is

$$\frac{1}{C_S} = \frac{1}{C_1} + \frac{1}{C_2} + \frac{1}{C_3} + \ldots \qquad (20.19)$$

Parallel Circuits
The **equivalent resistance** of a network of **resistors** connected in parallel is

$$\frac{1}{R_P} = \frac{1}{R_1} + \frac{1}{R_2} + \frac{1}{R_3} + \ldots \qquad (20.17)$$

The **equivalent capacitance** of a network of **capacitors** connected in parallel is

$$C_P = C_1 + C_2 + C_3 + \ldots \qquad (20.18)$$

RC Circuits
The **charge on the plates** of a **charging capacitor** is

$$q = q_0 [1 - e^{-t/(RC)}] \qquad (20.20)$$

The time constant of an RC circuit is

$$\tau = RC \qquad (20.21)$$

The **charge on the plates** of a **discharging capacitor** is

$$q = q_0 e^{-t/(RC)} \qquad (20.22)$$

Kirchhoff's Rules

Loop rule - the sum of potential drops equals the sum of potential rises around any closed loop.

Junction rule - the sum of the magnitudes of the currents entering a junction equals the sum of the magnitudes of currents leaving the junction.

DISCUSSION OF SELECTED SECTIONS

20.2 Ohm's Law

The current that a battery can "push" through a wire is analogous to the water flow through a pipe. Long and narrow pipes offer greater resistance to the moving water and lead to smaller flow rates. In the electrical case, the **resistance** (R) is defined as the ratio of the voltage V applied across a piece of the material to the current I through the material, or R = V/I. If the ratio V/I for a given material is constant over a wide range of voltages and currents, the resistance is constant. This relation is referred to as **Ohm's law**, after the German physicist Georg Simon Ohm, who discovered it.

Example 1
A resistor is connected to a 6.0-V battery. If the current flowing from the battery is 1.2 A, what is the resistance of the resistor?

The resistance of the resistor is given by equation (20.2),

$$R = \frac{V}{I} = \frac{6.0\,V}{1.2\,A} = 5.0\,\Omega$$

20.3 Resistance and Resistivity

In the case of the water pipe, the length and cross-sectional area of the pipe determine the resistance the pipe offers to the flow of water; the longer the pipe, and the smaller the cross-sectional area, the greater the resistance. In the electrical case, an analogous effect occurs. If L is the length of the material, and A its cross-sectional area, we have

$$R = \rho\frac{L}{A} \qquad (20.3)$$

where ρ is a proportionality constant known as the **resistivity** of the material. The resistivity is an inherent property of the material, while the resistance may vary depending on the geometry of the material.
     The resistivity of the material depends on temperature. In metals, the resistivity normally increases with increasing temperature according to the relation

$$\rho = \rho_0[1 + \alpha(T - T_0)] \qquad (20.4)$$

In equation (20.4), ρ and ρ_0 are the resistivities at temperatures T and T_0, respectively. The term α is known as the **temperature coefficient of resistivity**.

Example 2

A 4.0 m long piece of copper wire having a diameter of 2.0 mm is heated from 20.0 °C to 352 °C. If the resistivity of the copper at 20.0 °C is $\rho_0 = 1.72 \times 10^{-8} \, \Omega$ m, and the temperature coefficient of resistivity is $3.93 \times 10^{-3} (°C)^{-1}$, what is the resistance of the wire at 352 °C?

Using equation (20.4) we can find the resistivity of the copper at 352 °C. We have,

$$\rho = \rho_0[1 + \alpha(T - T_0)]$$

$$\rho = (1.72 \times 10^{-8} \, \Omega \text{ m})[1+(3.93 \times 10^{-3}/ °C)(352 °C - 20.0 °C)] = 3.96 \times 10^{-8} \, \Omega \text{ m}$$

The resistance of the wire can now be obtained using equation (20.3). Note that the cross-sectional area of the copper wire is $A = \pi r^2 = \pi(1.0 \times 10^{-3} \text{ m})^2 = 3.1 \times 10^{-6} \text{ m}^2$. Therefore, the resistance is

$$R = \rho \frac{L}{A} = \frac{(3.96 \times 10^{-8} \, \Omega \text{ m})(4.0 \text{ m})}{3.1 \times 10^{-6} \text{ m}^2} = 5.1 \times 10^{-2} \, \Omega$$

20.4 Electric Power

When there is a current I in a circuit as a result of a voltage V, the electric power delivered to the circuit is

$$P = IV \tag{20.6a}$$

Using $V = IR$ and substituting in equation (20.6), alternate expressions for P can be derived. These are

$$P = IV = I(IR) = I^2R \tag{20.6b}$$

$$P = IV = \left(\frac{V}{R}\right)V = \frac{V^2}{R} \tag{20.6c}$$

Example 3

A cigarette lighter in a car is just a resistor that, when activated, is connected across the 12-V battery. Suppose the lighter dissipates 33 W of power. Find (a) the resistance of the lighter and (b) the current that the lighter draws from the battery.

(a) Use equation (20.6c) to find the resistance,

$$R = V^2/P = (12 \text{ V})^2/(33 \text{ W}) = 4.4 \, \Omega.$$

(b) The current, from equation (20.6a), is

$$I = P/V = (33 \text{ W})/(12 \text{ V}) = 2.8 \text{ A}.$$

20.5 Alternating Current

While many electric circuits involve direct currents (dc), considerably more operate with alternating current (ac), in which the direction of charge flow reverses itself periodically. The most common kind of ac generator produces a voltage that fluctuates sinusoidally between positive and negative values as time passes. The voltage that exists at time t can be expressed as

$$V = V_0 \sin 2\pi ft \qquad (20.7)$$

where V_0 is the maximum or peak value of the voltage, and f is the frequency at which the voltage oscillates. In the United States, the voltage present at most home wall outlets has a peak value of approximately 170 volts and oscillates with a frequency of 60 Hz. The current in an ac circuit also varies sinusoidally according to

$$I = I_0 \sin 2\pi ft \qquad (20.8)$$

where I_0 is the peak current in the circuit. The power delivered to the circuit by an ac generator is $P = IV$, so

$$P = I_0 V_0 \sin^2 2\pi ft \qquad (20.9)$$

Of more significance in an ac circuit is the average power of the circuit, which is just one-half the peak power,

$$\overline{P} = \frac{1}{2} I_0 V_0 \qquad (20.10)$$

The relationship between the peak power, $I_0 V_0$, and the average power is shown in the following figure.

On the basis of the above expressions, we can rearrange terms to reveal that

$$\overline{P} = \left(\frac{I_0}{\sqrt{2}}\right)\left(\frac{V_0}{\sqrt{2}}\right) = I_{ms} V_{ms} \qquad (20.11)$$

where I_{rms} and V_{rms} are called the **root mean square** (rms) current and voltage, respectively. We see that

$$I_{ms} = \frac{I_0}{\sqrt{2}} \qquad (20.12) \qquad\qquad V_{ms} = \frac{V_0}{\sqrt{2}} \qquad (20.13)$$

So in the United States the maximum ac voltage is $V_0 = 170$ volts, and the corresponding rms voltage is $V_{rms} = 170$ volts$/\sqrt{2} = 120$ volts. Electrical appliances usually specify 120 V, so the rms voltage is indicated. In terms of the rms voltage, current, and power, the above equations may be written

$$V_{rms} = I_{rms} R \qquad (20.14)$$

$$\overline{P} = I_{rms} V_{rms} \qquad (25.15a)$$

$$\overline{P} = I_{rms}^2 R \qquad (25.15b)$$

$$\overline{P} = \frac{V_{rms}^2}{R} \qquad (25.15c)$$

Example 4

An ac generator applies a peak voltage of 75 V to an electrical device whose resistance is 22 Ω. Determine the (a) rms voltage, (b) rms current, (c) average power (d) and the peak power of this circuit.

(a) The peak value of the voltage is 75 V, so the corresponding rms value is

$$V_{rms} = \frac{V_0}{\sqrt{2}} = \frac{75 \text{ V}}{\sqrt{2}} = 53 \text{ V}$$

(b) The rms current is obtained from Ohm's law

$$I_{rms} = \frac{V_{rms}}{R} = \frac{53 \text{ V}}{22 \text{ }\Omega} = 2.4 \text{ A}$$

(c) The average power is therefore

$$\overline{P} = I_{rms} V_{rms} = (2.4 \text{ A})(53 \text{ V}) = 130 \text{ W}$$

(d) The peak power is

$$P = I_0 V_0 = (I_{rms} \sqrt{2})(V_{rms} \sqrt{2}) = 2 I_{rms} V_{rms} = 2(130 \text{ W}) = 260 \text{ W}$$

20.6 Series Wiring

The hallmark of devices connected in series is that the current through each device MUST be the same under all circumstances. Two or more devices in a circuit may happen to have the same current through them for certain values of resistance and applied voltage, but not be in series. In looking at a circuit and trying to determine whether the devices are in series or not, one may look for devices which are connected together at one end with no other devices being connected to that end. The entire current through one device has no other choice than to go through the second device.

These resistors are in series. These resistors are NOT in series.

The equivalent resistance of a network of series resistors may be found by simply adding the individual resistances according to equation (20.16). Note that the equivalent resistance is ALWAYS larger than the largest series resistor.

Example 5
What is the equivalent resistance of a 15.0 Ω resistor and a 26.0 Ω resistor connected in series? How much current would a 12.0 V battery deliver to the resistor network, and what is the voltage drop across each resistor?
The equivalent resistance of the two resistors is

$$R_S = 15.0\ \Omega + 26.0\ \Omega = 41.0\ \Omega.$$

The battery "sees" only the equivalent resistance. Ohm's law then requires that the battery send a current

$$I = (12.0\ V)/(41.0\ \Omega) = 0.293\ A.$$

The current through each resistor is the same and equal to the current supplied by the battery. Ohm's law gives for the 15 Ω resistor

$$V_{15} = (0.293\ A)(15.0\ w) = 4.4\ V$$

and for the 26 Ω resistor

$$V_{26} = (0.293\ A)(26.0\ \Omega) = 7.6\ V$$

Note that the voltage drops add to 12.0 V, the emf of the battery.

Example 6
The resistor bin in an electronics lab has a quantity of the following values of resistors: 47 Ω, 68 Ω and 225 Ω. A resistor is needed which has a value of 1500 Ω. Which of the resistors (all of the same value connected in series) could you use to do the job?

The equivalent resistance of N resistors each of resistance R connected in series is $R_S = NR$. We want $R_S = 1500\ \Omega$ to within two significant figures. We need

$$N = (1500\ \Omega)/(47\ \Omega) = 32$$

of the 47 Ω resistors and

$$N = (1500 \ \Omega)/(68 \ \Omega) = 22$$

of the 68 Ω resistors.

We cannot use the 225 Ω resistors at all since we need N = (1500 Ω)/(225 Ω) = 6.7 of them. Using 6 would give a resistance of 6(225 Ω) = 1350 Ω which is too little and using 7 would give 7(225 Ω) = 1575 Ω which is too much.

20.7 Parallel Wiring

The hallmark of parallel wiring is that the voltage across each of the devices connected in parallel MUST be the same under all circumstances. We recognize that devices may happen to have the same voltage across them under special circumstances, but not be wired in parallel.

In order to be in parallel BOTH ends of the devices must be connected directly together with no other devices inserted between them.

These resistors are in parallel.   NONE of these resistors are in parallel.

The equivalent resistance of a network of resistors connected in parallel can be found from equation (20.17). Note that the equivalent resistance is ALWAYS less than the smallest parallel resistor.

Example 7
A 47.0 Ω and a 33.0 Ω resistor are connected in parallel. What is the equivalent resistance of the resistors? How much current would a 12.0 V battery supply to the network and how much current would flow through each resistor?

Equation (20.17) gives

$$\frac{1}{R_p} = \frac{1}{47.0 \ \Omega} + \frac{1}{33.0 \ \Omega}$$

which gives

$$R_p = 19.4 \ \Omega.$$

Note that this is less than either of the resistor values.

The battery "sees" only the equivalent resistance and will supply a current consistent with Ohm's law of

$$I = (12.0 \ V)/(19.4 \ \Omega) = 0.619 \ A.$$

The voltage across each of the resistors is the same and equal to the emf supplied by the battery. Ohm's law applied to the 47.0 Ω resistor gives

$$I_{47} = (12.0 \ V)/(47.0 \ \Omega) = 0.255 \ A$$

and to the 33.0 Ω resistor gives

$$I_{33} = (12.0 \ V)/(33.0 \ \Omega) = 0.364 \ A.$$

The currents through each of the resistors total to equal the current supplied by the battery.

Example 8
You have on hand several resistors of each of the following values: 24 Ω, 36 Ω and 6.0 Ω. How could you use these (all of one value) in a parallel circuit to make an equivalent resistance of 8.0 Ω?

First we see that the 6.0 Ω resistors are useless in making an 8.0 Ω resistance since any parallel combination of resistors with a 6.0 Ω resistor will result in an equivalent resistance of LESS than 6.0 Ω.

If we wish to use a network of all one value, we would need N resistors each of value R to give an equivalent R_P = R/N. The number of resistors needed is then N = R/R_P. We would need N = (24 Ω)/(8.0 Ω) = 3 of the 8.0 Ω resistors or N = (36 Ω)/(8.0 Ω) = 4.5 of the 36 Ω resistors. Four of these gives an equivalent resistance of 9.0 Ω which is too much while five gives 7.2 Ω which is too little. We would have to use three 24 Ω resistors.

20.8   Circuits Wired Partially in Series and Partially in Parallel

Some circuits can be analyzed by reducing the circuit to an equivalent one by combining series resistors and parallel resistors using the previous methods. This is feasible ONLY if you can identify the devices as being wired in series and/or wired in parallel. Otherwise you will need to use the Kirchhoff rules given later in this chapter.

Example 9
Identify the series and parallel combinations of resistors in the circuit diagram. Reduce the circuit to an equivalent resistance and find the current supplied by the battery.

The two 4.0 Ω resistors are in series with each other and both together are in parallel with the 8.0 Ω resistor. The remaining resistors are in series with each other and with the network containing the two 4.0 Ω and the 8.0 Ω resistors.

First reduce the circuit by combining the two 4.0 Ω resistors into a 4.0 Ω + 4.0 Ω = 8.0 Ω equivalent.

Next reduce the resulting parallel combination of the two 8.0 Ω resistances to an equivalent of 4.0 Ω.

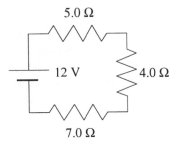

Now the three series resistances can be added to give the final equivalent $5.0\ \Omega + 4.0\ \Omega + 7.0\ \Omega = 16.0\ \Omega$

The battery "sees" only the final equivalent resistance and will provide a current consistent with Ohm's law.

$$I = (12\ \text{V})/(16.0\ \Omega) = 0.75\ \text{A}$$

20.9 Internal Resistance

All real voltage sources such as generators and batteries have their own "internal resistance" since they contain wires, devices, plates, etc which have resistance. Whenever a voltage source is required to deliver current, a voltage drop (due to Ohm's law) will appear across this internal resistance and reduce the terminal voltage or output of the source. A real voltage source can be thought of as an ideal voltage source in series with the internal resistance.

Example 10
A laboratory signal generator has an internal resistance of 600 Ω and delivers 0.050 A of current to a load circuit whose equivalent resistance is also 600 Ω. What is the emf of the generator? What is the voltage drop across the generator?

The emf of the generator is the terminal voltage plus the drop across its internal resistance, r. The terminal voltage is just the voltage drop across the load circuit since the generator and load circuit are connected in parallel.

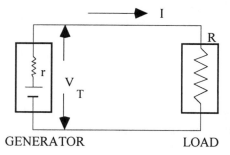

$$\text{emf} = V_T + Ir = I(R + r)$$

$$\text{emf} = (0.050\ \text{A})(600\ \Omega + 600\ \Omega) = 60\ \text{V}$$

The voltage drop across the generator is

$$Ir = (0.050\ \text{A})(600\ \Omega) = 30\ \text{V}$$

20.10 Kirchhoff's Rules

The Kirchhoff rules are basically statements of two conservation laws. The loop rule is simply a statement of the conservation of energy for charges going around a closed loop in a circuit, while the junction rule is a statement of the conservation of charge entering and leaving a junction.

Kirchhoff's rules are applicable to ALL circuits and must be used when the circuit in question cannot be reduced to an equivalent circuit by identifying parallel and series combinations of devices.

The method of applying Kirchhoff's rules to a circuit is basically the same for all circuits and involves the following steps.

1. Identify the independent loops and junctions in the circuit.
2. Choose current directions for the currents flowing in the circuit. Your choice is arbitrary. If you choose wrong, Kirchhoff's laws will tell you.
3. Pick a loop and proceed clockwise around the loop adding up potential drops and adding up potential rises. Set the drops equal to the rises. Repeat for each independent loop in the circuit. You should now have one equation for each loop.
4. Choose a junction and apply the junction rule by setting the total current into the junction equal to the total current out of the junction. Repeat for each independent junction to obtain an equation for each junction.
5. Solve the loop equations and junction equations simultaneously for the unknown quantities. You can find one unknown quantity for each of the equations that you have written.

Example 11
Find each of the three currents indicated in the circuit diagram.

This circuit is a good candidate for Kirchhoff's rules since none of the devices can be identified as being wired in series or parallel.

Step 1: The circuit has three possible loops ABEF, BCDE, and ACDF (only two will be independent) and two possible junctions B and E (only one will be independent).

Step 2: The current directions have already been chosen. We are not guaranteed that they are correct, however.

Step 3: Since the loop equations are somewhat complicated, we will suppress the units which normally accompany numbers. The loop rule applied to loop ABEF gives

$$(20.0)I_1 + 10.0 = (10.0)I_2 + 10.0$$

or

$$(20.0)I_1 - (10.0)I_2 = 0 \qquad (1)$$

The loop rule applied to the loop BCDE gives

$$(5.00)I_3 + (10.0)I_2 = 20.0 + 10.0$$

or

$$(10.0)I_2 + (5.00)I_3 = 30.0 \qquad (2)$$

The loop rule applied to loop ACDF gives

$$(20.0)I_1 + (5.00)I_3 = 20.0 + 10.0$$

or

$$(20.0)I_1 + (5.00)I_3 = 30.0$$

NOTE: The last equation can be derived by adding equations (1) and (2). Only two of the equations are independent and can be used. Which two? It's your choice.

Step 4: The junction rule applied at B gives

$$I_1 + I_2 = I_3$$

or

$$I_1 + I_2 - I_3 = 0 \qquad (3)$$

The junction rule applied at E gives

$$I_3 = I_1 + I_2$$

or

$$I_1 + I_2 - I_3 = 0$$

NOTE: The second junction equation is identical to the first, therefore only one junction equation can be used.

Step 5: This step is pure algebra. Use your favorite technique to solve the three equations simultaneously for the currents.
Add (1) and (2) to obtain

$$(20.0)I_1 + (5.00)I_3 = 30.0 \qquad (4)$$

Multiply (3) by 10 and add the result to (1) to get

$$(30.0)I_1 - (10.0)I_3 = 0 \qquad (5)$$

Multiply (4) by 2 and add the result to (5) and find

$$(70.0)I_1 = 60.0 \text{ or } I_1 = + 0.857 \text{ A}$$

Use this result in (5) to find

$$I_3 = + 2.57 \text{ A.}$$

Equation (3) then gives the last current to be

$$I_2 = + 1.71 \text{ A.}$$

All of the currents are positive, indicating that we guessed the correct directions.

20.11 The Measurement of Current and Voltage

The moving coil galvanometer is the heart of analog voltmeters and ammeters. The galvanometer is capable of measuring small currents flowing through its coil.

The key to constructing an ammeter is to allow only a small known fraction of the current to be measured to flow through the galvanometer. This is accomplished by "shunting" the bulk of the current around the galvanometer via a shunt resistor placed in parallel with the galvanometer. The ammeter constructed this way must be INSERTED INTO the circuit so that the current can flow through it. Since the ammeter contributes a series resistance to the circuit, its very presence changes the current in the circuit. It is desired that the equivalent resistance of the ammeter be as small as possible so that it affects the current to be measured in only a small way.

Example 12
A perfect 9.0 V battery is connected to a 1800 Ω load resistance so that 5.0 mA flows in the circuit. An ammeter with a resistance of 100 Ω is inserted in the circuit and used to measure the current. What error does the ammeter introduce in the measurement?

When the ammeter is inserted, the battery "sees" a new equivalent resistance of

$$1800 \ \Omega + 100 \ \Omega = 1900 \ \Omega.$$

Ohm's law then demands that the battery produce a current of

$$(9.0 \ \text{V})/(1900 \ \Omega) = 4.7 \ \text{mA}$$

which is the current that the ammeter will register. The error introduced by the ammeter is then

$$(100\%)(5.0 \ \text{mA} - 4.7 \ \text{mA})/(5.0 \ \text{mA}) = 6 \ \%.$$

The voltmeter is constructed by placing a large resistor value in series with the galvanometer. When the voltmeter is connected ACROSS the device whose voltage is to be measured, a small current will flow through the galvanometer and cause a deflection of the pointer. This current, hence the deflection of the pointer, is proportional to the voltage across the voltmeter. Again, the introduction of the meter changes the very thing it is supposed to measure. To ensure accuracy, the resistance of the voltmeter is usually very high so that little current is "drawn" by the voltmeter.

Example 13

A 1.0 MΩ voltmeter is used to measure the voltage drop across one of two 25 kΩ resistors connected in series to a perfect 12 V battery. How much error in the measured voltage is introduced by the voltmeter?

Before the voltmeter is connected the voltage drop across each resistor is just 6.0 V, since the resistors are identical and connected in series. After the voltmeter is connected in parallel with one resistor, the battery "sees" a new equivalent resistance of

$$R_S = 25 \ \text{k}\Omega + R_P$$

where R_P is given by

$$1/R_P = 1/(25 \ \text{X} \ 10^3 \ \Omega) + 1/(1.0 \ \text{x} \ 10^6 \ \Omega)$$

so $R_P = 24$ kΩ and $R_S = 49$ kΩ.

The battery now delivers a new current to the circuit of

$$(12 \ \text{V})/(49 \ \text{X} \ 10^3 \ \Omega) = 0.24 \ \text{mA}.$$

The new voltage drop across the resistor-voltmeter combination is then

$$IR_P = (0.24 \ \text{X} \ 10^{-3} \ \text{A})(24 \ \text{X} \ 10^3 \ \Omega) = 5.8 \ \text{V}.$$

The error is

$$(100 \ \%)(6.0 \ \text{V} - 5.8 \ \text{V})/6.0 \ \text{V}) = 3 \ \%.$$

20.12 Capacitors in Series and Parallel

The magnitude of the charges on the plates of capacitors connected in series MUST be the same due to the conservation of charge. The voltages across the capacitors may be different if the capacitors have different capacitances. An equivalent capacitance can be defined for series capacitors and is given by (20.19).

The voltage across capacitors connected in parallel MUST be the same. This allows different charges to be on the capacitors if their capacitances are not equal. An equivalent capacitance for parallel capacitors is given by (20.18).

Circuit analysis for capacitors is very similar to the analysis for resistors. Often you can reduce a circuit containing parallel and series capacitors to a circuit containing one equivalent capacitor. If this cannot be done, you may use the Kirchhoff loop rule along with the principle of conservation of charge.

Example 14
Find the equivalent capacitance, the charge separated by the battery and the charge on each capacitor in the circuit shown.

The 5.00 μF and the 4.00 μF capacitors are in parallel and have an equivalent capacitance of

$$C_P = 5.00 \text{ μF} + 4.00 \text{ μF} = 9.00 \text{ μF}.$$

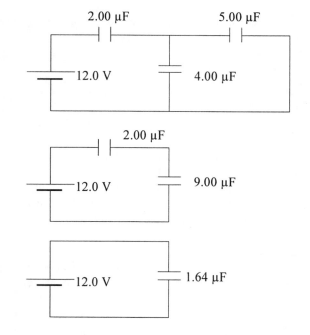

This combination is in series with the 2.00 μF capacitor so the final equivalent capacitance of the circuit is found from

$$1/C_S = 1/(2.00 \text{ μF}) + 1/(9.00 \text{ μF})$$

and is $C_S = 1.64$ μF.

The battery "sees" only the equivalent capacitance and separates an amount of charge

$$q = (1.64 \text{ X } 10^{-6} \text{ F})(12.0 \text{ V})$$
$$q = 1.97 \text{ X } 10^{-5} \text{ C}.$$

Now the 2.00 μF capacitor and the 9.00 μF equivalent capacitance are in series and must have the same charge on their plates. That is, $1.97 \text{ X } 10^{-5}$ C = 19.7 μC.

The 5.00 μF and 4.00 μF share the 19.7 μC in such a way that the voltage across each of them is the same.

$$V_4 = V_5$$

or

$$q_4/q_5 = C_4/C_5 = 0.800.$$

We also know that

$$q_4 + q_5 = 19.7 \text{ μC}$$

so

$$q_5 = 10.9 \text{ μC and } q_4 = 8.8 \text{ μC}.$$

20.13 RC Circuits

Example 15
What fraction of charge is lost by a discharging capacitor during the first five time constants?

Let the charge on the capacitor at time = 0 s be $q_0 = CV_0$. Equation (20.22) gives for t = 5τ

$$q = q_0 \, e^{-t/(RC)} = q_0 \, e^{-5} = q_0 \, (6.7 \text{ X } 10^{-3})$$

The fraction of the initial charge remaining after 5τ is q/q_0 so that the fraction lost is

$$1 - q/q_0 = 1 - (6.7 \text{ X } 10^{-3}) = 0.993.$$

PRACTICE PROBLEMS

1. A 75-W light bulb draws a current of 0.80 A. If the light bulb stays on for 3.0 hours (a) how much charge flows through the light bulb? (b) How many electrons have passed through the bulb?

2. A certain material used for a wire is 25 m long and has a diameter of 3.0 mm. The wire carries a 2.5 A current when a 12-V potential difference is applied between its ends. What is the resistivity of the wire?

3. At what temperature would the resistance of a silver conductor be double its resistance at 20.0 °C? Take the temperature coefficient of resistivity for silver to be 3.80×10^{-3} (C°)$^{-1}$.

4. A wire whose resistance is 12.0 Ω is melted down. From the melted material a new wire is made that is 1/4 the diameter of the original wire. What is the resistance of the new wire?

5. An electric alarm clock uses a 5.0-W motor and runs all day, every day. If electricity costs $0.10 per kWh, determine the yearly cost of running the clock.

6. An ac generator produces a voltage given by V = 283 sin 188t, where V is in volts and t in seconds. (a) What is the peak voltage produced by the generator? (b) What is its frequency? (c) If the voltage supplied by the generator produces an rms current of 2.0 A in a device in the circuit, what is the resistance of this device?

7. A device is rated at 750 watts when used with a 120-V ac line. (a) What peak current flows through it? (b) What peak current would flow through it if connected to a 240-V ac line? (c) In which case is more power utilized? (Assume the resistance of the device is constant).

8. Find the equivalent resistance of the circuit shown.

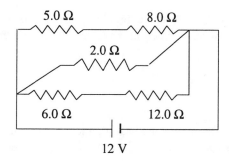

9. A battery with an emf of 12.0 V is used to turn the starter motor on a car. When the starter motor is engaged, the battery applies only 8.2 V to the motor while it delivers a current of 120 A. What is the internal resistance of the battery and the resistance of the starter motor?

10. A galvanometer whose full scale current is 150.0 μA and coil resistance is 75.0 Ω is used with a 1.50 Ω shunt resistor to make an ammeter. What is the largest current which can be measured by the ammeter?

11. An ammeter whose resistance is 9.0 Ω is inserted into a circuit containing a 95 Ω resistor and a 9.0 V battery. What percentage error does the ammeter introduce into the current it measures?

12. A galvanometer whose full scale current is 150 μA and coil resistance is 75 Ω is used with a 25 kΩ series resistor to make a voltmeter. What is the largest voltage which can be measured by the voltmeter?

13. Apply Kirchhoff's rules to find the currents in the circuit shown.

14. Find the current through each of the resistors in the diagram shown. Also, find the current delivered by the battery.

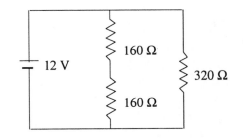

15. Find the equivalent capacitance of the circuit shown in the diagram. How much charge is separated by the battery?

16. A 1500 Ω resistor and a 25 μF capacitor are connected together to form a circuit. The capacitor has been previously charged to a voltage of 35 V. How long will it take for the voltage on the capacitor to drop to 15 V?

17. A 1500 μF capacitor is charged through a 25 kΩ resistor by a 12 V battery. How long does it take for the battery to charge the capacitor to a voltage of 8.0 V?

SOLUTIONS AND ANSWERS

Practice Problems

---

1. (a) Using equation (20.1),
    $$\Delta q = I\Delta t = (0.80 \text{ C/s})(3 \text{ h})(3600 \text{ s/1 h}) = \textbf{8.6 x } \textbf{10}^3 \textbf{ C}.$$
    (b) We know,
    $$N = (\Delta q)/e = (8.6 \text{ x } 10^3 \text{ C})/(1.6 \text{ x } 10^{-19} \text{ C}) = \textbf{5.4 x } \textbf{10}^{22} \textbf{ electrons}.$$

---

2. Equation (20.2) gives,
    $$R = V/I = (12 \text{ V})/(2.5 \text{ A}) = 4.8 \ \Omega$$
    The area is
    $$A = \pi r^2 = \pi(1.5 \text{ x } 10^{-3} \text{ m})^2 = 7.1 \text{ x } 10^{-6} \text{ m}^2.$$

    Using equation (20.3), $R = \rho L/A$ gives
    $$\rho = RA/L = (4.8 \ \Omega)(7.1 \text{ x } 10^{-6} \text{ m}^2)/(25 \text{ m}) = \textbf{1.4 x } \textbf{10}^{-6} \ \boldsymbol{\Omega} \textbf{ m}.$$

---

3. Using equation (20.5) we have,
    $$R = R_0[1 + \alpha(T - T_0)].$$
    In this case we have $R = 2R_0$, so that
    $$R/R_0 = 2 = 1 + \alpha(T - T_0)$$
    or
    $$T = T_0 + 1/\alpha = 20.0 \ ^\circ\text{C} + 1/(3.80 \text{ x } 10^{-3}/\text{C}^\circ) = \textbf{283} \ ^\circ\textbf{C}.$$

---

4. The volume of the wire is a constant so $A_1L_1 = A_2L_2$ gives
    $$L_2/L_1 = A_1/A_2.$$
    Now use equation (20.3) to get,
    $$R_2/R_1 = (\rho L_2/A_2)/(\rho L_1/A_1) = (L_2/L_1)(A_1/A_2) = A_1^2/A_2^2.$$
    Since $A = \pi d^2/4$, we have
    $$R_2 = R_1(A_1^2/A_2^2) = R_1(d_1/d_2)^4 = (12.0 \ \Omega)(4)^4 = \textbf{3070} \ \boldsymbol{\Omega}.$$

---

5. Since power is $P = E/t$, then
    $$E = Pt = (5.0 \text{ J/s})(365 \text{ days})(24 \text{ h/1 day})(3600 \text{ s/1 h}) = 1.6 \text{ x } 10^8 \text{ J}.$$
    So the cost is,
    $$\text{Cost} = (\$0.10/1 \text{ kWh})(1 \text{ kWh}/3.6 \text{ x } 10^6 \text{ J})(1.6 \text{ x } 10^8 \text{ J}) = \textbf{\$4.40}.$$

---

6. (a) The equation is in the form $V = V_0 \sin 2\pi ft$, where V_0 is the peak voltage and f is the frequency.
    Therefore, since $V = 283 \sin 188t$, we see that the peak voltage is **283 V**.

    (b) The frequency is obtained from $2\pi f = 188$ or
    $$f = 188/2\pi = \textbf{29.9 Hz}.$$

(c) We know that

$$R = V_{rms}/I_{rms} = (283 \text{ V}/\sqrt{2})/(2.0 \text{ A}) = \textbf{1.0 x 10}^2 \text{ } \Omega.$$

---

7. (a) The rms voltage is 120 V, and the average power is 750 W. The rms current is obtained from $P = I_{rms}V_{rms}$, so

$$I_{rms} = P/V_{rms} = (750 \text{ W})/(120 \text{ V}) = 6.3 \text{ A}.$$

The peak current is

$$I_0 = I_{rms} \sqrt{2} = \textbf{8.9 A}.$$

(b) If connected to a 240-V line, the power drawn is no longer 750 W. We need the resistance of the device,

$$R = V_{rms}/I_{rms} = (120 \text{ V})/(6.3 \text{ A}) = 19 \text{ } \Omega.$$

The new current is therefore,

$$I'_{rms} = V'_{rms}/R = (240 \text{ V})/(19 \text{ } \Omega) = 13 \text{ A}.$$

The peak current is therefore,

$$I'_0 = I'_{rms} \sqrt{2} = \textbf{18 A}.$$

(c) The power drawn from the 240-V line is, $P' = I'_{rms}V'_{rms} = (13 \text{ A})(240 \text{ V}) = 3100 \text{ W}.$

So the **power drawn from the 240-V line is 4 times as great** as that drawn from the 120-V line.

---

8. The 5.0 Ω and 8.0 Ω are in series with an equivalent resistance of 13.0 Ω. The 6.0 Ω and 12.0 Ω are also in series and have an equivalent resistance of 18.0 Ω. The 2.0 Ω resistor is in parallel with the 13.0 Ω and 18.0 Ω equivalents so the equivalent resistance of the entire circuit is

$$1/R = 1/(13.0 \text{ } \Omega) + 1/(18.0 \text{ } \Omega) + 1/(2.0 \text{ } \Omega) \text{ or } R = \textbf{1.6 } \Omega.$$

---

9. The terminal voltage of a battery is $V_T = \text{emf} - Ir$, so

$$r = (\text{emf} - V_T)/I = (12.0 \text{ V} - 8.2 \text{ V})/(120 \text{ A}) = \textbf{0.032 } \Omega.$$

The resistance of the starter motor is
$$R = V_T/I = (8.2 \text{ V})/(120 \text{ A}) = \textbf{0.068 } \Omega.$$

---

10. The shunt resistor and the galvanometer are in parallel so each will have the same voltage drop. The drop across the galvanometer is

$$V = I_g R_g = (150.0 \text{ X } 10^{-6} \text{ A})(75.0 \text{ } \Omega) = 1.13 \text{ X } 10^{-2} \text{ V}.$$

The equivalent resistance of the ammeter is

$$1/R = 1/(75.0 \text{ } \Omega) + 1/(1.50 \text{ } \Omega)$$

which gives

$$R = 1.47 \text{ } \Omega.$$

The current through the ammeter at full scale is

$$I = V/R = (1.13 \text{ X } 10^{-2} \text{ V})/(1.47 \text{ } \Omega) = \textbf{7.69 X 10}^{-3} \text{ A}.$$

11. The current in the circuit before the ammeter is inserted is the "true" value and is

$$I = (9.0 \text{ V})/(95 \text{ } \Omega) = 0.095 \text{ A}.$$

    After the ammeter is inserted, the current in the circuit is reduced to

$$I' = (9.0 \text{ V})/(104 \text{ } \Omega) = 0.087 \text{ A}.$$

    The percentage error introduced by the ammeter is

$$\% \text{ error} = (0.095 \text{ A} - 0.087 \text{ A})/(0.095 \text{ A}) \text{ X } 100 \% = \textbf{8 \%}.$$

---

12. A full scale voltage across the voltmeter produces a current of $I_g = 150 \text{ } \mu\text{A}$ through the voltmeter. The equivalent resistance of the voltmeter is $R = 75 \text{ } \Omega + 25 \text{ k}\Omega$, so

$$V = (150 \text{ X } 10^{-6} \text{ A})(75 \text{ } \Omega + 25 \text{ X } 10^3 \text{ } \Omega) = \textbf{3.8 V}.$$

---

13. Label the currents and choose the current directions. It is not necessary to "guess" the true current directions. Let I_1 be up through the 12 V battery, I_2 be down through the 15 Ω resistor, and I_3 be down through the 25 Ω resistor. Apply the loop rule (drops = rises) to the left loop starting at the upper left corner and proceeding clockwise.

$$I_3(25 \text{ } \Omega) = 12 \text{ V},$$

    then $\textbf{I}_3 = \textbf{0.48 A}$. Repeat for the outside loop

$$I_2(15 \text{ } \Omega) = 6.0 \text{ V} + 12 \text{ V}, \text{ then } \textbf{I}_2 = \textbf{1.2 A}.$$

    Apply the junction rule (currents in = currents out) at the top junction.

$$I_1 = I_2 + I_3$$

    which gives $\textbf{I}_1 = \textbf{1.7 A}$.

    Note that all of the currents are positive, indicating that we "guessed" their directions correctly.

---

14. The two 160 Ω resistors are in series and in parallel with the battery. Their equivalent resistance is

$$160 \text{ } \Omega + 160 \text{ } \Omega = 320 \text{ } \Omega$$

    and the voltage across them is 12 V, so the current through them is

$$I_{160} = (12 \text{ V})/(320 \text{ } \Omega) = \textbf{3.8 X } \textbf{10}^{-2} \textbf{ A}.$$

    The 320 Ω resistor is in parallel with the battery, hence "draws" a current

$$I_{320} = (12 \text{ V})/(320 \text{ } \Omega) = \textbf{3.8 X } \textbf{10}^{-2} \textbf{ A}.$$

    The total current delivered by the battery is $I = I_{160} + I_{320} = \textbf{7.6 X } \textbf{10}^{-2} \textbf{ A}.$

15. The 8.0 μF and 22 μF capacitors are in series and have an equivalent capacitance given by

$$1/C_1 = 1/(8.0 \text{ μF}) + 1/(22 \text{ μF})$$

or

$$C_1 = 5.9 \text{ μF}.$$

The equivalent capacitance, C_1, is in parallel with the 16 μF capacitor with a new equivalent capacitance of

$$C_2 = 5.9 \text{ μF} + 16 \text{ μF} = 22 \text{ μF}.$$

The equivalent capacitance is in series with the 6.0 μF capacitor, so the equivalent capacitance of the circuit is found from

$$1/C_3 = 1/(22 \text{ μF}) + 1/(6.0 \text{ μF})$$

Then

$$\mathbf{C_3 = 4.7 \ μF.}$$

The charge separated by the battery is

$$q = C_3 V = (4.7 \times 10^{-6} \text{ F})(12 \text{ V}) = \mathbf{5.6 \times 10^{-5} \ C.}$$

---

16. The charge on the capacitor plates at any time, t, is given by equation (20.22)

$$q = CV_0 e^{-t/(RC)}.$$

The charge on the plates of the capacitor is $q = CV$ so

$$V/V_0 = e^{-t/(RC)}.$$

Taking the natural logarithm of both sides and solving for t yields

$$t = -(RC) \ln (V/V_0) = -(1500 \text{ Ω})(25 \times 10^{-6} \text{ F}) \ln (15 \text{ V}/35 \text{ V}) = \mathbf{3.2 \times 10^{-2} \ s.}$$

---

17. Equation (20.20) applies to the charging of a capacitor in a series RC circuit: $q = CV_0 (1 - e^{-t/(RC)})$.

The charge on the plates of the capacitor is $q = CV$, so

$$V = V_0 (1 - e^{-t/(RC)}).$$

Rearranging, taking the natural log of both sides, and solving for t yields

$$t = -(RC) \ln (1 - V/V_0) = -(25 \times 10^3 \text{ Ω})(1500 \times 10^{-6} \text{ F}) \ln (1 - 8.0 \text{ V}/12.0 \text{ V})$$
$$t = \mathbf{41 \ s.}$$

CHAPTER 21
Magnetic Forces and Magnetic Fields

PREVIEW

In this chapter you will study the properties of magnetic fields. You will learn how a magnetic field exerts a force on moving charges and currents, how electric currents produce magnetic fields, and about the properties of magnetic materials. You will see how magnetic fields play important roles in such devices as mass spectrometers, stereo speakers and tape recorders, galvanometers, and electric motors.

QUICK REFERENCE

Important Terms

Magnetic field
A vector field which exists in the region around a magnet. The direction of the field at any point is the direction indicated by the north pole of a small compass needle placed at that point. The SI unit is the tesla (T).

Magnetic force
The force exerted on a **moving** charge due to a magnetic field. The magnetic force is always perpendicular to both the direction of the magnetic field and the velocity vector.

Mass spectrometer
An instrument which uses magnetic fields to determine the relative masses of atoms and molecules.

Hall effect
An effect where moving charges in a current-carrying metal or semiconductor are deflected by an external magnetic field. The deflected charge carriers create an emf, the polarity of which can be used to determine whether the charge carriers are positive or negative.

Ferromagnetic materials
Materials, such as iron, which can be permanently magnetized. These materials are made up of tiny regions that behave like small magnets. In a permanent magnet, these regions are aligned, and a high degree of magnetism results.

Equations

The **magnitude** B of the magnetic field at any point in space is defined as

$$B = \frac{F}{q_0 \, (v \sin \theta)} \tag{21.1}$$

The **radius** of the circular **path of a charged particle** moving in a **magnetic field** is given by

$$r = \frac{mv}{qB} \tag{21.2}$$

The **magnetic force** on a **current-carrying wire** of length L is

$$F = ILB \sin \theta \tag{21.3}$$

The **torque** τ **on a current-carrying loop of wire in a magnetic field** can be written as

$$\tau = NIAB \sin \phi \qquad (21.4)$$

The **magnetic field** produced by a **long straight wire** has a magnitude

$$B = \frac{\mu_0 I}{2\pi r} \qquad (21.5)$$

The **magnetic field** at the **center of a current-carrying loop** of wire has a magnitude

$$B = N \frac{\mu_0 I}{2R} \qquad (21.6)$$

The **magnetic field in the interior of a long solenoid** has a magnitude

$$B = \mu_0 nI \qquad (21.7)$$

DISCUSSION OF SELECTED SECTIONS

21.2 The Force That a Magnetic Field Exerts on a Moving Charge

We are all familiar with simple magnets, and we know that magnets attract certain metallic objects (like iron). In fact, magnets can exert forces on each other. A magnet has two poles, a **north pole** and a **south pole**. When two magnets are placed near one another it is found that like poles repel each other, and unlike poles attract each other. This behavior is similar to that of like and unlike electric charges.

We know that an electric field exists in the space around electric charges. Similarly, a **magnetic field** exists in the region around a magnet. The magnetic field is a vector that has both magnitude and direction. The direction of the magnetic field at any point in space is the direction indicated by the north pole of a small compass needle placed at that point.

When a charge is placed in a magnetic field, does it experience a magnetic force? Yes, if two conditions are met:

  1. The charge must be moving, for no magnetic force acts on a stationary charge.
  2. The velocity of the moving charge must have a component that is perpendicular to the direction of the field.

Consider a test charge $+q_0$ moving with a velocity **v** through a magnetic field **B**. If the charge is moving at an angle θ with respect to the magnetic field, the field will exert a force **F** on the test charge. The magnitude B of the magnetic field at any given point in space is then given by

$$B = \frac{F}{q_0 (v \sin \theta)} \qquad (21.1)$$

where F and v are the magnitudes of the force and velocity, respectively. The SI unit of magnetic field is:

$$\frac{\text{newton second}}{\text{coulomb meter}} = 1 \text{ tesla (T)}$$

The direction of the magnetic force **F** is perpendicular to both **v** and **B**; in other words, **F** is perpendicular to the plane defined by **v** and **B**. The directions of the force, magnetic field, and velocity are related by a convenient rule called **Right-Hand Rule No. 1** (RHR-1). This can be stated as

> **Right-Hand Rule No. 1.** Extend the right hand so the fingers point along the direction of the magnetic field **B** and the thumb points along the velocity **v** of the charge. The palm of the hand then faces in the direction of the magnetic force **F** that acts on a positive charge.

If the moving charge is negative instead of positive, the direction of the magnetic force is opposite to that predicted by RHR-1. So, to find the force on a negative charge, use RHR-1 as if the charge were positive, and find the direction of the force. Then, reverse this direction to find the direction of the force that acts on the negative charge.

Example 1

An electron traveling at a velocity of 5.0 x 10^6 m/s west enters a region of uniform magnetic field of magnitude 2.0 x 10^{-3} T directed upward. What are the magnitude and direction of the force that acts on this electron?

We can use equation (21.1) to find the magnitude of the force. We have

$$F = qvB \sin \theta = (1.6 \times 10^{-19} \text{ C})(5.0 \times 10^6 \text{ m/s})(2.0 \times 10^{-3} \text{ T}) \sin 90° = 1.6 \times 10^{-15} \text{ N.}$$

RHR-1 tells us that the force on a positive charge would be north (place your fingers upward with your thumb to the west, your palm then faces north). Therefore, the force on the electron (which is negative) would be south.

21.3 The Motion of a Charged Particle in a Magnetic Field

Consider the case in which a particle has a velocity that is perpendicular to a uniform magnetic field. As shown in the figure below, the magnetic force serves to move the particle in a circular path.

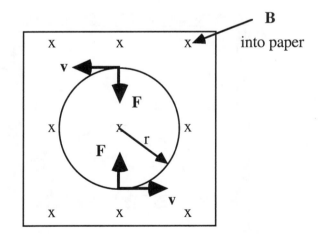

Note that with the magnetic field pointing into the paper (x's represent **B** into the page) and with the velocity **v** at each position shown, RHR-1 predicts the force **F** at each point to be directed in toward the center of a circle of radius r. That is, the magnetic force always remains perpendicular to the velocity and is directed toward the center of the circular path shown. The radius of the circular path is obtained by equating the magnetic force to the centripetal force.

We can therefore write,

$$F_c = \frac{mv^2}{r} = qvB \implies r = \frac{mv}{qB} \qquad (21.2)$$

Example 2

In a magnetic field of B = 0.75 T, for what path radius will an electron circulate at a speed of 3.0×10^7 m/s? What will its kinetic energy be?

Using equation (21.2), we can find the radius of the orbit for $v = 3.0 \times 10^7$ m/s.

$$r = \frac{mv}{qB} = \frac{(9.11 \times 10^{-31} \text{ kg})(3.0 \times 10^7 \text{ m/s})}{(1.6 \times 10^{-19} \text{ C})(0.75 \text{ T})} = 2.3 \times 10^{-4} \text{ m}$$

The kinetic energy of the electron is

$$KE = \frac{1}{2}mv^2 = \frac{1}{2}(9.11 \times 10^{-31} \text{ kg})(3.0 \times 10^7 \text{ m/s})^2 = 4.1 \times 10^{-16} \text{ J}$$

21.4 The Mass Spectrometer

A mass spectrometer is a device for determining the relative masses and abundances of atoms and molecules. In the type illustrated below, the atoms or molecules are vaporized and ionized by the ion source, S. The charged ion is given a charge +e by the removal of one electron from the atom/molecule. The positive ions are then accelerated through a potential difference V and are injected into a region of constant magnetic field **B**, where they are deflected in semicircular paths of radius r.

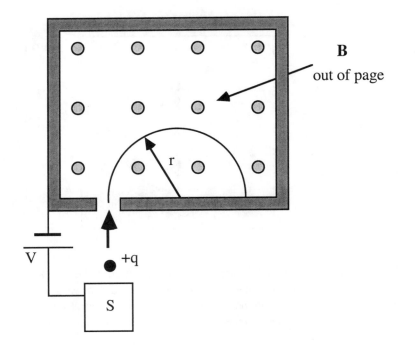

The ion has a kinetic energy KE = $(1/2)mv^2$ = eV, and since r = mv/qB we can combine equations to obtain

$$m = \left(\frac{er^2}{2V}\right) B^2$$

Example 3

A mass spectrometer uses a potential difference of 2.00 kV to accelerate a singly ionized ion to the proper speed. A 0.400-T magnetic field then bends the ion into a circular path of radius 0.226 m. What is the mass of the ion?

Using the above equation we obtain

$$m = \frac{er^2 B^2}{2V} = \frac{(1.60 \times 10^{-19} \text{ C})(0.226 \text{ m})^2(0.400 \text{ T})^2}{2(2.00 \times 10^3 \text{ V})} = 3.27 \times 10^{-25} \text{ kg}$$

21.5 The Force on a Current in a Magnetic Field

We have seen that a moving charge in a magnetic field experiences a magnetic force. Since a current is nothing more than a stream of moving charges, it is natural to expect that a current-carrying wire will experience a magnetic force when placed in a magnetic field. For a wire of length L carrying a current I placed in a magnetic field B, the force exerted is

$$F = ILB \sin \theta \qquad\qquad (21.3)$$

where θ is the angle between the direction of the current and the direction of the magnetic field.

Example 4

An electric power line carries a current of 1400 A in a location where the earth's magnetic field is 0.50×10^{-4} T. The line makes an angle of 75° with respect to the field. Determine the magnitude of the magnetic force on a 120-m length of line.

Using equation (21.3) we have

$$F = ILB \sin \theta = (1400 \text{ A})(120 \text{ m})(0.50 \times 10^{-4} \text{ T}) \sin 75° = 8.1 \text{ N}.$$

21.7 Magnetic Fields Produced by Currents

In the previous section we saw that a current-carrying wire experiences a magnetic force when placed in a magnetic field. We now consider a phenomena in which **a current-carrying wire produces a magnetic field**.

Consider a long, straight wire carrying a current. If we place compass needles at different points near the wire, we find that the magnetic field lines form a circular pattern about the wire. If the direction of the current is reversed, we find that the direction of the magnetic field has also reversed. The direction of the magnetic field can be obtained by using **Right-Hand Rule No. 2** (RHR-2). It can be stated as follows:

Right-Hand Rule No. 2. Curl the fingers of the right hand into the shape of a half-circle. Point the thumb in the direction of the conventional current I, and the tips of the fingers will point in the direction of the magnetic field **B**.

Right-Hand Rule No. 2 is illustrated in the following diagram.

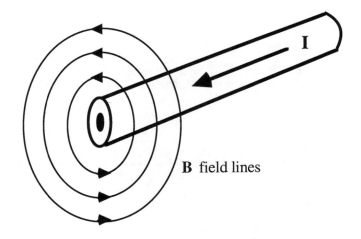

B field lines

The magnitude of the magnetic field at a distance r from a long, straight wire is given by the expression:

$$B = \frac{\mu_0 I}{2\pi r} \qquad (21.5)$$

where μ_0 is a constant known as the **permeability of free space**, and its value is $\mu_0 = 4\pi \times 10^{-7}$ T m/A.

Example 5

A 6.00-μC charge is moving with a velocity of 7.50×10^6 m/s parallel to a long, straight wire. The wire carries a current of 67.0 A in the same direction as the moving charge, and is 5.00×10^{-2} m away from the charge. Find the magnitude and direction of the force on the charge.

First calculate the magnetic field produced by the wire and then the force that this field exerts on the moving charge.

$$B = \frac{\mu_0 I}{2\pi r} = \frac{(4\pi \times 10^{-7} \text{ T m/A})(67.0 \text{ A})}{2\pi(5.00 \times 10^{-2} \text{ m})} = 2.68 \times 10^{-4} \text{ T}$$

The magnitude of the force on the charge is given by equation (21.1) with $\theta = 90°$. Therefore,

$$F = qvB \sin \theta = (6.00 \times 10^{-6} \text{ C})(7.50 \times 10^6 \text{ m/s})(2.68 \times 10^{-4} \text{ T}) \sin 90° = 1.21 \times 10^{-2} \text{ N}$$

The direction of the force on the moving charge is obtained by applying **each** of the right-hand rules. As seen in the following diagram, if the current in the wire is directed OUT of the page, the magnetic field produced by this wire circles counterclockwise about the wire, according to RHR-2. The charge is moving in the same direction as the current, i.e., OUT of the page. At the site of the moving charge, the magnetic field is directed UP. By applying RHR1, our fingers point upward (in the direction of the B-field), our thumb points out of the page (in the direction of the velocity) and the palm of our hand (which represents the direction of the magnetic force) points toward the wire. The force on the moving charge is therefore directed toward the wire.

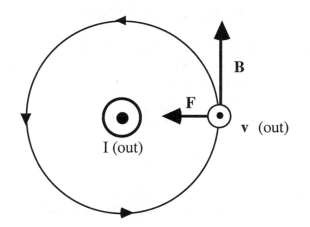

If we take a wire and bend it in the shape of a circular loop, the magnetic field at the center of the loop is given by

$$B = N \frac{\mu_0 I}{2R} \qquad\qquad (21.6)$$

where the loop contains N turns of wire.

Another device of interest is the **solenoid**. It is made up of a long coil of wire in the shape of a helix. If the wire is wound so the turns are packed close to each other and the solenoid is long compared to its diameter, the magnetic field inside the solenoid is nearly constant in magnitude and directed parallel to the axis. The magnitude of the field in the interior of a long solenoid is

$$B = \mu_0 n I \qquad\qquad (21.7)$$

where n is the number of turns per unit length of the solenoid and I is the current. If, for example, the solenoid contains 500 turns and has a length of 0.1 m, the number of turns per unit length is n = 5000 turns/m.

Example 6
A solenoid is 0.50 m long, has three layers of windings of 750 turns each, and carries a current of 4.0 A. What is the magnetic field at the center of the solenoid?

Since there are three layers of windings, there are (3 x 750) turns in 0.50 m. So, n = (3 x 750)/0.50 = 4500 turns/m. The magnetic field at the center of the solenoid is obtained using equation (21.7),

$$B = \mu_0 \, n \, I = (4\pi \times 10^{-7} \text{ T m/A})(4500 \text{ turns/m})(4.0 \text{ A}) = 2.3 \times 10^{-2} \text{ T}.$$

PRACTICE PROBLEMS

1. The electrons in the beam of a television tube have an energy of 15 keV. The tube is oriented so that the electrons move horizontally from east to west. The earth's magnetic field points down and has a magnitude $B = 4.5 \times 10^{-5}$ T. (a) In what direction will the beam deflect? (b) What is the acceleration of a given electron?

2. Protons in a magnetic field of 0.80 T follow a circular trajectory of 75-cm radius. (a) What is the speed of the protons? (b) If electrons traveled at the same speed in this field, what would the radius of their trajectory be?

3. Two isotopes of carbon, carbon-12 and carbon-13, have masses of 19.92×10^{-27} kg and 21.59×10^{-27} kg, respectively. These two isotopes are singly ionized and each is given a speed of 6.667×10^5 m/s. The ions then enter the bending region of a mass spectrometer where the magnetic field is 0.8500 T. Determine the spatial separation between the two isotopes after they have traveled through a half-circle.

4. A weighing scale supports a battery, to which a rigid wire loop is attached, as shown in the figure. The lower part of the loop is in a magnetic field of 0.20 T. If the combined mass of the battery and wire loop is 0.25 kg, what current in the wire is necessary for the scale to indicate zero weight? Which pole of the battery is positive?

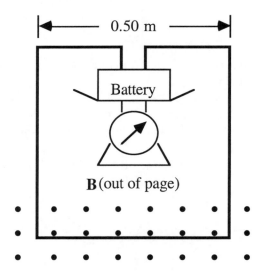

5. Two wires parallel to one another are separated by 1.0 m, as shown in the figure. Each carries a current of 3.0 A, but in opposite directions. Find the magnitude and direction of the magnetic field at a point P midway between the wires.

6. Two parallel straight wires are separated by a distance of 0.75 m and carry currents $I_1 = 25$ A and $I_2 = 35$ A.

   Find the force (magnitude and direction) that each wire exerts on a 2.5 m length of the other wire if the currents are (a) in the same direction and (b) in opposite directions.

7. You are given 251 m of copper wire that can carry a current of 0.500 A without excessive heating. This wire is used to make a solenoid of 25.0 cm length and 3.00 cm diameter. Determine the magnetic field that can be achieved at the center of the solenoid.

8. Four long, parallel conductors all carry currents of 5.0 A. An end view of the conductors is shown in the figure below. The current directions for wires A and D are into the page (indicated by crosses) while the current directions for B and C are out of the page (indicated by dots). Calculate the magnitude and direction of the magnetic field at point P, located at the center of the square of edge length 0.40 m.

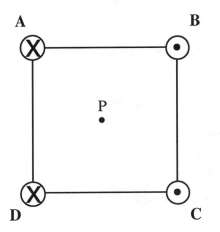

SOLUTIONS AND ANSWERS

Practice Problems

---

1. (a) The direction of the beam is obtained using RHR-1. With **B** down (fingers point down), and the velocity to the west (thumb to the west), the palm of the hand (**F**) is facing south. For an electron, force is directed **north**.
    (b) The kinetic energy is

$$KE = (1/2)mv^2 = (15\ 000\ eV)(1.6 \times 10^{-19}\ J/1\ eV) = 2.4 \times 10^{-15}\ J.$$

Solving for v yields,

$$v^2 = 2KE/m = 2(2.4 \times 10^{-15}\ J)/(9.11 \times 10^{-31}\ kg) \Rightarrow v = 7.3 \times 10^7\ m/s.$$

The magnitude of the force is $F = qvB \sin \theta$ ($\theta = 90°$). The acceleration can therefore be obtained using

$$a = F/m = (qvB \sin 90°)/m = (1.6 \times 10^{-19}\ C)(7.3 \times 10^7\ m/s)(4.5 \times 10^{-5}\ T)/(9.11 \times 10^{-31}\ kg) = \mathbf{5.8 \times 10^{14}\ m/s^2}.$$

---

2. (a) Solving equation (21.2) for the velocity yields

$$v = qrB/m = (1.6 \times 10^{-19}\ C)(0.75\ m)(0.80\ T)/(1.67 \times 10^{-27}\ kg) = \mathbf{5.7 \times 10^7\ m/s}.$$

(b) For an electron,

$$r = mv/qB = (9.11 \times 10^{-31}\ kg)(5.7 \times 10^7\ m/s)/[(1.6 \times 10^{-19}\ C)(0.80\ T)] = \mathbf{4.1 \times 10^{-4}\ m}.$$

---

3. The spatial separation of the two isotopes is the difference between the diameter of their trajectories. Therefore,

$$r_{12} = m_{12}v/qB = (19.92 \times 10^{-27}\ kg)(6.667 \times 10^5\ m/s)/[(1.602 \times 10^{-19}\ C)(0.8500\ T)] = 9.753 \times 10^{-2}\ m.$$

$$r_{13} = m_{13}v/qB = (21.59 \times 10^{-27}\ kg)(6.667 \times 10^5\ m/s)/[(1.602 \times 10^{-19}\ C)(0.8500\ T)] = 10.57 \times 10^{-2}\ m.$$

The spatial separation is therefore, $s = 2(r_{13} - r_{12}) = \mathbf{1.6 \times 10^{-2}\ m}.$

---

4. To support the weight of the battery, the magnetic force on the wire must equal the weight of the battery. That is $F = ILB \sin \theta = mg$. Using $\theta = 90°$ we can solve for the current to obtain,

$$I = mg/LB = (0.25\ kg)(9.80\ m/s^2)/[(0.50\ m)(0.20\ T) = \mathbf{25\ A}.$$

In order to balance the weight, the magnetic force must be directed UP. Since the field is OUT of the page, and the force is up, RHR-1 tells us I is to the left. Therefore, the **right-hand terminal is positive**.

---

5. At point P, the field due to each wire is pointing DOWN, as determined from RHR-2. Since each wire carries the same current and is at the same distance from P, the total field is

$$B_{TOT} = 2B = 2(\mu_0 I/2\pi r) = \mu_0 I/\pi r = (4\pi \times 10^{-7}\ T\ m/A)(3.0\ A)/\pi(0.50\ m) = \mathbf{2.4 \times 10^{-6}\ T,\ DOWN}.$$

---

6. The magnitude of the force between the wires (see example 8 in the text) is

$$F = \mu_0 I_1 I_2 L/2\pi r = (4\pi \times 10^{-7} \text{ T m/A})(25 \text{ A})(35 \text{ A})(2.5 \text{ m})/2\pi(0.75 \text{ m}) = \mathbf{5.8 \times 10^{-4} \text{ N}}.$$

(a) If the currents are in the same direction, by RHR-2 and RHR-1, the force is one of **attraction**.
(b) If the currents flow in opposite directions through each wire, the force is one of **repulsion**.

---

7. To find the number of turns of wire, divide the total length of the wire (251 m) by the length per turn, which is just $l = 2\pi r = 2\pi(1.50 \times 10^{-2} \text{ m}) = 0.0942$ m per turn. Thus, $N = (251 \text{ m})/(0.0942 \text{ m/turn}) = 2660$ turns. Since the solenoid is 25.0 cm in length, $n = 2660$ turns/0.250 m = 10 600 turns/m. The magnetic field at the center is

$$B = \mu_0 n I = (4\pi \times 10^{-7} \text{ T m/A})(10\ 600 \text{ m}^{-1})(0.500 \text{ A}) = \mathbf{6.66 \times 10^{-3} \text{ T}}.$$

---

8. The diagram showing the directions of the fields due to each wire is shown below.

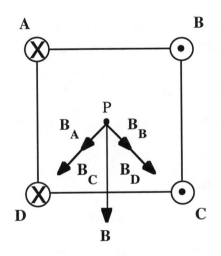

Since each wire carries the same current, and each is equidistant from the center, $r^2 = (0.20 \text{ m})^2 + (0.20 \text{ m})^2$, so $r = 0.28$ m. The field due to each wire is

$$B = \mu_0 I/2\pi r = (4\pi \times 10^{-7} \text{ T m/A})(5.0 \text{ A})/2\pi(0.28 \text{ m}) = 3.6 \times 10^{-6} \text{ T}.$$

Note that the x-components of the fields vanish, while the y-components add together. The total field is thus,

$$B_{tot} = 4B \sin 45° = 4(3.6 \times 10^{-6} \text{ T}) \sin 45° = \mathbf{1.0 \times 10^{-5} \text{ T, DOWN}}.$$

CHAPTER 22
Electromagnetic Induction

PREVIEW

Whenever the magnetic flux changes through a loop, an emf is generated in the loop. This observation is given substance in the form of Faraday's Law of Electromagnetic Induction which is presented in this chapter. Electromagnetic induction is used to explain the operation of electric generators and useful circuit devices called inductors along with the operation of the common transformer.

QUICK REFERENCE

Important Terms

Induced current
The current produced in a conducting coil when it is exposed to a changing magnetic field or its area or orientation changes relative to a magnetic field.

Induced emf
The emf in a coil or loop when it is exposed to a changing magnetic field or its area or its orientation changes in a magnetic field. The induced emf MAY produce an induced current if the coil or loop is a conductor.

Motional emf
The emf produced across a conductor due to its motion through a magnetic field.

Magnetic flux
The product of the area of a loop, the magnetic field through the loop and the cosine of the angle that the magnetic field makes with a line drawn normal to the plane of the loop. It is the change of magnetic flux through a loop which produces the induced emf.

Induced magnetic field
The magnetic field which is produced by an induced current.

Back emf
The induced emf produced in the rotating coils of a motor whose polarity is opposite the polarity of the emf applied to the motor.

Mutual induction
The process by which a changing current in one circuit produces a changing magnetic field which induces a current in another circuit.

Self-induction
The process by which a changing current in a circuit induces an emf in the same circuit.

Inductors
Coils which are used as circuit elements to exploit their property of self-induction.

Transformer
A device which uses mutual inductance between primary and secondary coils to increase or decrease an ac voltage.

Induced emfs

The emf induced in a **rod moving perpendicularly with respect to a magnetic field** is

$$emf = vBL \qquad (22.1)$$

The emf induced in a **coil of N loops** (Faraday's Law)

$$\text{emf} = -N\frac{\Delta\Phi}{\Delta t} \tag{22.3}$$

The emf produced by a **generator**

$$\text{emf} = NAB\omega\sin\omega t \tag{22.4}$$

The emf due to **mutual induction**

$$(\text{emf})_s = -M\frac{\Delta I_p}{\Delta t} \tag{22.7}$$

The emf due to **self-induction**

$$\text{emf} = -L\frac{\Delta I}{\Delta t} \tag{22.9}$$

Magnetic Flux

The magnetic flux through a **plane area or coil**

$$\Phi = BA\cos\phi \tag{22.2}$$

Inductance

Mutual inductance of **two coils**

$$M = \frac{N_s\Phi_s}{I_p} \tag{22.6}$$

Self-induction of a coil

$$L = \frac{N\Phi}{I} \tag{22.8}$$

Motors

The **current** drawn by a motor (emf is the "back emf" of the motor)

$$I = \frac{V - \text{emf}}{R} \tag{22.5}$$

Energy in Inductors

The **energy** stored in an inductor due to **self inductance** is

$$\text{Energy} = \frac{1}{2}LI^2 \tag{22.10}$$

The **energy density** in a **magnetic field**

$$\text{Energy density} = \frac{1}{2\mu_0}B^2 \tag{22.11}$$

Transformers

The **transformer equation**

$$\frac{V_s}{V_p} = \frac{N_s}{N_p} \tag{22.12}$$

DISCUSSION OF SELECTED SECTIONS

22.1 Induced Emf and Induced Current

It can be demonstrated that relative motion between a magnet and a coil which forms part of a circuit will lead to an induced current flowing in the circuit containing the coil. The induced current is seen to change direction when the direction of the relative motion is reversed. (Perhaps your instructor will demonstrate this.) Further investigation indicates that two additional actions can lead to an induced current in the coil: A change in the area of the coil and a change in the orientation of the coil with respect to the magnetic field. The emf which is produced in the coil which drives the induced current is called the "induced emf". The induced emf exists whether or not the coil is part of a closed circuit. However, a closed circuit is necessary for the induced current to flow.

22.2 Motional Emf

Motional emf is an emf induced in a conductor moving through a magnetic field. The emf arises from a charge separation in the conductor due to the magnetic forces exerted on charges in the conductor. It depends on the component of the magnetic field perpendicular to the motion of the conductor, the speed and length of the conductor. This emf is present whether or not the conductor is part of a closed circuit.

Example 1

The horizontal component of the earth's magnetic field at a certain location is 2.6×10^{-5} T and points due north. What motional emf appears in the 0.80 m vertical radio antenna of a car traveling east with a speed of 95 km/h? What motional emf appears in the antenna if the car turns north?

Equation (22.1) applies to the emf induced in a conductor moving perpendicular to a magnetic field. If the car is going east, then its vertical antenna is moving perpendicular to the horizontal component of the earth's magnetic field. The induced emf is

$$\text{emf} = vBL = (95 \times 10^3 \text{ m/3600 s})(2.6 \times 10^{-5} \text{ T})(0.80 \text{ m}) = 5.5 \times 10^{-4} \text{ V}.$$

If the car is moving north, then the vertical antenna is moving parallel to the magnetic field and no emf will be generated.

Example 2

A rectangular loop of wire is moving at 22 m/s as shown through a region of constant magnetic field of magnitude 0.55 T. What emf is induced in the top and bottom wires of the loop? What emf is induced in the side wires? If the loop has a resistance of 0.015 Ω, how much current flows through it?

The top (ab) and bottom (cd) wires are moving perpendicular to the magnetic field, so equation (22.1) gives the emf across each of them to be

$$\text{emf} = vBL$$
$$\text{emf} = (22 \text{ m/s})(0.55 \text{ T})(0.50 \text{ m}) = 6.1 \text{ V}.$$

The end wires are oriented along the direction of the magnetic field so no emf will be produced in them. The current in the loop is driven by the NET emf induced in it. The polarity of the emf in the top wire is (a = +) the same as the polarity of the emf induced in the bottom wire (d = +). The net emf is the sum of the emfs around the loop.

$$\text{net emf} = + 6.0 \text{ V} - 6.0 \text{ V} + 0 \text{ V} + 0 \text{ V}$$
$$\text{net emf} = 0 \text{ V}.$$

Since the net emf is zero, there can be no current in the loop.

22.3 Magnetic Flux

As indicated previously, an emf can be produced in a coil by changing the magnetic field through the coil, the area of the coil and/or the orientation of the coil. These three methods of producing an emf are summarized by introducing the concept of magnetic flux $\Phi = BA \cos \phi$, where B is the magnitude of the magnetic field penetrating the coil, A is the area of the coil and ϕ is the angle between the magnetic field vector and the normal to the surface. An emf can be produced in a coil by changing the magnetic flux through the coil. (Faraday's Law).

Example 3
A long solenoid of 150 turns per cm and radius 1.5 cm carries a dc current of 0.50 A. What is the total magnetic flux through a 0.25 m length of the solenoid due to its OWN magnetic field?

Previously, it was found that the magnetic field produced inside a long current-carrying solenoid is constant in magnitude and directed along the axis of the solenoid: $B = \mu_0 nI$.

Each turn of the solenoid has a flux $\Phi = BA \cos 0° = BA$ through it . The total flux through the solenoid is then

$$\Phi_T = NBA = N\mu_0 nIA$$
$$\Phi_T = \mu_0 n^2 LIA$$
$$\Phi_T = (4\pi \times 10^{-7} \text{ T m/A})(150 \times 10^2 \text{ turns/m})^2(0.25 \text{ m})(0.50 \text{ A})(\pi)(1.5 \times 10^{-2} \text{ m})^2$$
$$\Phi_T = 2.5 \times 10^{-2} \text{ Wb}$$

22.4 Faraday's Law of Electromagnetic Induction

A change in the magnetic flux through a coil produces an induced emf in the coil given by (22.3).

Example 4
The current in the solenoid of example 3 changes from 0.50 A to 0 A in 1.5 ms. What is the magnitude of the average emf induced in the solenoid?

Faraday's law applied to the solenoid gives the magnitude of the emf induced in N turns:

$$\text{emf} = N(\Delta\Phi/\Delta t)$$

The flux changes through the solenoid because the decreasing current causes the magnetic field inside the solenoid to decrease to zero. The flux change through one turn is $\Delta\Phi = \mu_0 nIA$.

The magnitude of the emf induced in the solenoid is then

$$\text{emf} = (N\mu_0 nIA)/\Delta t = (2.5 \times 10^{-2} \text{ Wb})/(1.5 \times 10^{-3} \text{ s}) = 17 \text{ V}$$

22.5 Lenz's Law

Faraday's Law gives the emf induced in a coil by a changing magnetic flux. Lenz's Law gives the polarity of the induced emf. The induced emf ALWAYS has a polarity which WOULD produce a current whose magnetic field WOULD oppose the change in flux that caused the induced emf in the first place. A key word here is WOULD. It is not necessary that a current actually flow in a closed circuit. The emf will be produced and its polarity will be given by Lenz's Law even if no closed circuit exists.

Example 5

An electromagnet generates a magnetic field which "cuts" through a coil as shown. What is the polarity of the emf generated in the coil if the applied field, B (a) points to the right and is increasing? (b) points to the right and is decreasing? (c) is pointing to the left and increasing? (d) is pointing to the left and decreasing?

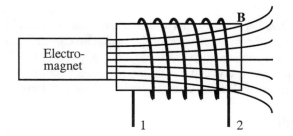

(a) If the magnetic field through the coil points to the right and is increasing, then the flux through the coil is increasing. Lenz's Law demands that if a current could flow then it would produce an "induced magnetic field", B_I which would oppose the increase in flux. The action of B_I is to produce a new flux through the coil to attempt to increase the total flux. B_I must point to the left. Now imagine that a resistor has been placed in series with the coil to provide a closed loop. What current direction through the resistor is needed to produce B_I in the coil? RHR-2 gives the current to be in the direction shown in the drawing. The resistor "sees" the coil as a source of emf much like a battery. Current leaves a source of emf at the + terminal and enters at the - terminal, so point 1 is POSITIVE while point 2 is NEGATIVE.

(b) If **B** points to the right and is decreasing, then B_I will attempt to produce a flux to increase the total flux through the coil. In this case B_I must be directed to the right and will be produced by a current flowing to the left through the resistor. This corresponds to point 1 NEGATIVE and point 2 POSITIVE.

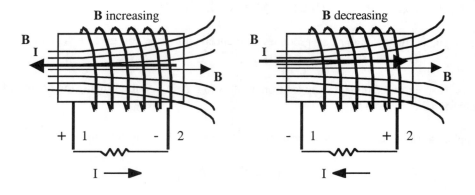

(c) and (d) The analysis of the situation where the applied field, **B**, points to the left is very similar to the analysis in parts (a) and (b). The drawing below gives the results.

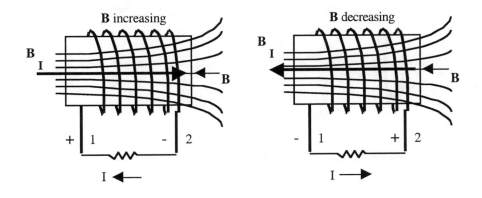

Sometimes Lenz's Law can be interpreted in a much more general way which may be useful. The following example illustrates one such interpretation.

Example 6

A metal hoop is suspended from a support by a string and a magnet is moved relative to the hoop as shown in the drawing. What is the direction of the induced current in the hoop if the magnet is moved (a) north end first toward the hoop? (b) south end first away from the hoop (c) south end first toward the hoop? (d) north end first away from the hoop?

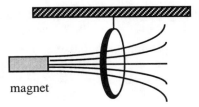

(a) Refer to the drawing shown below. If the north end of the magnet is moved toward the hoop, then the action of the induced current is to attempt to decrease the flux through the hoop. The flux through the hoop is increasing ultimately because the magnet is moving. The induced current will produce a north pole on the left of the loop to repel the magnet in an attempt to stop the relative motion. The left side of the coil is a north pole, hence the induced magnetic field, **B**$_I$, is directed from right to left through the hoop. The current direction needed to produce this magnetic field is, according to RHR-2, CW as seen by the "eye".

(b) the situation here is analyzed in the same way as in part (a). As the magnet moves away from the hoop, the hoop develops a south pole on its left to attempt to stop the relative motion by attracting the magnet. The induced field is then from left to right through the hoop. The current as seen by the "eye" is CCW.

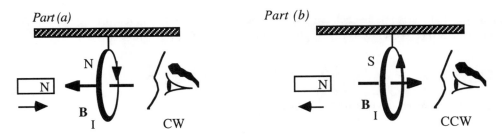

(c) Repeating the above analysis for the magnet moving south end first toward the hoop results in a current flowing CCW. See the drawing below.

(d) Similarly, if the magnet moves away from the hoop with its south end toward the hoop a CW current will result as shown in the drawing.

22.7 The Electric Generator

Electromagnetic induction is used to produce an emf in a generator which can be used to power external devices such as motors. The simplest generator is described extensively in the text. It produces an emf given by (22.4).

Example 7

The maximum strength of the earth's magnetic field is about 7.0×10^{-5} T near the south magnetic pole. In principle, this field could be used with a rotating coil to generate 60.0 Hz ac electricity. What is the minimum number of turns (area per unit turn = 0.016 m^2) that the coil must have so as to produce an rms voltage of 120 V?

The peak emf produced by the generator is given by (22.4) to be NABω. The rms voltage is then $V = (NAB\omega)/\sqrt{2}$. The number of turns needed is

$$N = \frac{\sqrt{2}\,V}{BA\omega} = \frac{\sqrt{2}\,(120\text{ V})}{(7.0 \times 10^{-5}\text{ T})(0.016\text{ m}^2)(2\pi)(60.0\text{ Hz})} = 4.0 \times 10^5$$

The very desirable induced emf which appears in a generator becomes a "back emf" in a motor. The back emf retards the motion of the armature in a motor by acting with a polarity opposite that of the voltage which is applied to run the motor. The net current "drawn" by an electric motor is given by (22.5).

Example 8

A motor is designed to operate on 117 V and draws a current of 37.5 A when it first starts up. At its normal operating speed, the motor draws a current of 4.10 A. Obtain (a) the resistance of the armature coil, (b) the back emf developed at normal speed, (c) the back emf at one third normal speed.

(a) When the motor first starts up, its back emf is zero since its armature is not yet turning, hence no induced emf can appear in it. Under these conditions (22.5) gives the resistance of the armature coil to be

$$R = V/I = (117\text{ V})/(37.5\text{ A}) = 3.12\ \Omega.$$

(b) At normal operating speed the rms back emf is given by (22.5) to be

$$\text{emf} = V - IR = 117\text{ V} - (4.10\text{ A})(3.12\ \Omega) = 104\text{ V}.$$

(c) The peak emf at any speed is given by (22.4). Applying this equation to normal speed and one third speed and dividing the resulting equations gives the rms back emf at one third speed.

$$\text{emf} = (104 \text{ V})(1/3) = 34.7 \text{ V}$$

22.8 Mutual Inductance and Self-Inductance

Mutual Inductance
When two coils are placed close together and one of the coils (the primary coil) carries a changing current, the second coil (the secondary coil) will develop an induced emf as the magnetic flux produced by the primary coil changes through it. In many cases, the flux through the secondary coil is proportional to the current flowing in the primary coil. The constant of proportionality is called the mutual inductance of the coils, M. The emf produced in the secondary coil by a change in the current in the primary coil is given by (22.7). The mutual inductance of two coils depends on the geometry of the arrangement and any magnetic material which may be placed in the coils.

Example 9
A current in one coil changes uniformly from 2.0 A to 0.50 A in 45 ms. A voltage whose magnitude is 5.5 V is seen to develop in a second coil located close to the first. What is the mutual inductance of the two coils?

The magnitude of the voltage is given by (22.7) without the minus sign. Therefore, the mutual inductance is

$$M = \frac{(\text{emf})_s}{\dfrac{\Delta I_p}{\Delta t}} = \frac{5.5 \text{ V}}{\dfrac{1.5 \text{ A}}{(45 \times 10^{-3} \text{ s})}} = 0.17 \text{ H}$$

Self-Inductance
Self-induction occurs when a coil carrying a changing current induces an emf in itself. Of course, this comes about because the time changing current produces a changing magnetic field which "cuts" through the coil. The associated changing flux through the coil produces a self-induced emf in the coil. Often, the flux through a coil is proportional to the current through the coil. The constant of proportionality in this case is called the self-inductance of the coil. The emf developed in a coil due to self-induction is given by (22.9).

Example 10
A solenoid of 550 turns has a self-inductance of 5.2 mH. (a) What is the flux per turn through the solenoid when it carries a current of 25 mA? (b) What is the magnitude of the emf induced in the solenoid if the current changes uniformly to 0 A in 2.0 ms?

(a) The magnetic flux per turn through the solenoid is given by (22.8).

$$\Phi = LI/N = (5.2 \times 10^{-3} \text{ H})(25 \times 10^{-3} \text{ A})/(550 \text{ turns}) = 2.4 \times 10^{-7} \text{ Wb/turn}$$

(b) The magnitude of the induced emf is given by (22.9) without the minus sign.

$$\text{emf} = L\frac{\Delta I}{\Delta t} = (5.2 \text{ mH})\frac{25 \text{ mA}}{2.0 \text{ ms}} = 65 \text{ mV}.$$

Current-carrying inductors store energy in their magnetic fields much like charged capacitors store energy in their electric fields. As a matter of fact, the mathematical expression for the energy stored in inductors is similar in form to the expression for energy stored in capacitors. The storage of energy in these devices is of great importance in ac circuits which are studied in the next chapter.

22.9 Transformers

The transformer is probably the single most important application of mutual inductance in our technological world. It allows the "stepping up" and "stepping down" of time changing voltages. This is very important in the transmission of ac electricity. The relationship between the primary and secondary voltages and the "turns ratio" of the ideal transformer is given by (22.12) while the relationships between the primary and secondary currents and the turns ratio is given by (22.13).

Example 11

A particular "ac adapter" for a calculator is a transformer designed to be plugged into a wall socket and to operate on 120 V. It produces an output secondary voltage of 5.7 V and an output current of 240 mA. (a) What is the turns ratio of the adapter? (b) If the adapter were an ideal transformer, how much current would it draw from the wall socket? (c) If the adapter really draws 50.0 mA from the wall socket, what percentage of the power is lost in the transformer?

(a) The turns ratio can be found from (22.12).

$$\frac{N_s}{N_p} = \frac{V_s}{V_p} = \frac{5.7 \text{ V}}{120 \text{ V}} = 0.048$$

This can also be written as 1:21.

(b) The current drawn from the wall socket is the primary current in the transformer. Equation (22.13) gives

$$I_p = \left(\frac{V_s}{V_p}\right) I_s = \left(\frac{5.7 \text{ V}}{120 \text{ V}}\right)(240 \text{ mA}) = 11 \text{ mA}$$

(c) The power input to the primary coil is $P_p = I_p V_p = (50.0 \times 10^{-3} \text{ A})(120 \text{ V}) = 6.0 \text{ W}$. The power available at the secondary coil is $P_s = I_s V_s = (240 \times 10^{-3} \text{ A})(5.7 \text{ V}) = 1.4 \text{ W}$.

The percentage of power lost is

$$\% \text{ power lost} = \frac{6.0 \text{ W} - 1.4 \text{ W}}{6.0 \text{ W}} \times 100 \% = 77 \%$$

PRACTICE PROBLEMS

1. A 1.0 m long conducting rod is moving perpendicular to a 0.35 T magnetic field. How fast must the rod move in order to produce an emf of 12 V across its ends?

2. A conducting bar such as that in Figure 22.4b in the text moves perpendicular to a 1.5 T magnetic field with a constant speed of 150 m/s. The conducting bar delivers a current of 5.5 A to a 6.0 Ω load. How long is the bar? How much force is necessary to keep the bar moving at constant speed?

3. A cube, 0.50 m on a side, is placed in a uniform magnetic field of strength 0.45 T oriented in the direction of a diagonal of the front face of the cube. Find the magnetic flux through each of the faces of the cube. What is the total magnetic flux through the cube?

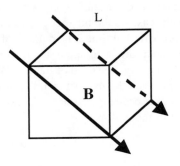

4. Use Faraday's Law to show that an emf = vBL is induced across a conducting bar of length, L, moving with a constant speed, v, perpendicular to a magnetic field of magnitude, B.

5. A 0.25 m long coil consists of 560 square turns 6.5 cm on a side. The coil is placed in a uniform magnetic field of strength 1.2 T. Initially the coil is oriented so that its axis coincides with the magnetic field direction. The coil is then rotated 90 ° about an axis perpendicular to the magnetic field direction in 0.20 ms. What is the magnitude of the emf induced in the coil?

6. The speed sensor for an automotive cruise control consists of a magnet mounted on the drive shaft along with a coil of wire mounted on the chassis. As the magnet passes the coil, an emf is induced for a short period of time. The number of induced emfs occurring per second allows a computer to determine the drive shaft speed and ultimately the speed of the car. What is the polarity of the emf when the magnet is at the three positions A, B, and C shown in the drawing? Assume the north pole of the magnet is closest to the coil.

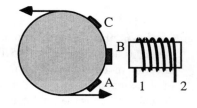

7. **CHALLENGE PROBLEM** A ring of aluminum is placed over the iron core of an electromagnet as shown in the drawing. The current in the coil of the electromagnet is 60.0 Hz ac. Use Lenz's Law to show that the flux change due only to the time changing magnetic field through the ring is insufficient to explain why the ring is repelled from the electromagnet.

ring

8. A generator consists of 150 rectangular turns of wire. Each turn has an area of 0.050 m^2. The generator is rotated in a 1.6 T magnetic field and produces an rms emf of 25 V. What is the frequency of the emf produced by the generator?

9. An experimenter needs an inductor with a self-inductance of 0.45 mH. The coil must be no longer than 2.5 cm and no more than 0.50 cm in diameter. How many turns of wire will the experimenter need if the coil can be assumed to be a perfect solenoid?

10. A 1:150 transformer has a primary voltage of 4500 V. (a) What is the voltage in the secondary coil? (b) If the transformer consumes 45 W, what are the currents in the primary and secondary coils?

SOLUTIONS AND ANSWERS

Practice Problems

---

1. Equation (22.1) applies to a rod moving through a magnetic field.

$$v = (emf)/(BL) = (12\ V)/[(0.35\ T)(1.0\ m)] = \textbf{34 m/s.}$$

---

2. The emf produced by the moving bar is given by equation (22.1). This emf provides the current in the load. According to Ohm's law, emf = IR. The length of the rod is

$$L = IR/vB = [(5.5\ A)(6.0\ \Omega)]/[(150\ m/s)(1.5\ T)] = \textbf{0.15 m.}$$

The force which must be exerted to keep the rod moving at a constant speed is

$$F = ILB = (5.5\ A)(1.5\ m)(0.15\ T) = \textbf{1.2 N.}$$

---

3. The flux through an area is given by equation (22.2). For the

front face: $\Phi_1 = BL^2 \cos 90° = \textbf{0 Wb.}$ **back face**: $\Phi_2 = BL^2 \cos 90° = \textbf{0 Wb.}$

top face: $\Phi_3 = BL^2 \cos 135° = \textbf{- 0.080 Wb.}$ **bottom face**: $\Phi_4 = BL^2 \cos 45° = \textbf{0.080 Wb.}$

right face: $\Phi_5 = BL^2 \cos 135° = \textbf{- 0.080 Wb.}$ **left face**: $\Phi_6 = BL^2 \cos 45° = \textbf{0.080 Wb.}$

The total flux through the cube is the sum of the fluxes through the faces, so $\Phi = \textbf{0 Wb.}$

---

4. Faraday's law can be used with real or imaginary loops. Construct an imaginary loop containing the moving bar as shown in the drawing.

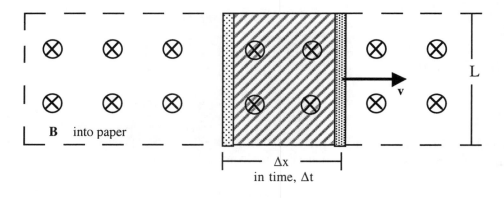

In time, Δt, the bar moves a distance, Δx, through the magnetic field. The flux change through the loop is entirely due to the change in area of the loop which is $\Delta A = L\ \Delta x$. The flux change through the loop is

$$\Delta \Phi = B\ \Delta A = BL\ \Delta x.$$

Faraday's law gives the magnitude of the induced emf to be

$$emf = \Delta\Phi/\Delta t = BL\ (\Delta x/\Delta t).$$

Now $\Delta x/\Delta t$ is the speed of the bar, so

$$\text{emf} = vBL.$$

---

5. The flux change through the coil is due to a change in the orientation of the coil. The flux through the coil before it is rotated is

$$\Phi_0 = BNA \cos 0° = BNA.$$

The flux through the coil after it is rotated is

$$\Phi = BNA \cos 90° = 0 \text{ Wb}.$$

The change in flux during the 0.20 ms time period is then

$$\Delta\Phi = BNA$$

and the magnitude of the emf induced in the coil is

$$\text{emf} = BNA/\Delta t = (1.2 \text{ T})(560)(6.5 \times 10^{-2} \text{ m})^2/(0.20 \times 10^{-3} \text{ s})$$
$$\text{emf} = \mathbf{1.4 \times 10^4 \ V}.$$

---

6. Mentally connect a resistor between points 1 and 2 of the coil.

When the magnet is at **position A**, it is moving toward the coil and causing a change in the magnetic flux through the coil. A generalized interpretation of Lenz's law requires that the coil react to try to stop this motion. Therefore, the left end of the coil will become a north pole to repel the north pole of the magnet. The current in the coil needed to produce a north pole at the left end flows through the coil from 2 to 1. The current flows from the coil at 1, through the imaginary resistor, and back into the coil at 2. The **polarity of 1 is** then **POSITIVE** and **the polarity of 2 is NEGATIVE.**

When the magnet is at **position C**, then it is moving away from the coil. The coil will produce a south pole at its left end to attract the magnet in an attempt to stop its motion, hence the flux change. The current in the coil must flow through the coil from 1 to 2. The current in the imaginary resistor then flows from 2 to 1 which indicates that **point 2 is POSITIVE** and **point 1 is NEGATIVE.**

When the magnet is at **position B**, the left end of the coil is changing from a north to a south pole. At this position the flux through the coil is momentarily constant and **no emf is induced in the coil** at all.

---

7. An ac current in the electromagnet produces the following magnetic fields: **B** is up and increasing, **B** is up and decreasing, **B** is down and increasing, **B** is down and decreasing. The induced current in the ring produces the following induced fields, $\mathbf{B_I}$.

B is up and increasing $\Rightarrow \mathbf{B_I}$ is down $\Rightarrow$ The electromagnet and ring repel since they are north to north.

B is up and decreasing $\Rightarrow \mathbf{B_I}$ is up $\Rightarrow$ The electromagnet and ring attract since they are north to south..

B is down and increasing $\Rightarrow \mathbf{B_I}$ is up $\Rightarrow$ The electromagnet and the ring are south to south and repel.

B is down and decreasing $\Rightarrow \mathbf{B_I}$ is down $\Rightarrow$ The electromagnet and ring are south to north and attract.

The net force on the ring over one cycle of the ac current is obviously zero. The force of gravity over one cycle of the ac current is not zero, so if the time changing magnetic field through the ring is the only way the flux can change, the net force on the ring could not be zero and the ring will not float.

In the actual case, the flux through the ring changes as the ring moves up and down in the magnetic field since the magnetic field is not uniform. This flux change gives rise to an additional induced magnetic field which generates an average repulsive force on the ring.

8. The peak emf produced by the generator is given by equation (22.4) to be $NAB\omega$, so the rms emf produced is $NAB\omega/\sqrt{2}$. Now the frequency of the emf is

$$f = \frac{\omega}{2\pi} = \frac{\sqrt{2}\ (\text{rms})}{2\pi NAB} = \frac{\sqrt{2}\ (25\ \text{V})}{2\pi(150)(0.050\ \text{m}^2)(1.6\ \text{T})} = \mathbf{0.47\ Hz}$$

---

9. The self-inductance of a solenoid is given in Example 10 in the text to be $L = \mu_0 n^2 Ax$. Here x is the length of the solenoid. Now

$$n = \sqrt{\frac{0.45 \times 10^{-3}\ \text{H}}{(4\pi \times 10^{-7}\ \text{T m/A})(\pi)(0.25 \times 10^{-2}\ \text{m})^2(2.5 \times 10^{-2}\ \text{m})}} = 27\ 000\ \text{m}^{-1}$$

The total number of turns needed is then $N = nx = \mathbf{680}$.

---

10. (a) Equation (22.12) gives

$$V_s = (N_s/N_p)V_p = (1/150)(4500\ \text{V}) = \mathbf{3.0 \times 10^1\ V}.$$

(b) The current delivered to the primary is

$$I_p = P/V_p = (45\ \text{W})/(4500\ \text{V}) = \mathbf{1.0 \times 10^{-2}\ A}.$$

If no power is lost in the transformer, the current in the secondary will be

$$I_s = P/V_s = (45\ \text{W})/(3.0 \times 10^1\ \text{V}) = \mathbf{1.5\ A}.$$

CHAPTER 23
Alternating Current Circuits

PREVIEW

In this chapter you will learn the basics of alternating current (ac) circuits. You will learn how capacitors, resistors, and inductors are combined in series with ac voltage sources to produce alternating currents. Phasor diagrams are used to illustrate the relationships between current and voltage in ac circuits. Resonance circuits and their applications are discussed. An introduction to semiconductor devices is included as well.

QUICK REFERENCE

Important Terms

Capacitive reactance
A quantity which determines how much rms current exists in a capacitor in response to a given rms voltage across the capacitor. The units of reactance are ohms (Ω).

Inductive reactance
A quantity which determines how much rms current exists in an inductor in response to a given rms voltage across the inductor. The units of reactance are ohms (Ω).

Impedance
The total opposition to the flow of charge in a series RCL circuit. The resistance, capacitive reactance, and inductive reactance all contribute to the impedance.

Resonance
A condition which occurs when the frequency of the oscillating voltage supplied by a generator matches the natural frequency of the components in the circuit. At resonance the impedance of the series RCL circuit has a minimum value equal to the resistance R, and the rms current has a maximum value.

n-type semiconductors
Materials such as silicon and germanium that are "doped" by adding impurity atoms which contribute mobile electrons to the material.

p-type semiconductors
Materials such as silicon and germanium that are "doped" by adding impurity atoms which contribute mobile positive holes to the material.

p-n junction diode
A device formed from an n-type and a p-type semiconductor. When placed in a circuit it will allow current to flow in only one direction in that circuit.

Equations

The **rms voltage across a capacitor** can be expressed as,

$$V_{rms} = I_{rms} X_C \qquad (23.1)$$

where X_C is known as the **capacitive reactance** and is given by

$$X_C = \frac{1}{2\pi f C} \qquad (23.2)$$

Here, f is the frequency in Hz, C is the capacitance in farads, and X_C is measured in ohms.

The **rms voltage across an inductor** can be expressed as,

$$V_{rms} = I_{rms} X_L \tag{23.3}$$

where X_L is known as the **inductive reactance** and is given by

$$X_L = 2\pi f L \tag{23.4}$$

The **total rms voltage in a series RCL circuit** is given by,

$$V_{rms}^2 = V_R^2 + (V_L - V_C)^2 \tag{23.5}$$

The **total rms voltage** is **related** to **the rms current** through the following expressions:

$$V_{rms} = I_{rms} Z \tag{23.6}$$

$$Z = \sqrt{R^2 + (X_L - X_C)^2} \tag{23.7}$$

where Z is known as the **impedance** of the RCL circuit. The **phase angle** of an RCL circuit is defined as

$$\tan \phi = \frac{X_L - X_C}{R} \tag{23.8}$$

The **average power dissipated in an RCL circuit** can be written as

$$\overline{P} = I_{rms} V_{rms} \cos \phi \tag{23.9}$$

where $\cos \phi$ is called the **power factor** of the circuit, defined as $\cos \phi = R/Z$. When the impedance of a circuit is minimized, i.e., when $X_L = X_C$ in equation (23.7) and thus $Z = R$, the circuit will **resonate** with

$$f_0 = \frac{1}{2\pi\sqrt{LC}} \tag{23.10}$$

where f_0 is known as the **resonance frequency** of the circuit.

DISCUSSION OF SELECTED SECTIONS

23.1 Capacitors and Capacitive Reactance

The rms voltage across a resistor is $V_{rms} = I_{rms}R$. For a capacitor, notice that the term X_C replaces the resistance in equation (23.1). The **capacitive reactance** X_C, like resistance, is measured in ohms and determines how much rms current exists in a capacitor in response to a given rms voltage across the capacitor.

The capacitive reactance depends on the capacitance of the capacitor and the frequency of the ac voltage applied to the capacitor according to

$$X_C = \frac{1}{2\pi f C} \qquad (23.2)$$

Thus, according to equation (23.2), the capacitive reactance is large for small frequencies f. When X_C is large, the current is small since $I_{rms} = V_{rms}/X_C$. When the frequency is high, X_C is small, and I_{rms} is large.

Example 1
A 5.00 μF capacitor is in series with an ac voltage source which supplies an rms voltage of 32.0 V. Determine the rms current in the circuit when the frequency of the source is (a) 60.0 Hz and (b) 6.00 x 10^3 Hz.

(a) At 60.0 Hz,

$$X_C = \frac{1}{2\pi f C} = \frac{1}{2\pi(60.0 \text{ Hz})(5.00 \text{ x } 10^{-6} \text{ F})} = 531 \text{ }\Omega$$

$$I_{rms} = \frac{V_{rms}}{X_C} = \frac{32.0 \text{ V}}{531 \text{ }\Omega} = 0.0603 \text{ A}$$

(b) For 6.00 x 10^3 Hz,

$$X_C = \frac{1}{2\pi f C} = \frac{1}{2\pi(6.00 \text{ x } 10^3 \text{ Hz})(5.00 \text{ x } 10^{-6} \text{ F})} = 5.31 \text{ }\Omega$$

$$I_{rms} = \frac{V_{rms}}{X_C} = \frac{32.0 \text{ V}}{5.31 \text{ }\Omega} = 6.03 \text{ A}$$

Let's compare the effects that resistors and capacitors have on circuits. Consider the behavior of the **instantaneous** (not rms) voltage or current for a resistor as a function of time, as in the following diagram.

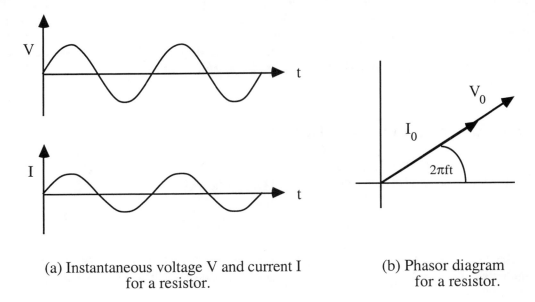

(a) Instantaneous voltage V and current I
for a resistor.

(b) Phasor diagram
for a resistor.

The graphs in figure (a) shown above indicate that, when only resistance is present, the voltage and current are proportional to each other at every moment. As the voltage across the resistor increases, the current increases proportionately. As the voltage drops, the current follows suit. For this reason, the current in a resistance R is said to be **in phase** with the voltage across the resistor. This aspect is depicted in figure (b) above through the use of a **phasor diagram**. In this model, voltage and current are represented by rotating arrows, called phasors, whose lengths correspond to the maximum voltage V_0 and the maximum current I_0. The **phasors rotate counterclockwise** at a frequency f. For a resistor, the phasors are colinear as they rotate, because the voltage and current are in phase. Now consider the situation for capacitors, as shown in the following diagrams.

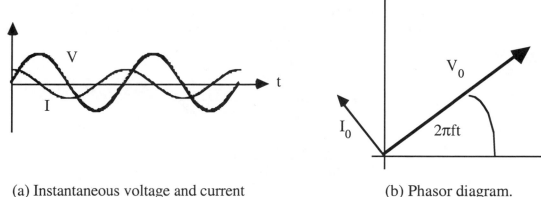

(a) Instantaneous voltage and current in a capacitive circuit.

(b) Phasor diagram.

For a capacitor, the current and voltage are **not** in phase. As the voltage increases at the beginning of the cycle, the charge on the capacitor begins to build up. The current, or rate of flow of charge, has a maximum positive value at the start of the charging process, when there is no charge on the capacitor. As charge builds up, and the voltage across the capacitor increases, the flow of charge is impeded, and hence the current drops. When the capacitor is fully charged, no further charge can flow, so the instantaneous current is zero. The capacitor then begins to discharge, and current begins to flow in the opposite (negative) direction. When all the charge is gone from the capacitor, the voltage across the capacitor goes to zero, and the current has its maximum negative value. The voltage and current are therefore one-quarter cycle out of step, or 90° out of phase. From the figure it is seen that the current through a capacitor **leads** the voltage across the capacitor by 90°. In figure (b) note that the phasors remain perpendicular while rotating, because the phase angle between the current and voltage is 90°. Since the current leads the voltage for a capacitor, the current phasor is ahead of the voltage phasor in the direction of rotation (counterclockwise). The instantaneous voltage and current are given by the vertical components of the phasors.

23.2   Inductors and Inductive Reactance

An inductor, as discussed in Section 22.8, is usually a coil of wire which develops a voltage that opposes a change in the current. The rms voltage across an inductor is given by the expression,

$$V_{rms} = I_{rms} X_L \qquad (23.3)$$

which is analogous to $V_{rms} = I_{rms} R$ for a resistor. The term that appears in place of R in equation (23.3), X_L, is called the **inductive reactance**. It depends on the inductance of the inductor and the frequency of the generator according to the relation

$$X_L = 2\pi fL \qquad (23.4)$$

Equation (23.4) indicates that the larger the frequency of the voltage generator, the larger the inductive reactance, and hence the smaller the rms current. Low frequencies mean small inductive reactances and large currents.

Example 2

A 7.50-mH inductor is in a circuit with a generator which provides an rms voltage of 32.0 V. Find X_L and the rms current at frequencies of (a) 60.0 Hz and (b) 6.00 x 10^3 Hz.

(a) At 60.0 Hz,

$$X_L = 2\pi fL = 2\pi(60.0 \text{ Hz})(7.50 \text{ x } 10^{-3} \text{ H}) = 2.83 \text{ } \Omega$$

$$I_{rms} = \frac{V_{rms}}{X_L} = \frac{32.0 \text{ V}}{2.83 \text{ } \Omega} = 11.3 \text{ A}$$

(b) At 6.00 x 10^3 Hz,

$$X_L = 2\pi fL = 2\pi(6.00 \text{ x } 10^3 \text{ Hz})(7.50 \text{ x } 10^{-3} \text{ H}) = 283 \text{ } \Omega$$

$$I_{rms} = \frac{V_{rms}}{X_L} = \frac{32.0 \text{ V}}{283 \text{ } \Omega} = 0.113 \text{ A}$$

By virtue of its inductive reactance, an inductor affects the amount of current that exists in an ac circuit. Consider the following figure (a) in which we plot a graph of the instantaneous voltage and current versus time for a circuit containing only an inductor.

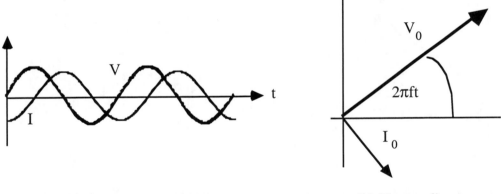

(a) Instantaneous voltage and current   (b) Phasor diagram.
       in an inductive circuit.

At the points where the current is a maximum or a minimum, the current doesn't change much with time, and hence the voltage generated by the inductor to oppose the ac current is zero. At points where the current is zero, the graph is at its steepest, and the current has its largest rate of change. Correspondingly, the voltage generated by the inductor to oppose the ac current has its largest value (positive or negative). Since the current reaches its maximum after the voltage does, it is said that the current **lags** behind the voltage by a phase angle of 90°.

In figure (b) above, the current and voltage phasors remain perpendicular as they rotate, for there is a 90° phase angle between them. The current phasor lags the voltage phasor, relative to the direction of rotation (counterclockwise), in contrast to the situation for the capacitor. Once again, the instantaneous voltage and current are given by the vertical components of the phasors.

23.3 Circuits Containing Resistance, Capacitance, and Inductance

Capacitors, inductors, and resistors can be combined in a single circuit, as shown in the following figure.

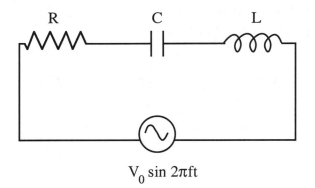

$$V_0 \sin 2\pi ft$$

In a series RCL circuit the total opposition to the flow of charge is called the **impedance** of the circuit. In order to determine how the maximum voltages across the resistor, V_R, the capacitor, V_C, and the inductor, V_L, combine to yield the maximum voltage, V_0, we need to study the phasor diagram for the combination. Kirchhoff's loop rule indicates that the phasors add together to give the total voltage V_0 that is supplied to the circuit by the generator. The addition, however, must be a vector addition to take into account the different directions. Consider the phasor diagram for the RCL circuit shown above.

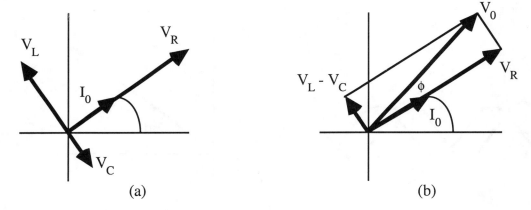

(a)   (b)

In figure (a) the phasors for V_R, V_C, and V_L are shown relative to the maximum current I_0. Notice that the current is in phase with V_R, leads V_C by 90°, and lags behind V_L by 90°. Figure (b) shows how to add the voltages together vectorially to produce V_0, the resultant voltage. From the drawing we see that the total voltage is given by:

$$V_0^2 = V_R^2 + (V_L - V_C)^2$$

The rms voltage of the circuit is obtained by dividing this equation by $(\sqrt{2})^2$ so that

$$V_{rms}^2 = V_R^2 + (V_L - V_C)^2 \qquad (23.5)$$

where the symbols on the right now refer to rms quantities.

Finally, since $V_R = I_{rms}R$, $V_C = I_{rms}X_C$, and $V_L = I_{rms}X_L$, for the entire RCL circuit it follows that

$$V_{rms} = I_{rms} Z \qquad (23.6)$$

$$Z = \sqrt{R^2 + (X_L - X_C)^2} \qquad (23.7)$$

where the quantity Z is known as the **impedance** of the circuit. The **phase angle** ϕ between the current phasor I_0 and the voltage phasor V_0, shown in part (b) of the above diagram, is given by

$$\tan \phi = \frac{X_L - X_C}{R} \qquad (23.8)$$

The average power dissipated by the circuit is given by

$$\overline{P} = I_{rms} V_{rms} \cos \phi \qquad (23.9)$$

where the term $\cos \phi = R/Z$ is called the **power factor** of the circuit.

Example 3
A series RCL circuit consists of a 65.0-Ω resistor, a 3.00-μF capacitor, and a 65.0-mH inductor. The generator has a frequency of 444 Hz and an rms voltage of 18.0 V. Obtain (a) the rms voltage across and current through each element and, (b) the electrical power consumed by the circuit.

(a) First determine the impedance of the circuit,

$$X_C = \frac{1}{2\pi fC} = \frac{1}{2\pi(444 \text{ Hz})(3.00 \times 10^{-6} \text{ F})} = 119 \ \Omega$$

$$X_L = 2\pi fL = 2\pi(444 \text{ Hz})(6.50 \times 10^{-2} \text{ H}) = 181 \ \Omega$$

$$Z = \sqrt{R^2 + (X_L - X_C)^2} = \sqrt{(65.0 \ \Omega)^2 + (181 \ \Omega - 119 \ \Omega)^2} = 9.0 \times 10^1 \ \Omega$$

The rms current in the circuit is therefore,

$$I_{rms} = \frac{V_{rms}}{Z} = \frac{18.0 \text{ V}}{9.0 \times 10^1 \ \Omega} = 0.20 \text{ A}$$

The rms voltages across each device are therefore

$$V_R = I_{rms}R = (0.20 \text{ A})(65.0 \ \Omega) = 13 \text{ V}$$
$$V_C = I_{rms}X_C = (0.20 \text{ A})(119 \ \Omega) = 24 \text{ V}$$
$$V_L = I_{rms}X_L = (0.20 \text{ A})(181 \ \Omega) = 36 \text{ V}$$

(b) To find the power dissipated by the circuit we need to first find the power factor. We know that

$$\cos \phi = \frac{R}{Z} = \frac{65.0 \ \Omega}{9.0 \times 10^1 \ \Omega} = 0.72$$

Therefore, the power dissipated by the circuit is

$$\overline{P} = I_{rms} V_{rms} \cos \phi = (0.20 \text{ A})(18.0 \text{ V})(0.72) = 2.6 \text{ W}$$

23.4   Resonance in Electric Circuits

Under certain conditions in a series RCL circuit a situation known as **resonance** can occur. Resonance occurs when the frequency of the vibrating force exactly matches the natural frequency of the object to which the force is applied. In the electric case the vibrating force is provided by the oscillating electric field that is related to the voltage supplied by the generator. In a resonant series RCL circuit the current reaches its largest possible value. The largest value of I_{rms} occurs when the impedance is at its minimum value. From equation (23.7) we see that this occurs when $X_L = X_C$. Using the definitions for reactance, equations (23.2) and (23.4), the **resonant frequency** f_0 occurs when $2\pi f_0 L = 1/(2\pi f_0 C)$, that is, when

$$f_0 = \frac{1}{2\pi\sqrt{LC}} \tag{23.10}$$

Note that the resonant frequency depends on the inductance and capacitance, but not the resistance.

Example 4
A series RCL circuit consists of a 150-Ω resistor, a 5.5-nF capacitor, and a 48-mH inductor. The circuit is connected to the terminals of an ac oscillator whose output is 75 V and whose frequency is adjustable over a wide range.
(a) What is the resonant frequency of the circuit? (b) How much current flows in the circuit at resonance?
(c) What are the voltages across each of the circuit components?

(a) The resonant frequency is,

$$f_0 = \frac{1}{2\pi\sqrt{LC}} = \frac{1}{2\pi\sqrt{(48 \times 10^{-3} \text{ H})(5.5 \times 10^{-9} \text{ F})}} = 9800 \text{ Hz}$$

(b) At resonance, we see from equation (23.7) that $Z = R$. Therefore,

$$I_{rms} = \frac{V_{rms}}{R} = \frac{75 \text{ V}}{150 \ \Omega} = 0.50 \text{ A}$$

(c) The voltages are: $\quad V_R = I_{rms} R = (0.50 \text{ A})(150 \ \Omega) = 75 \text{ V}$

$$X_C = 1/(2\pi f C) = 1/[2\pi(9800 \text{ Hz})(5.5 \times 10^{-9})] = 3.0 \times 10^3 \ \Omega,$$

$$V_C = (0.50 \text{ A})(3.0 \times 10^3 \text{ W}) = 1500 \text{ V}$$

$$X_L = 2\pi f L = 2\pi(9800 \text{ Hz})(48 \times 10^{-3} \text{ H}) = 3.0 \times 10^3 \ \Omega; \ V_L = (0.50 \text{ A})(3.0 \times 10^3 \ \Omega) = 1500 \text{ V}$$

PRACTICE PROBLEMS

1. A capacitor has a capacitance of 1.90 μF. What is the reactance of this capacitor at frequencies of (a) 333 Hz, (b) 165 Hz, and (c) 66.6 Hz?

2. An inductor has an inductance of 121 mH. What is the reactance of this inductor at frequencies of (a) 333 Hz, (b) 165 Hz, and (c) 66.6 Hz?

3. A 5.0-mH inductor is connected to an ac voltage source whose rms voltage is 24 V. The rms current flowing in the circuit is 1.2 A. What is the frequency of the voltage source?

4. A 2700-Ω resistor and a 1.1 μF capacitor are connected in series across a generator (60.0 Hz, 120 V). Determine the power dissipated in the circuit.

5. A circuit consists of a 215-Ω resistor and a 0.200-H inductor. These two elements are connected in series across a generator that has a frequency of 106 Hz and a voltage of 234 V. (a) What is the current in the circuit? (b) Determine the phase angle between the current and the voltage.

6. A 66 Hz ac generator operates in a series RCL circuit which has an rms current of 1.0 A. The following rms voltages are found across the resistor, capacitor, and inductor, respectively:

$$V_R = 55 \text{ V}, \quad V_C = 32 \text{ V}, \quad V_L = 48 \text{ V}.$$

(a) Determine the source voltage and the phase angle between the current and the voltage. (b) Determine the values of R, C, and L.

7. The elements in a series RCL circuit are a 106-Ω resistor, a 3.30-μF capacitor, and a 31.0-mH inductor. What is the impedance of the circuit and the phase angle between the current and the voltage when the frequency is (a) 215 Hz and (b) 609 Hz? In each case interpret the algebraic sign of the phase angle, remembering that a positive sign means the voltage "leads" the current and the circuit is more inductive than capacitive.

8. A series RCL circuit has a resonance frequency of 55 kHz. If the value of the inductance is 3.5 mH, what is the value of the capacitance?

9. A series RCL circuit contains a 240-V, 120-Hz ac voltage, a 2200-Ω resistor, a coil of 3.0-H inductance, and a capacitance C that can be varied. (a) What value should C have in order for the voltage to lead the current by 45°? (b) At this value of C, find the rms current in the circuit. (c) What is the voltage across the coil?

10. A series RCL circuit has R = 25 Ω, C = 1.8 μF, and L = 6.3 mH. (a) At what frequency of an ac generator will the circuit resonate? (b) What is the rms current at resonance if the ac voltage is 25 V?

SOLUTIONS AND ANSWERS

Practice Problems

1. The capacitive reactance is calculated from equation (23.2), $X_C = 1/(2\pi fC)$. Therefore,

   (a) $X_C = 1/[2\pi(333 \text{ Hz})(1.90 \times 10^{-6} \text{ F})] = \textbf{252 } \Omega$.

   (b) $X_C = 1/[2\pi(165 \text{ Hz})(1.90 \times 10^{-6} \text{ F})] = \textbf{508 } \Omega$.

   (c) $X_C = 1/[2\pi(66.6 \text{ Hz})(1.90 \times 10^{-6} \text{ F})] = \textbf{1260 } \Omega$.

2. (a) $X_L = 2\pi fL = 2\pi(333 \text{ Hz})(0.121 \text{ H}) = \textbf{253 } \Omega$.

   (b) $X_L = 2\pi(165 \text{ Hz})(0.121 \text{ H}) = \textbf{125 } \Omega$.

   (c) $X_L = 2\pi(66.6 \text{ Hz})(0.121 \text{ H}) = \textbf{50.6 } \Omega$.

3. We know

   $$V_{rms} = I_{rms}X_L = I_{rms}(2\pi fL) \Rightarrow f = V_{rms}/(2\pi I_{rms}L) = (24 \text{ V})/[2\pi(1.2 \text{ A})(5.0 \times 10^{-3} \text{ H})] = \textbf{640 Hz}.$$

4. Find the impedance using $Z^2 = R^2 + X_C^2$ where

   $$X_C = 1/(2\pi fC) = 1/[2\pi(60.0 \text{ Hz})(1.1 \times 10^{-6} \text{ F})] = 2400 \text{ } \Omega.$$

   Thus,
   $$Z = \sqrt{(2700 \text{ } \Omega)^2 + (2400 \text{ } \Omega)^2} = 3600 \text{ } \Omega.$$

   The power factor is
   $$\cos \phi = R/Z = (2700 \text{ } \Omega)/(3600 \text{ } \Omega) = 0.75.$$

   The power is
   $$P = I_{rms}V_{rms} \cos \phi = (V_{rms}^2/Z) \cos \phi = (120 \text{ V})^2 (0.75)/(3600 \text{ } \Omega) = \textbf{3.0 W}.$$

5. Find the impedance of the circuit using $Z^2 = R^2 + X_L^2$ where

   $$X_L = 2\pi fL = 2\pi(106 \text{ Hz})(0.200 \text{ H}) = 133 \text{ } \Omega.$$

   Thus,
   $$Z = \sqrt{(215 \text{ } \Omega)^2 + (133 \text{ } \Omega)^2} = 253 \text{ } \Omega.$$

   We can now find the current and phase angle:

   (a) The rms current is therefore,

   $$I_{rms} = V_{rms}/Z = (234 \text{ V})/(253 \text{ } \Omega) = \textbf{0.925 A}.$$

   (b) Since $\cos \phi = R/Z$,
   $$\phi = \cos^{-1} [(215 \text{ } \Omega)/(253 \text{ } \Omega)] = \textbf{31.8°}.$$

6.   Find the reactances using $V_{rms} = I_{rms}X$. So,

$$R = V_R/I = (55 \text{ V})/(1.0 \text{ A}) = 55 \ \Omega;$$

$$X_C = V_C/I = (32 \text{ V})/(1.0 \text{ A}) = 32 \ \Omega; X_L = V_L/I = (48 \text{ V})/(1.0 \text{ A}) = 48 \ \Omega.$$

(a) Equation (23.5),

$$V_{rms}^2 = V_R^2 + (V_L - V_C)^2 = (55 \text{ V})^2 + (48 \text{ V} - 32 \text{ V})^2 \Rightarrow V_{rms} = \textbf{57 V}.$$

Equation (23.8),

$$\tan \phi = (X_L - X_C)/R = (16 \ \Omega)/(55 \ \Omega) = 0.29 \Rightarrow \phi = \tan^{-1}(0.29) = \textbf{16}°.$$

(b) We know

$$X_C = 1/(2\pi f C) \Rightarrow C = 1/(2\pi f X_C) = 1/[2\pi(66 \text{ Hz})(32 \ \Omega)] = \textbf{75 }\mu\textbf{F}.$$

$$X_L = 2\pi f L \Rightarrow L = X_L/(2\pi f) = (48 \ \Omega)/[2\pi(66 \text{ Hz})] = \textbf{0.12 H}. \text{ Of course, R} = \textbf{55 }\Omega.$$

---

7.   (a) At 215 Hz; $X_L = 2\pi f L = 41.9 \ \Omega$, $X_C = 1/(2\pi f C) = 224 \ \Omega$, R = 106 Ω, and

$$Z^2 = R^2 + (X_L - X_C)^2 = (106 \ \Omega)^2 + (41.9 \ \Omega - 224 \ \Omega)^2 \Rightarrow Z = \textbf{211 }\Omega.$$

We also know,

$$\tan \phi = (X_L - X_C)/R \Rightarrow \phi = \tan^{-1}(-182 \ \Omega/106 \ \Omega) = \textbf{-59.8}°.$$

Since the voltage LAGS the current by 59.8°, the circuit is **capacitive**.

(b) At 609 Hz; $X_L = 119 \ \Omega$, $X_C = 79.2 \ \Omega$, so that

$$Z^2 = (106 \ \Omega)^2 + (119 \ \Omega - 79.2 \ \Omega)^2 \Rightarrow Z = \textbf{113 }\Omega.$$

$$\phi = \tan^{-1} (39.8 \ \Omega/106 \ \Omega) = \textbf{+20.6}°.$$

So the voltage leads the current and the circuit is **inductive**.

---

8.   Using equation (23.10),

$$f_0 = 1/[2\pi\sqrt{(LC)}] \Rightarrow C = 1/(4\pi^2 f_0^2 L) = 1/[4\pi^2(55 \times 10^3 \text{ Hz})^2 (3.5 \times 10^{-3} \text{ H})] = \textbf{2.4 nF}.$$

---

9.   (a) Since $\tan \phi = (X_L - X_C)/R$ and since $\phi = 45°$, we have

$$X_C = X_L - R \tan 45° \Rightarrow 1/(2\pi f C) = 2\pi f L - R \tan 45°.$$

Solving,

$$C = 1/(4\pi^2 f^2 L - 2\pi f R \tan 45°) = 1/[4\pi^2(120 \text{ Hz})^2(3.0 \text{ H}) - 2\pi(120 \text{ Hz})(2200 \ \Omega) \tan 45°] = \textbf{21 }\mu\textbf{F}.$$

(b) The impedance of the circuit is given by

$$Z^2 = (2200 \ \Omega)^2 + (2200 \ \Omega)^2 \Rightarrow Z = 3100 \ \Omega.$$

Therefore,

$$I_{rms} = V_{rms}/Z = (240 \text{ V})/(3100 \text{ } \Omega) = \mathbf{0.077 \text{ A}}.$$

(c) We know

$$V_L = I_{rms}X_L = (0.077 \text{ A})(2300 \text{ } \Omega) = \mathbf{180 \text{ V}}.$$

---

10.   (a) The resonance frequency is

$$f_0^2 = 1/(4\pi^2 LC) = 1/[4\pi^2(6.3 \times 10^{-3} \text{ H})(1.8 \times 10^{-6} \text{ F})] \Rightarrow f_0 = \mathbf{1500 \text{ Hz}}.$$

(b) At resonance, $X_L = X_C$ so that $Z = R$. Therefore, the rms current is given by

$$I_{rms} = V_{rms}/Z = (25 \text{ V})/(25 \text{ } \Omega) = \mathbf{1.0 \text{ A}}.$$

CHAPTER 24
Electromagnetic Waves

PREVIEW

In this chapter you are introduced to electromagnetic waves, of which radio waves, infrared radiation, visible light, ultraviolet radiation, X-rays and gamma rays are familiar examples. Electromagnetic waves as oscillating electric and magnetic fields moving through space are investigated. Evidence that electromagnetic waves are transverse waves is demonstrated by an analysis of the polarization of the waves.

QUICK REFERENCE

Important Terms

Electromagnetic wave
Fluctuating electric and magnetic fields which propagate together.

Near field
The electric and magnetic fields which exist in the region close to an antenna.

Radiation field
The electric and magnetic fields formed far from an antenna which propagate as an electromagnetic wave.

Electromagnetic spectrum
Electromagnetic waves arranged in order of their frequency.

Intensity
The intensity of an electromagnetic wave is the total energy per unit time per unit area carried by an electromagnetic wave.

Unpolarized wave
A wave in which the vibrations can occur randomly in any direction.

Linearly polarized wave
A wave in which the vibrations always occur along one direction.

Polarizing material
A material which produces a linearly polarized wave.

Transmission axis
The transmission axis of a polarizing material is the preferred transmission direction of an electric field passing through the material.

Equations

The **speed** of an **electromagnetic wave**

$$c = \frac{1}{\sqrt{\varepsilon_0 \mu_0}} \qquad (24.1)$$

The **total energy density** of an **electromagnetic wave**:

$$u = \frac{1}{2}\varepsilon_0 E^2 + \frac{1}{2\mu_0}B^2 \qquad (24.2a)$$

$$u = \varepsilon_0 E^2 \qquad (24.2b)$$

$$u = \frac{1}{\mu_0}B^2 \qquad (24.2c)$$

The **relationship** between the magnitudes of the **electric field** and the **magnetic field** in an electromagnetic wave:

$$E = cB \qquad (24.3)$$

The **intensity** of an **electromagnetic wave**:

$$S = cu \qquad (24.4)$$

$$S = cu = \frac{1}{2}c\varepsilon_0 E^2 + \frac{c}{2\mu_0}B^2 \qquad (24.5a)$$

$$S = c\varepsilon_0 E^2 \qquad (24.5b)$$

$$S = \frac{c}{\mu_0}B^2 \qquad (24.5c)$$

Malus' Law gives the **average intensity** of light **through an analyzer**:

$$\bar{S} = \bar{S_0}\cos^2\theta \qquad (24.7)$$

DISCUSSION OF SELECTED SECTIONS

24.1 The Nature of Electromagnetic Waves

Oscillating currents in wires produce oscillating magnetic fields in the region about the wire. Faraday's Law requires that a changing magnetic flux through a surface, not necessarily a "real" surface, induce an emf around a loop enclosing the surface. An emf is a potential difference and is associated with an electric field. Hence, an oscillating current results in an oscillating magnetic field which produces an oscillating electric field. Oscillating electric fields can also produce oscillating magnetic fields. Once the process is started the oscillating magnetic and electric fields produce each other and can become detached from the original source (the current carrying wire) and travel through space as an independent wave - an electromagnetic wave.

24.3 The Speed of Light

Example 1
Radio waves are a form of electromagnetic radiation. If a spacecraft leaves the earth and travels a great distance, some time is required for radio transmissions from the earth to reach the spacecraft and for a reply by the spacecraft to

return to the earth. What is the minimum delay in the reply to a message sent from earth if the spacecraft is close to Mars and is 3.78×10^{11} m from the earth?

The time that the radio waves take to travel from the earth to the spacecraft is

$$t = s/c = (3.78 \times 10^{11} \text{ m})/(3.00 \times 10^8 \text{ m/s}) = 1.26 \times 10^3 \text{ s.}$$

The time taken for the reply to travel from the spacecraft to the earth is the same. The total time that the radio waves spend in transit is then 2.52×10^3 s which is 42 min. This is the total time required for the reply to reach the earth even if the spacecraft broadcasts the reply with no delay.

Example 2
A distance is to be measured by reflecting a pulse of laser light from a mirror and measuring the total time taken for the pulse to return to a detector at the site of the laser. How much time elapses for a distance of 152 m?

The total time required for the pulse of laser light to return to the detector is

$$t = 2(152 \text{ m})/(3.00 \times 10^8 \text{ m/s}) = 1.01 \times 10^{-6} \text{ s.}$$

This time, though short, is well within the measurement capabilities of modern electronics.

24.4 The Energy Carried by Electromagnetic Waves

The energy density in an electromagnetic wave is just the sum of the energy densities of the electric and magnetic fields. The magnitudes of the electric and magnetic fields are related by $E = cB$. This gives equations (24.2) which can be used to calculate the energy density of an electromagnetic wave. Note that E and B in these equations are sinusoidal functions of time, hence u is also a function of time. The equations can be interpreted as giving the average energy density if the rms values of E and B are used.

Example 3
What is the average energy density in an electromagnetic wave whose maximum magnetic field is 9.33×10^{-7} T? We must use the rms values of E and B to get the average energy density in the wave.
$B_{rms} = B_0/\sqrt{2} = 6.60 \times 10^{-7}$ T. Then equation (24.2c) gives the

$$\text{average energy density} = B^2_{rms}/\mu_0 = (6.60 \times 10^{-7} \text{ T})^2/(4\pi \times 10^{-7} \text{ T m/A})$$
$$= 3.47 \times 10^{-7} \text{ J/m}^3.$$

The intensity of an electromagnetic wave is the energy that it carries per second per square meter. Equations (24.5) give the time varying intensity. Again these equations may be used to find average intensities if the rms values of E and B are used.

Example 4
Sunlight falling on the ground on a certain day has an average intensity of 1100 W/m². What are the maximum values of the electric and magnetic fields in the sunlight?

Equation (24.5b) can be used to find the rms value of E if S is interpreted as being the average intensity of the light.

$$E_{rms} = \sqrt{\frac{\overline{S}}{c\varepsilon_0}} = \sqrt{\frac{1100 \text{ W/m}^2}{(3.00 \times 10^8 \text{ m/s})(8.85 \times 10^{-12} \text{ C}^2/\text{N m}^2)}} = 640 \text{ N/C}$$

The maximum value of the electric field is then

$$E_{max} = \sqrt{2} \ E_{rms} = 910 \text{ N/C}$$

Similarly, equation (24.5c) is used to find the rms value of the magnetic field.

$$B_{rms} = \sqrt{\frac{\mu_0 \overline{S}}{c}} = \sqrt{\frac{(4\pi \times 10^{-7} \text{ T m/A})(1100 \text{ W/m}^2)}{3.00 \times 10^8 \text{ m/s}}} = 2.1 \times 10^{-6} \text{ T}$$

The maximum magnetic field is then

$$B_{max} = \sqrt{2} \ B_{rms} = 3.0 \times 10^{-6} \text{ T}$$

24.6 Polarization

Only transverse waves may be polarized. Many sources of electromagnetic waves produce a random polarization. That is, the direction of the electric field vector may change randomly. If all possible electric field directions are filtered out except for one, then the electromagnetic wave is linearly polarized. A polarizing material in essence does this filtering. It limits the electric field oscillations to a single direction by selectively transmitting it and reflecting or absorbing all others.

Example 5
A device made from polarizing material is often called a polarizer. The intensity of unpolarized light reaching a polarizer is S. Show that the intensity of the light transmitted through the polarizer is 1/2 S.

The randomly polarized light has an electric field vector which oscillates in all possible directions. At any given time the electric field vector of the light wave can be resolved into a component along the transmission axis of the polarizer and a component perpendicular to the transmission axis. Since the light is randomly polarized, these two components will, on the average be equal. The component parallel to the transmission axis is

$$E \cos 45° = \frac{1}{\sqrt{2}} \ E.$$

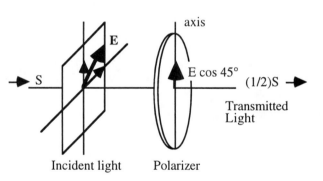

This is the only component transmitted and corresponds to an intensity, given by equation (24.5b), of $(1/2)c\varepsilon_0 \ E^2$. The intensity of the light incident on the polarizer is $S = c\varepsilon_0 \ E^2$. Comparing, we see that the intensity of the light transmitted through the polarizer is $(1/2) \ S$.

Malus' Law, equation (24.7), gives the average intensity of previously polarized light transmitted through a sheet of polarizing material whose transmission axis is an angle, θ, to the electric field direction of the light. If the light were originally polarized by a sheet of polarizer, the second sheet of polarizing material is referred to as an analyzer.

Example 6
When the axes of two sheets of polarizing material are perpendicular, or crossed (see drawing), no light is transmitted. However, when a third sheet is placed between the other two and has its axis tilted with respect to both of them, then light is transmitted. Explain.

When two sheets of polarizing materials are crossed, the first transmits only the electric field component along its transmission axis, hence no component of electric field exists along the axis of the second sheet. No electric field is transmitted through the pair, so the average intensity of the transmitted light is zero.

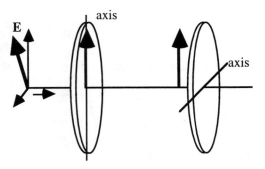

When the third sheet is introduced between the crossed pair, it transmits the component of the electric field along its axis. This electric field now has a component along the axis of the last sheet and some light will be transmitted.

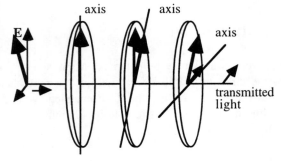

PRACTICE PROBLEMS

1. A certain star is located 2.93×10^{15} miles from the earth. How long ago did the light reaching the earth leave the star?

2. Television transmissions leaving the earth could possibly be received by inhabitants of another planet in a galaxy far far away. Imagine that somewhere there are intelligent creatures watching an episode of *I Love Lucy* which was broadcast in 1952. In 1988 how far in miles is the planet from the earth? Ignore the extra days in leap years.

3. A certain laser produces a cylindrical beam of light with a diameter of 1.5 mm and an average power of 1.5 W. The laser beam can be focused by a lens to a circular spot 2.0 μm in radius. (a) What is the average intensity of the focused beam? (b) What are the maximum values of the electric and magnetic fields in the focused beam?

4. Unpolarized light of average intensity, I_0, is incident on polarizing sheets with crossed axes. A third sheet is placed between the other two and has its transmission axis oriented at an angle, θ, to the transmission axis of the first polarizer. Find an expression for the average intensity of light, I, transmitted through the system as a function of the angle, θ.

5. The arrangement in Figure 24.6 can be used as a simple radio transmitter if an energy source is connected across the transformer in place of the amplifier circuit. The energy source excites the LC circuit to produce a current in the coil which oscillates at the natural frequency of the circuit. This oscillating current produces an electromagnetic wave which travels away from the antenna. What frequency of radio wave can be produced by a circuit in which L = 0.15 mH and C = 4.7 pF?

6. Linearly polarized light falls on a sheet of polarizing material whose transmission axis is oriented at an angle, θ, to the direction of polarization of the light. What fraction of the incident intensity is transmitted through the sheet for $\theta = 15.0°$, $30.0°$, $60.0°$ and $75.0°$?

7. A vertical radio antenna broadcasts vertically polarized radio waves. What fraction of the radio wave intensity arriving at a receiving antenna is available to produce a current if the receiving antenna is oriented 35° to the vertical?

8. A flat coil of wire is used with an LC-tuned circuit as a receiving antenna. The coil has a radius of 25 cm and consists of 450 turns. The transmitted radio wave has a frequency of 1.2 MHz. The magnetic field of the wave is parallel to the normal of the coil and has a maximum value of 2.0×10^{-13} T. Using Faraday's law of induction estimate the magnitude of the average emf induced in the antenna during one quarter of a wave period.

SOLUTIONS AND ANSWERS

Practice Problems

---

1. The time taken for the light to travel from the star to the earth is

$$t = s/c = (2.93 \times 10^{15} \text{ mi})(1.609 \times 10^3 \text{ m/1 mi})/(3.00 \times 10^8 \text{ m/s})$$
$$t = \mathbf{1.57 \times 10^{10}} \text{ s} = \mathbf{498} \text{ y.}$$

---

2. The distance traveled by the light in its trip from the earth to the planet is

$$s = ct = (3.00 \times 10^8 \text{ m/s})(6.22 \times 10^{-4} \text{ mi/1 m})(36 \text{ y})(365 \text{ d/1 y})(24 \text{ h/1 d})(3600 \text{ s/1 h})$$
$$s = \mathbf{2.12 \times 10^{14}} \text{ mi.}$$

---

3. (a) The average intensity of the concentrated beam is

$$\overline{S} = \frac{P}{A} = \frac{1.5 \text{ W}}{\pi (2.0 \times 10^{-6} \text{ m})^2} = \mathbf{1.2 \times 10^{11}} \text{ W/m}^2.$$

(b) Equation (24.5b) written for the average intensity gives

$$E_m = \sqrt{2}E_{rms} = \sqrt{\frac{2\overline{S}}{c\varepsilon_0}} = \sqrt{\frac{2 (1.2 \times 10^{11} \text{ W/m}^2)}{(3.00 \times 10^8 \text{ m/s})(8.85 \times 10^{-12} \text{ C}^2/\text{N m}^2)}} = \mathbf{9.5 \times 10^6} \text{ N/C}$$

Equation (24.3) then gives
$$B_m = E_m/c = \mathbf{0.032} \text{ T.}$$

---

4. The unpolarized incident light has an electric field E_0. Only 1/2 S_0 of the light is transmitted through the first polarizing sheet and its electric field is $1/\sqrt{2}$ E_0. Only the component of the electric field along the second polarizing sheet's axis is transmitted, that is, $1/\sqrt{2}$ $E_0 \cos \theta$. The transmission axis of the last polarizing sheet is at an angle of 90° - θ to the electric field, so the component of the field which is transmitted through the last polarizing sheet is $1/\sqrt{2}$ $E_0 \cos \theta \sin \theta$. The intensity of the transmitted light is then $S = cE^2$ or

$$S = 1/2 \ c_0E_0^2 \cos^2 \theta \sin^2 \theta = \tfrac{1}{2}I_0\cos^2 \theta \ \sin^2 \theta.$$

---

5. The frequency of the radio wave produced by the oscillating current has the same frequency as the current.

$$f = \frac{\omega}{2\pi} = \frac{1}{2\pi}\sqrt{\frac{1}{LC}} = \frac{1}{2\pi}\sqrt{\frac{1}{(0.15 \times 10^{-3} \text{ H})(4.7 \times 10^{-12} \text{ F})}} = \mathbf{6.0 \times 10^6} \text{ Hz}$$

---

6. Malus' law, equation (24.7) gives

$$\frac{\overline{S}}{S_0} = \cos^2 \theta$$

which gives for: $\theta = 15.0°$, **0.933**; $\theta = 30.0°$, **0.750**; $\theta = 60.0°$, **0.250**; $\theta = 75.0°$, **0.0670**.

---

7. Malus's law applies since the radio waves are already linearly polarized. The fraction of the radio wave intensity available to produce a current is $\cos^2 35° = $ **0.67**.

---

8. The flux through the coil changes from 0 Wb to $B_m NA$ in 1/4 of a wave period, T. The magnitude of the average emf induced in the coil during this time is, according to Faraday's law,

$$emf = B_m NA/(1/4 \ T)$$

or

$$emf = 4B_m NAf = 4(2.0 \times 10^{-13} \ T)(450)(\pi)(25 \times 10^{-2} \ m)^2 (1.2 \times 10^6 \ Hz)$$

$$emf = \textbf{8.5} \times \textbf{10}^{-5} \ \textbf{V}.$$

CHAPTER 25
The Reflection of Light: Mirrors

PREVIEW

In this chapter the reflection of light from mirrors is discussed. You will see how reflections are formed by plane and spherical mirrors, how to construct ray diagrams for reflecting surfaces, and how to determine magnification.

QUICK REFERENCE

Important Terms

Wave fronts
Surfaces on which all points of a wave are in the same phase of motion.

Rays
Lines that are perpendicular to the wave fronts and point in the direction of the velocity of the wave.

Law of reflection
For light that reflects off a smooth surface, (a) the incident ray, the reflected ray, and the normal to the surface all lie in the same plane, and (b) the angle of reflection equals the angle of incidence.

Virtual image
An image from which rays of light do not actually come, but only appear to do so.

Real image
An image from which rays of light actually emanate.

Plane mirror
A flat reflecting surface which forms an upright, virtual image that is located as far behind the mirror as the object is in front of the mirror. In addition, the heights of the image and the object are equal.

Spherical mirror
A reflecting surface that has the shape of a section from the surface of a sphere.

Principal axis
The straight line drawn through the center of curvature and the middle of the mirror's surface.

Paraxial rays
Rays that lie close to the principal axis.

Radius of curvature
The distance from the center of curvature to the mirror.

Focal point
A point on the principal axis where paraxial rays that are parallel to the principal axis converge after being reflected from a concave mirror. For a convex mirror, it is the point on the principal axis from which such rays appear to emanate.

Focal length
The distance from the focal point to the middle of the mirror.

Magnification
The ratio of the image height to the object height.

Equations

The **focal length** of a **concave mirror**:

$$f = \frac{1}{2}R \qquad\qquad (25.1)$$

The **focal length** of a **convex mirror**:

$$f = -\frac{1}{2}R \qquad (25.2)$$

The **mirror equation**:

$$\frac{1}{d_o} + \frac{1}{d_i} = \frac{1}{f} \qquad (25.3)$$

The **magnification equation**:

$$m = \frac{\text{Image height}}{\text{Object height}} = -\frac{d_i}{d_o} \qquad (25.4)$$

DISCUSSION OF SELECTED SECTIONS

25.2 The Reflection of Light

Suppose a ray of light is incident on a flat, reflecting surface, such as the mirror shown in the diagram. The angle of incidence θ_i is the angle that the incident ray makes with respect to the normal to the surface. The angle of reflection θ_r is the angle that the reflected ray makes with the normal. The **law of reflection** states that the incident ray, the reflected ray, and the normal to the surface all lie in the same plane, and the angle of reflection equals the angle of incidence.

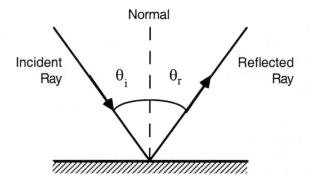

Example 1

Two plane mirrors are separated by 120°, as the drawing illustrates. If a ray strikes mirror M_1, at a 65° angle of incidence, at what angle θ does it leave mirror M_2?

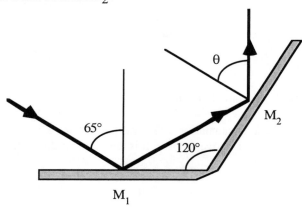

The ray leaves M_1 at an angle of 65° with respect to the normal, or 90° - 65° = 25° with respect to the plane of M_1. The first reflected ray therefore strikes M_2 at an angle of 180° - (120° + 25°) = 35° with respect to the plane of M_2. The second angle of incidence is 90° - 35° = 55°. Finally, the angle of reflection from M_2, θ, is equal to the angle of incidence for mirror M_2, which is just $\theta = 55°$.

25.3 The Formation of Images by a Plane Mirror

When you look into a plane mirror, you see an image of yourself which is upright, has left-right reversal, is located as far behind the mirror as you are in front of it, and is the same size as you are. To illustrate how this image is formed, consider the following **ray diagram**.

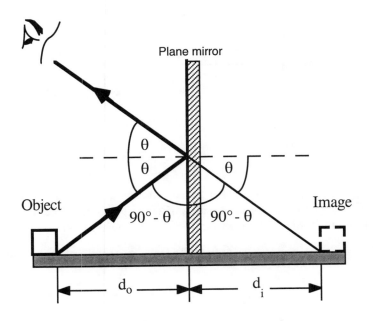

A ray of light leaving the bottom of the object reflects from the mirror and enters the eye. To the eye, it seems that the ray originates from behind the mirror, as shown. However, rays of light do not actually emanate from behind the mirror, and we refer to this as a **virtual image**. All plane mirrors produce virtual images.

From the diagram, it can be demonstrated that the image is located as far behind a plane mirror as the object is in front of it. That is, the object distance d_o equals the image distance d_i. In addition, by drawing a ray of light from the top of the object, rather than from the bottom, we find that the height of the image equals the height of the object.

25.4 Spherical Mirrors

A spherical mirror has the shape of a section from the surface of a sphere. If the inside or concave surface of the mirror is polished, we have a **concave mirror**, shown below.

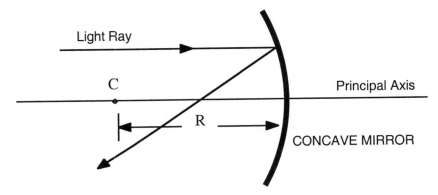

If the outside or convex surface is polished, we have a **convex mirror**, as in the figure. Each type of mirror has a radius of curvature R, where the center of curvature is located at point C. The **principal axis** of the mirror is a straight line drawn through C and the midpoint of the mirror.

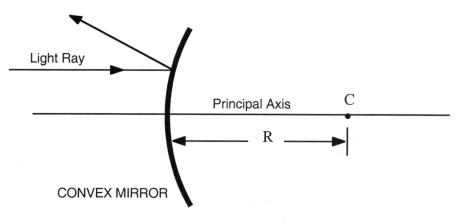

If light rays incident on a spherical mirror are parallel to the principal axis, and lie close to this axis, the reflected rays all pass through a common point, called the **focal point** F. The distance between this focal point and the mirror is called the **focal length** f of the mirror, as shown in the following figure.

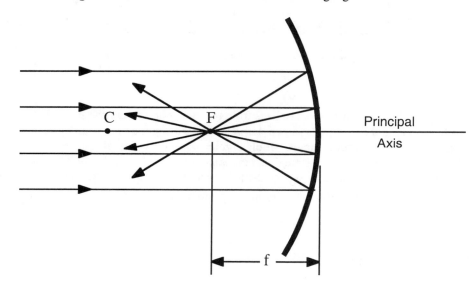

The focal length f and radius of curvature R for spherical mirrors are related. The relationships are as follows:
For a concave mirror:

$$f = \frac{1}{2}R \tag{25.1}$$

For a convex mirror:

$$f = -\frac{1}{2}R \tag{25.2}$$

Thus, the magnitude of the focal length is one-half the radius of curvature. It should be pointed out, however, that equations (25.1) and (25.2) are only valid for rays that lie close to the principal axis, that is, for **paraxial rays**. For spherical mirrors, rays which lie too far from the principal axis will not be reflected through the focal point, and this produces blurred images. This effect is known as **spherical aberration**. Also note that the focal length for convex mirrors is taken by convention to be a negative number.

25.5 The Formation of Images by Spherical Mirrors

When locating the position and size of an image formed by a **concave** mirror, three paraxial rays are useful.

Ray 1. A ray initially parallel to the principal axis, which will pass through the focal point F upon reflection.

Ray 2. A ray passing through the focal point, which will be reflected parallel to the principal axis, according to the principle of reversibility.

Ray 3. A ray traveling through the center of curvature C, which strikes the mirror perpendicularly, and is therefore reflected back on itself.

The procedure used to locate the image is as follows: Draw each of the three rays discussed above from a single point on the object (for example the top of the object). Upon reflection, these rays will converge at a point. This point will then represent the position of the top of the image, as shown in the following figure.

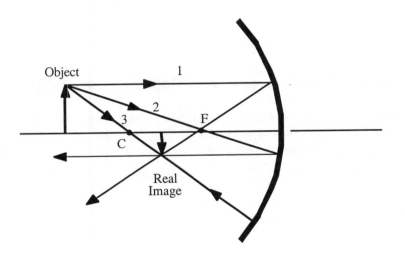

Notice that the image formed by the concave mirror above is a real image, is inverted, and smaller than the object. The image is real because the light rays actually pass through the image. It is possible, under certain conditions, to obtain a virtual image with a concave mirror. This occurs when the object is placed between the focal point F and the mirror.

The procedure for determining the location and size of an image for a **convex** mirror uses the same three paraxial rays discussed above. However, the focal point and center of curvature lie behind the mirror. Therefore, we have:

Ray 1. A ray initially parallel to the principal axis, which appears to originate from the focal point F after reflection from the mirror.

Ray 2. This ray heads toward F, emerging parallel to the principal axis after reflection.

Ray 3. A ray initially traveling toward the center of curvature C, which strikes the mirror perpendicularly and reflects back on itself.

The three rays discussed above appear to come from a point on a virtual image that is behind the mirror. The virtual image is upright, and smaller that the object, as shown in the following ray diagram.

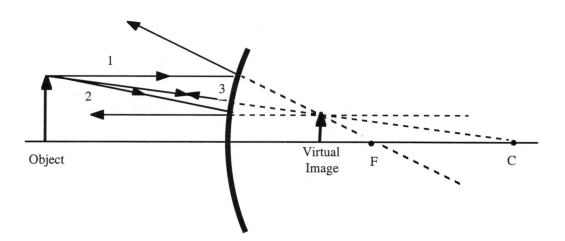

25.6   The Mirror Equation and the Magnification Equation

Ray diagrams are useful in determining the properties of the image formed by spherical mirrors. However, a more precise description of the image is available. Consider an object located a distance d_o from a spherical mirror, having a height h_o. The image formed by the mirror is located a distance d_i from the mirror and has a height h_i. If f is the focal length of the mirror, the object distance, image distance, and focal length are related by the **mirror equation**:

$$\frac{1}{d_o} + \frac{1}{d_i} = \frac{1}{f} \qquad (25.3)$$

The **magnification** m of a mirror is defined as the ratio of the image height to the object height: $m = h_i/h_o$. It can be shown (see proof in text) that the ratio of the image and object heights is equal to the ratio of the image and object distances. That is, $h_i/h_o = d_i/d_o$. The **magnification equation** can therefore be written as

$$m = \frac{\text{Image height}}{\text{Object height}} = - \frac{d_i}{d_o} \qquad (25.4)$$

By convention, the minus sign is inserted in equation (25.4) to help determine whether the object is upright or inverted. The magnification is positive if the image is upright, and negative if the image is inverted with respect to the object.

Example 2
An object is placed 40.0 cm in front of a concave mirror whose radius of curvature is 30.0 cm. Find the position of the image and the magnification.

First find the focal length of the concave mirror using equation (25.1):

$$f = \frac{1}{2}R = \frac{1}{2}(30.0 \text{ cm}) = 15.0 \text{ cm}$$

We can now determine the position of the image using the mirror equation,

$$\frac{1}{d_i} = \frac{1}{f} - \frac{1}{d_o} = \frac{1}{15.0 \text{ cm}} - \frac{1}{40.0 \text{ cm}} = 0.0417 \text{ cm}^{-1} \implies d_i = 24.0 \text{ cm}$$

Since d_i is a positive quantity, the image is **real**. To find the magnification use equation (25.4),

$$m = -\frac{d_i}{d_o} = -\frac{24.0 \text{ cm}}{40.0 \text{ cm}} = -0.600$$

So the image is only 60.0% as large as the object. The minus sign indicates that the image is **inverted** with respect to the object.

Example 3
An object is placed 10.0 cm in front of a concave mirror of focal length 15.0 cm. Find the image distance and magnification.

The mirror equation gives us:

$$\frac{1}{d_i} = \frac{1}{f} - \frac{1}{d_o} = \frac{1}{15.0 \text{ cm}} - \frac{1}{10.0 \text{ cm}} = -0.033 \text{ cm}^{-1} \implies d_i = -3.0 \times 10^1 \text{ cm}$$

So the image is a **virtual image** as the minus sign for d_i indicates. The magnification is

$$m = -\frac{d_i}{d_o} = -\frac{(-3.0 \times 10^1 \text{ cm})}{10.0 \text{ cm}} = 3.0$$

So the image is 3.0 times larger than the object. The fact that the magnification is positive in this case means the image is upright with respect to the object.

Example 4
An object is placed 75 cm in front of a convex mirror of focal length - 55 cm. Find the properties of the image.

The image distance is:

$$\frac{1}{d_i} = \frac{1}{f} - \frac{1}{d_o} = \frac{1}{-55 \text{ cm}} - \frac{1}{75 \text{ cm}} = -0.032 \text{ cm}^{-1} \implies d_i = -31 \text{ cm}$$

The image is of course virtual, and the magnification is

$$m = -\frac{d_i}{d_o} = -\frac{(-31 \text{ cm})}{75 \text{ cm}} = 0.41.$$

So the image is upright (m is positive) and smaller than the object (m is less than unity).

PRACTICE PROBLEMS

1. A 179-cm tall boy wants to buy a mirror that is tall enough so he can see himself full length in the mirror. What is the minimum height of such a mirror?

2. You hold a plane mirror 1.0 m in front of your eyes and are able to see a 15 m high tree behind you. If the mirror is 20.0 cm high, and the tree image completely fills the mirror, how far are you standing from the tree?

3. The image of an object viewed in a concave mirror of focal length 25.0 cm appears 75.0 cm in front of the mirror. Find the location of the object and the magnification.

4. An object is placed 35 cm in front of a convex mirror of focal length - 25 cm. Find the location and magnification of the image.

5. A concave makeup mirror is designed so the virtual image is twice the size of the object, when the distance between the object and mirror is 15 cm. (a) Determine the radius of curvature of the mirror. (b) Draw a ray diagram to scale, showing this situation.

6. Complete the following table. As an example, the first row of the table shows a completed set of data.

Type of Mirror	Radius of Curvature	Focal Length	Object Distance	Image Distance	Real?	Inverted?	Magnif.
Concave	20.0 cm	+10.0 cm	+6.67 cm	-20.0 cm	NO	NO	+ 3.0
Plane	——	——	+45 cm				
	50.0 cm		+5.0 cm				+ 0.84
				+75 cm			- 3.0
		+20.0 cm		-40.0 cm			
Convex			-10.0 cm				0.33 (sign?)
Concave	30.0 cm			+150 cm			

SOLUTIONS AND ANSWERS

Practice Problems

---

1.   Referring to example 1 in the textbook it is shown that a mirror needs to be one-half the height of the person in order to view his entire height in the mirror. Therefore, the height of the mirror is **89.5 cm**.

---

2.   The ray diagram can be drawn in the following way:

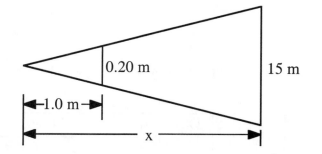

To find the distance x, use the two similar triangles to obtain the following ratio:

$$\frac{x}{1.0 \text{ m}} = \frac{15 \text{ m}}{0.20 \text{ m}} \implies x = \frac{(15 \text{ m})(1.0 \text{ m})}{0.20 \text{ m}} = 75 \text{ m}$$

Therefore, the tree is 75 - 1.0 m = **74 m** behind you.

---

3.   We have
$$1/d_o = 1/f - 1/d_i = 1/(25.0 \text{ cm}) - 1/(75.0 \text{ cm}) = 0.0267 \text{ cm}^{-1} \implies d_o = \textbf{37.5 cm.}$$

The magnification is
$$m = - d_i/d_o = - (75.0 \text{ cm})/(37.5 \text{ cm}) = \textbf{- 2.00.}$$

---

4.   We have
$$1/d_i = 1/f - 1/d_o = 1/(- 25 \text{ cm}) - 1/(35 \text{ cm}) = - 0.069 \text{ cm}^{-1} \implies d_i = \textbf{- 15 cm.}$$

The magnification is
$$m = - d_i/d_o = - (- 15 \text{ cm})/(35 \text{ cm}) = \textbf{+ 0.43.}$$

---

5.   The magnification is $m = - d_i/d_o = + 2$ where d_o is 15 cm. Therefore, $d_i = - 2d_o = - 3.0 \times 10^1$ cm. Therefore, the focal length is found from $1/f = 1/d_o + 1/d_i$ which yields f = + 30 cm. Therefore, R = 2f = **60 cm**.

---

6. The completed table is shown below.

Type of Mirror	Radius of Curvature	Focal Length	Object Distance	Image Distance	Real?	Inverted?	Magnif.
Concave	20.0 cm	+10.0 cm	+6.67 cm	-20.0 cm	NO	NO	+ 3.0
Plane	——	——	+45 cm	-45 cm	NO	NO	+ 1.0
Convex	50.0 cm	- 25.0 cm	+5.0 cm	-4.2 cm	NO	NO	+ 0.84
Concave	38 cm	+19 cm	+25 cm	+75 cm	YES	YES	- 3.0
Concave	40.0 cm	+20.0 cm	+13.3 cm	-40.0 cm	NO	NO	+ 3.0
Convex	30.0 cm	-15.0 cm	+30.0 cm	-10.0 cm	NO	NO	+0.33
Concave	30.0 cm	+15.0 cm	+16.7 cm	+150 cm	YES	YES	- 9.0

CHAPTER 26
The Refraction of Light: Lenses and Optical Instruments

PREVIEW

In this chapter you will be introduced to one of the most useful properties of light: refraction at an interface between two transparent materials. Snell's law governs how much refraction occurs at an interface. You will learn to apply Snell's law to find how much the path of a light ray is bent in passing through an interface and how much the path of a light ray is displaced when passing through a pane of glass.

A special case of Snell's law when the angle of incidence is sufficiently large leads to the phenomenon of total internal reflection at the interface. This phenomenon has importance in the transmission of light through optical fibers and is responsible for the brilliance of the light reflected from a diamond.

The polarization of light by reflection from a boundary is discussed as well as the dispersion of light into its rainbow of colors when it is refracted through a prism.

Many applications of refraction involve lenses. You will learn how lenses refract light to form images of the object from which the light originated. In particular you will learn to calculate where the images will be formed, what kind of image is formed and the size of the image for the special case of thin lenses. After treating thin lenses separately, you will study thin lenses in combinations. Two familiar applications of lenses in combination are the microscope and the telescope which are also discussed along with the camera, the human eye, and the magnifying glass.

QUICK REFERENCE

Important Terms

Interface
The boundary between two materials.

Refraction
The change in the direction of travel of light as it passes from one material to another.

Index of refraction
The ratio of the speed of light in a vacuum to the speed of light in a particular material.

Apparent depth
The depth of a submerged object as seen by someone not under the water. The apparent depth is smaller than the actual depth of the object due to the refraction of light as it travels from the water into the air.

Critical angle
The angle of incidence at which the angle of refraction is 90°.

Total internal reflection
The phenomenon in which a light ray incident on an interface is totally reflected and no refraction occurs at an interface.

Brewster angle
The angle of incidence of a light ray at which the reflected ray is completely polarized. At the Brewster angle the reflected and refracted rays are 90° to each other.

Dispersion
The spreading of light into its color components.

Focal point
The point where a lens focuses light rays that are near the principle axis and parallel to it.

Focal length
The distance between a lens and its focal point.

Converging lens
A lens which causes incident parallel rays to converge at the focal point.

Diverging lens
A lens which causes incident parallel rays to diverge after passing through the lens, the rays then appearing to come from the focal point.

Camera
A device which uses a converging lens system to produce real, inverted images on film.

Human eye
It is very much like a camera. It forms a real, inverted image on a light-sensitive surface, called the retina.

Accommodation
The process by which the focal length of the eye is automatically adjusted, so that objects at different distances can be made to produce focused images on the retina.

Near point
The point nearest the eye at which an object can be placed and still produce a sharp image on the retina.

Far point
The location of the farthest object on which the fully relaxed eye can focus.

Nearsightedness (myopia)
A condition in which the eye can focus on nearby objects, but not on distant ones. The condition can be corrected through the use of glasses made from diverging lenses.

Farsightedness (hyperopia)
A condition in which the eye can focus on distant objects, but not on nearby ones. The condition can be corrected through the use of glasses made from converging lenses.

Refractive power
A figure of merit of a lens measured in diopters and given by 1/f, where f is the focal length of the lens in meters.

Angular size
The angle an object subtends at the eye of the viewer.

Angular magnification
The ratio of the final angular size produced by an optical instrument to the reference angular size of the object.

Magnifying glass
A single converging lens that forms an enlarged, upright, and virtual image of an object which is placed at or inside the focal point of the lens.

Compound microscope
A device consisting of an objective lens and an eyepiece lens which produces an enlarged, inverted, and virtual image of an object.

Astronomical telescope
A device which magnifies distant objects with the aid of objective and eyepiece lenses. It produces a final image which is inverted and virtual.

Spherical aberration
An effect in which rays from the outer edge of a spherical lens are not focused at the same point as rays that pass through the center of the lens, thus causing images to be blurred.

Chromatic aberration
An undesirable lens effect arising when a lens focuses different colors at different points.

Equations

The **index of refraction of a material** is

$$n = \frac{\text{Speed of light in a vacuum}}{\text{Speed of light in the material}} = \frac{c}{v} \qquad (26.1)$$

Snell's law of refraction

$$n_1 \sin \theta_1 = n_2 \sin \theta_2 \qquad (26.2)$$

The **apparent depth** of an object with the **observer directly above the object** is

$$d' = d\left(\frac{n_2}{n_1}\right) \tag{26.3}$$

The **critical angle for total internal reflection** is given by

$$\sin\theta_c = \frac{n_2 \sin 90°}{n_1} = \frac{n_2}{n_1} \qquad (n_1 > n_2) \tag{26.4}$$

Brewster's Law

$$\tan\theta_B = \frac{n_2}{n_1} \tag{26.5}$$

The **thin lens equation**

$$\frac{1}{d_o} + \frac{1}{d_i} = \frac{1}{f} \tag{26.6}$$

The **magnification equation**

$$m = \frac{\text{Image height}}{\text{Object height}} = \frac{h_i}{h_o} = -\frac{d_i}{d_o} \tag{26.7}$$

The **refractive power** of a lens is given by:

$$\text{Refractive power (in diopters)} = \frac{1}{f \text{ (in meters)}} \tag{26.8}$$

The **angular magnification of an optical instrument** is:

$$M = \frac{\text{Angular size produced by instrument}}{\text{Reference angular size}} = \frac{\theta'}{\theta} \tag{26.9}$$

The **angular magnification of a magnifying glass** is:

$$M = \frac{\theta'}{\theta} \cong \left(\frac{1}{f} - \frac{1}{d_i}\right) N \tag{26.10}$$

The **angular magnification of a compound microscope** is given by:

$$M \cong -\frac{(L - f_e)N}{f_o f_e} \tag{26.11}$$

The **angular magnification of an astronomical telescope** is:

$$M = \frac{\theta'}{\theta} \cong -\frac{f_o}{f_e} \qquad (26.12)$$

DISCUSSION OF SELECTED SECTIONS

26.2 Snell's Law and the Refraction of Light

When light travels across an interface from one medium to another, it undergoes a change in its speed as indicated by the different indices of refraction of the materials. This speed change results in a bending (refraction) of the path of the light unless it strikes the interface at normal incidence (See figure 26.2 in the text.). Snell's law (equation 26.2) can be used to calculate the amount of bending.

Example 1
Light strikes an interface between two materials of refractive indices, n_1 and n_2, at an angle θ_1 to the normal to the surface. Show that a ray of the light is bent towards the normal if $n_1 < n_2$ and that a ray is bent away from the normal if $n_1 > n_2$.

Snell's law gives for any light ray

$$\sin \theta_2 = \frac{n_1}{n_2} \sin \theta_1$$

If $n_1 < n_2$ then $\sin \theta_2 < \sin \theta_1$ and $\theta_2 < \theta_1$. This means that the ray is bent toward the normal.

If $n_1 > n_2$ then $\sin \theta_2 > \sin \theta_1$ and $\theta_2 > \theta_1$. This means that the ray is bent away from the normal.

Example 2
A corner reflector is placed on the bottom of a lake. Light is shined from a boat at an angle of 35° to the surface of the water, strikes the corner reflector, and passes through the water and back into the air. At what angle to the surface does the light leave the water? Note: It is a property of a corner reflector that it reflects light back along a path that is parallel to its original path.

Original ray   Returning ray

Snell's law applied to the light as it travels from the air to the water gives the angle of refraction into the water

$$\theta_2 = \sin^{-1}\left(\frac{n_1}{n_2} \sin \theta_1\right) = \sin^{-1}\left(\frac{1.00}{1.33} \sin 65°\right) = 43°$$

Similarly, the angle of refraction into the air for the returning light is

$$\theta_2 = \sin^{-1}\left(\frac{n_1}{n_2}\sin\theta_1\right) = \sin^{-1}\left(\frac{1.33}{1.00}\sin 43°\right) = 65°$$

The angle of refraction from the water to the air is the same as the original incident angle. This illustrates a useful property of light rays. A correct ray diagram is still valid if the ray reverses direction.

Example 3
An airborne pelican looks directly down and sees a fish which is actually 0.50 m below the surface of the water. At what apparent depth does the pelican see the fish?

The light leaving the fish and arriving at the pelican's eye travels along the normal to the water-air interface, so equation (26.3) may be used to find the apparent depth of the fish.

$$d' = d\left(\frac{n_2}{n_1}\right) = (0.50 \text{ m})\left(\frac{1.00}{1.33}\right) = 0.38 \text{ m}$$

Example 4
The windshields on old automobiles were flat instead of curved. What is the displacement, as seen by the driver, of objects whose light strikes a 0.81 cm thick flat windshield of index of refraction 1.50 at an angle of 45°? Do you think that the displacement of the objects as seen by the driver will hinder his driving ability?

Refer to the drawing. A light ray from the object is displaced by an amount d as it travels through the windshield. The ray strikes the air-glass interface of the windshield at an angle of $\theta_1 = 45°$ to the normal. Snell's law requires that the angle of refraction be

$$\sin\theta_2 = (n_1/n_2)\sin\theta_1 = (1.00/1.33)\sin 45°$$

$$\theta_2 = 32°$$

The right triangles (123 and 134) shown in the drawing have a common hypotenuse, hence

$$\frac{t}{\cos\theta_2} = \frac{d}{\sin(\theta_1 - \theta_2)}$$

$$d = \left[\frac{\sin(\theta_1 - \theta_2)}{\cos\theta_2}\right]t = \left[\frac{\sin(45° - 32°)}{\cos 32°}\right](0.81 \text{ cm})$$

$$d = 0.21 \text{ cm}.$$

This represents a very small displacement of the objects as seen by the driver and should not hinder his driving ability.

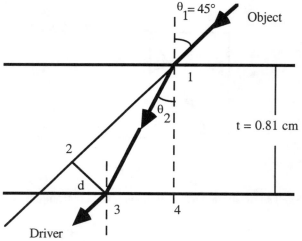

26.4 Polarization and the Reflection and Refraction of Light

When light strikes a nonmetallic surface at an angle of incidence equal to the Brewster angle, then the reflected light will be totally 100 % polarized parallel to the surface. The light transmitted through the surface, that is the refracted light, is then partially polarized perpendicular to the surface.

Example 5
The light produced inside of a laser tube is unpolarized. The light emitted by many lasers is polarized, however, because special windows, called Brewster windows are used at the ends of the tube. The windows are inclined at the Brewster angle to the axis of the tube. (a) Explain how these windows can completely polarize the laser light and determine the direction of polarization of the emitted light. (b) What is the necessary angle of inclination of the windows if the gas inside the tube has n = 1.00 and the windows are made of fused quartz?

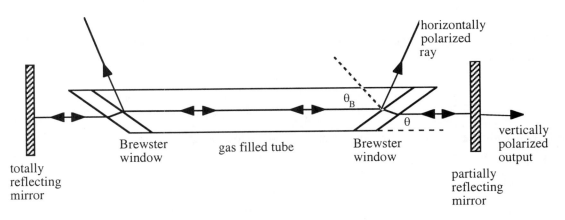

(a) Refer to the drawing. Not all of the laser light produced inside the tube is emitted as output light. Some is reflected back into the tube by a partially reflecting output mirror. This light has already passed through the Brewster window and is partially polarized in the vertical direction. It is this light that is amplified in the laser tube and passes through another window before being reflected back by a totally reflecting mirror. After several passes through the tube, the light becomes totally polarized in the vertical direction since the horizontal component is reflected out of the tube, hence the output beam is vertically polarized.

(b) The angle of inclination of the window to the axis of the tube, θ, is $90° - \theta_B$. Now

$$\theta_B = \tan^{-1} (1.458/1.00) = 55.6°$$

so

$$\theta = 34.4°.$$

26.7 The Formation of Images by Lenses
26.8 The Thin Lens Equation and the Magnification Equation

The same equation applies to image formation by THIN lenses as applied to image formation by mirrors.

$$\frac{1}{d_o} + \frac{1}{d_i} = \frac{1}{f}$$

For a single **lens** the following conventions are used. These are the same conventions as used in the text, but are stated in a somewhat different manner.

1. The object is placed to the left of the lens.
2. Real images fall to the right of the lens.

3. Virtual images fall to the left of the lens.

4. The object distance, d_o, is always positive.

5. The image distance, d_i, is positive for real images and negative for virtual images.

6. The focal length, f, is positive for a converging lens and negative for a diverging lens.

7. The magnification, m, is positive for an upright image and negative for an inverted image.

Example 6

A converging lens can form both real and virtual images depending on where the object is placed relative to the focal point of the lens. (a) For what object distances will it produce a real image? Virtual image? (b) For what object distances will it produce an upright image? Inverted image? (c) For what object distances will the image be the same size as the object? Larger than? Smaller than?

(a) Whether the image is real or virtual is determined by the sign of the image distance, d_i. The thin lens equation gives the image distance to be

$$d_i = \frac{d_o f}{d_o - f}$$

Now the object distance, d_o, is positive for real objects and the focal length of a converging lens is positive, so the sign of d_i is determined by whether d_o or f has the larger value.

d_i is positive and the image is real if $d_o > f$.

d_i is negative and the image is virtual if $d_o < f$.

(b) Whether or not the image is upright or inverted is determined by the sign of the magnification:

$$m = -\frac{d_i}{d_o}$$

If $d_o < f$, then d_i is negative and m is positive so the image is upright.

If $d_o > f$, then d_i is positive and m is negative so the image is inverted.

(c) Solving the thin lens equation for d_i and substituting into the magnification equation yields

$$m = -\frac{f}{d_o - f}$$

Remember that both d_o (real object) and f are positive for a converging lens.

The magnitude of the magnification is equal to one if f - d_o = f, that is, d_o = 2f.

The magnitude of the magnification is greater than one if d_o - f < f, that is, d_o < 2f.

The magnitude of the magnification is less than one if d_o - f > f, that is d_o > 2f.

The magnitude of the magnification is INFINITE if d_o = f. This is the case where the image rays are parallel.

The above results are summarized in the following diagram.

IMAGE CHARACTERISTICS FOR VARIOUS OBJECT DISTANCES

Example 7
Repeat example 6 for a diverging lens.

(a) The thin lens equation gives

$$d_i = \frac{d_o f}{d_o - f}$$

This is the same as for the converging lens except that f is negative for a diverging lens. This requires that d_i be NEGATIVE for ALL object distances, hence the image formed by a diverging lens is ALWAYS VIRTUAL for real objects.

(b) The magnification is again

$$m = -\frac{d_i}{d_o}$$

In part (a) we found that d_i is always negative for a diverging lens, so the magnification is ALWAYS POSITIVE for real objects and corresponds to an UPRIGHT image.

(c) The magnification is also

$$m = -\frac{f}{d_o - f}$$

where f is negative for real objects. The magnitude of the magnification is ALWAYS LESS THAN ONE, since the denominator is always larger than the numerator in the above fraction. These results are summarized in the following diagram.

IMAGE CHARACTERISTICS FOR VARIOUS OBJECT DISTANCES

Image - virtual, upright and reduced
for ALL object distances.

26.9 Lenses in Combination

Finding the final image location for a combination of lenses involves successive applications of the thin lens equation. Find the image location for each lens always using the image of the previous lens as the object for the following lens.

Example 8

An object is placed 15.0 cm to the left of a diverging lens of focal length - 15.0 cm, which is 25 cm to the left of a converging lens with a focal length of 10.0 cm. Find the location of the final image of the combination. Describe the image.

First, ignore the converging lens and find the image location of the diverging lens alone. The thin lens equation gives

$$\frac{1}{d_{i1}} = \frac{1}{f} - \frac{1}{d_{o1}} = \frac{1}{-15.0 \text{ cm}} - \frac{1}{15.0 \text{ cm}}$$

$$d_{i1} = -7.50 \text{ cm}$$

This image serves as the object for the second lens located a distance $d_{o2} = 25.0$ cm + 7.50 cm = 32.5 cm away. The thin lens equation gives for the second lens

$$\frac{1}{d_{i2}} = \frac{1}{f} - \frac{1}{d_{o2}} = \frac{1}{10.0 \text{ cm}} - \frac{1}{32.5 \text{ cm}}$$

$$d_{i2} = 14.4 \text{ cm}$$

The final image distance is POSITIVE, so the image is REAL. The overall magnification of the combination is

$$m' = \left(-\frac{d_{i1}}{d_{o1}}\right)\left(-\frac{d_{i2}}{d_{o2}}\right) = \left(\frac{7.50 \text{ cm}}{15.0 \text{ cm}}\right)\left(\frac{-14.4 \text{ cm}}{32.5 \text{ cm}}\right) = -0.222$$

The final image is SMALLER than the object and is INVERTED since the magnification is negative.

In the above example, the image of the first lens was located on the left of the second lens, hence the object was real and the object distance was positive. If the image of the first lens happens to fall to the right of the second lens, then it forms a virtual object for the second lens. The object distance in this case is considered to be negative.

Example 9

Consider an object lying 15.0 cm to the left of a converging lens of focal length 10.0 cm, which is located 12.0 cm to the left of a diverging lens of focal length - 15.0 cm. Where is the final image located? Describe the image.

Again, ignore the second lens and find the location of the image for the first lens.

$$\frac{1}{d_{i1}} = \frac{1}{f} - \frac{1}{d_{o1}} = \frac{1}{10.0 \text{ cm}} - \frac{1}{15.0 \text{ cm}}$$

$$d_{i1} = 3.0 \times 10^1 \text{ cm}$$

This image serves as the object for the second lens, but it lies to the right of the lens.

Now, with $d_{o2} = 12.0$ cm - 3.0×10^1 cm = - 18 cm,

$$\frac{1}{d_{i2}} = \frac{1}{f} - \frac{1}{d_{o2}} = \frac{1}{-15.0 \text{ cm}} - \frac{1}{-18 \text{ cm}}$$

$$d_{i2} = -9.0 \times 10^1 \text{ cm}$$

The final image lies 9.0×10^1 cm to the left of the second lens and is a VIRTUAL image. The overall magnification is

$$m' = \left(-\frac{d_{i1}}{d_{o1}}\right)\left(-\frac{d_{i2}}{d_{o2}}\right) = \left(\frac{-30.0 \text{ cm}}{15.0 \text{ cm}}\right)\left(\frac{9.0 \times 10^1 \text{ cm}}{-18 \text{ cm}}\right) = +1.0 \times 10^1$$

The final image is ten times larger than the object and is upright as indicated by the positive magnification.

26.10   The Human Eye

The anatomy of the human eye is discussed in the textbook, so we will concentrate our discussion on the optics of the eye. The human eye is very similar to a camera; both have a lens system and a variable-aperture diaphragm (the iris and pupil fulfill this role in the eye). In the eye, the image formed on the retina is real, inverted, and smaller than the object, as is the case for the camera.

Unlike the camera, however, the eye has a fixed image distance, that is, the distance between the lens and the retina is constant. In order to focus on objects at different distances, therefore, the focal length of the lens must be adjustable. This adjustment is produced by physically changing the shape of the lens, thus changing the focal length. The ciliary muscle is responsible for "tensing" or "relaxing" the eye's lens. As an object moves closer to the eye, the ciliary muscle automatically tenses, literally squeezing the lens, increasing its curvature, and thus shortening its focal length. This permits a sharp image to be formed on the retina of this nearby object. For distant objects, the ciliary muscle relaxes, the curvature of the lens decreases, and the focal length increases to provide a sharp focus on the retina. The process in which the focal length of the lens changes for different object distances is referred to as **accommodation**. The eye cannot accommodate perfectly for all object distances, however. If an object is placed too close to the eye, a clear focus cannot be attained. The point nearest the eye at which an object can be placed and still produce a sharp image on the retina is called the **near point**. The near point for the normal eye is about 25 cm. Similarly, the **far point** of the eye is the location of the farthest object on which the fully relaxed eye can focus. For normal vision, the far point is located nearly at infinity.

A **nearsighted (myopic)** person can only focus clearly on nearby objects but not on distant ones. For such a person, the far point of the fully relaxed eye is not at infinity, but may be only a few meters away. The image of a distant object falls in front of the retina, as shown in the figure at the top of the next page, thus causing the image to be out of focus. To correct this, a diverging lens is used to refocus the light back on the retina.

Nearsightedness (myopia) can be corrected using a diverging lens as shown in the figure below. The diverging lens acts to produce a virtual image at the far point of the unaided myopic eye. Thus, instead of the eye trying to focus at infinity, it will focus on a virtual image located at the far point of the eye.

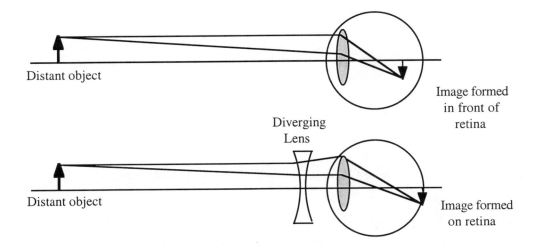

Example 10 *Use of a diverging lens to correct nearsightedness*

A nearsighted person is diagnosed to have a far point located only 350 cm from his eyes. Determine the focal length of contact lenses that will enable him to see distant objects clearly.

The virtual image formed by the diverging lens is located at the far point of the eye. Since contact lenses are placed right against the eyes, $d_i = -350$ cm, the negative sign indicating a virtual image. Since the object is far away, the object distance is taken to be infinity, that is, $d_o = \infty$. The thin-lens equation (26.6) then gives

$$\frac{1}{f} = \frac{1}{d_o} + \frac{1}{d_i} = \frac{1}{\infty} + \frac{1}{-350 \text{ cm}} \implies f = -350 \text{ cm}.$$

Note that the focal length is negative since it is a diverging lens.

A **farsighted (hyperopic)** person can usually see distant objects clearly, but cannot focus on those nearby. Whereas the near point of a "normal" eye is located about 25 cm from the eye, the near point of the farsighted eye may be much farther away. As shown in the figure below, an object located inside the near point of a hyperopic eye will be focused behind the retina, thus producing a blurred image. To correct for this effect, a converging lens is used to refocus the light onto the retina, so that a clear image may be formed. The converging lens forms a virtual image located at the near point of the hyperopic eye. Thus, instead of the eye trying to focus on the nearby object, it will focus on the virtual image located at the near point of the unaided eye. As shown in the figure below, farsightedness (hyperopia) can be corrected using a converging lens.

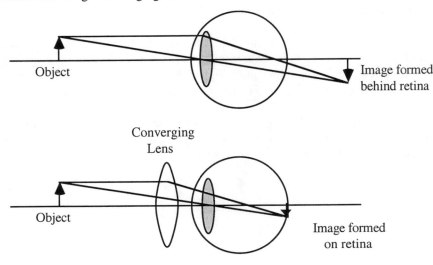

Example 11 *Use of a converging lens to correct farsightedness*
A farsighted person has a near point located 350 cm from the eyes. Obtain the focal length of the converging lens used in contact lenses so that this person can read a book held 25 cm from the eyes.

Since contact lenses are worn directly against the eyes, the object distance is $d_o = 25$ cm. Likewise, the lens forms an image at the near point, so $d_i = -350$ cm. Therefore,

$$\frac{1}{f} = \frac{1}{d_o} + \frac{1}{d_i} = \frac{1}{25 \text{ cm}} + \frac{1}{-350 \text{ cm}} = 0.037 \text{ cm}^{-1} \implies f = 27 \text{ cm}.$$

Notice the focal length is positive since it is a converging lens.

The **refractive power** of a lens is used to describe the degree to which a lens refracts light. It is defined as:

$$\text{Refractive power (in diopters)} = \frac{1}{f \text{ (in meters)}} \qquad (26.8)$$

The unit of refractive power is the diopter. One diopter is 1 m^{-1}.

Example 12
What is the refractive power of the lens described in example 11?

The focal length (in meters) of the lens in example 11 is $f = 27$ cm $= 0.27$ m. Therefore, the refractive power is

$$\text{Refractive power (in diopters)} = \frac{1}{f \text{ (meters)}} = \frac{1}{0.27 \text{ m}} = 3.7 \text{ diopters}.$$

Note that the refractive power of a converging lens is positive, while the refractive power of a diverging lens would be negative.

26.11 Angular Magnification and the Magnifying Glass

As an object is moved farther and farther away from the eye, it appears to get smaller. As far as the brain is concerned, the size of the image on the retina determines how large the object appears to be. Usually, however, the image size is not discussed directly, but rather, the angle θ subtended by the image can be used as a measure of the size. The angle θ is called the **angular size** of the object or image. This angle can be expressed in terms of the height h_o and distance d_o of the object according to the relation

$$\theta \text{ (in radians)} = \text{Angular size} \cong \frac{h_o}{d_o}$$

The approximation for angular size is demonstrated in the following figure.

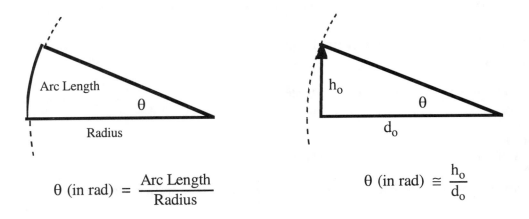

$$\theta \text{ (in rad)} = \frac{\text{Arc Length}}{\text{Radius}} \qquad\qquad \theta \text{ (in rad)} \cong \frac{h_o}{d_o}$$

Example 13

Determine the angular size of the sun as viewed from the earth. The sun has a diameter of 1.39×10^6 km and is at a distance of 1.50×10^8 km from the earth.

The angle subtended by the sun is

$$\theta \cong \frac{h_o}{d_o} = \frac{1.39 \times 10^6 \text{ km}}{1.50 \times 10^8 \text{ km}} = 9.27 \times 10^{-3} \text{ rad} = 0.531°$$

An optical instrument, such as a magnifying glass, allows us to view small or distant objects, because it produces a larger image on the retina than would be possible otherwise. If θ' is the angular size of the final image produced by the instrument, the **angular magnification** (or **magnifying power**) M is defined as

$$M = \frac{\text{Angular size produced by instrument}}{\text{Reference angular size}} = \frac{\theta'}{\theta} \qquad\qquad (26.9)$$

The simplest type of magnifying device is a magnifying glass. This is usually a single converging lens that forms an upright, virtual, enlarged image of an object placed at or inside the focal point of the lens.

For a magnifying glass held close to the eye, the angular magnification is approximately

$$M = \frac{\theta'}{\theta} \cong \left(\frac{1}{f} - \frac{1}{d_i}\right) N \qquad (26.10)$$

where N is the distance from the eye to the near point.

Example 14
A converging lens whose focal length is 11.0 cm is being used to examine a small object by a person whose near point is located 45 cm from his eyes. What is (a) the maximum angular magnification of the lens and (b) the angular magnification when the lens is positioned so the eye is fully relaxed?

(a) The maximum magnification occurs when the image is placed at the near point, so $d_i = -N$ and thus,

$$M \cong \frac{N}{f} + 1 = \frac{45\ cm}{11.0\ cm} + 1 = 5.1$$

(b) When the eye is relaxed, $d_i = -\infty$, so equation (26.10) becomes

$$M \cong \frac{N}{f} = \frac{45\ cm}{11.0\ cm} = 4.1$$

26.12   The Compound Microscope

To increase the angular magnification beyond that possible with a magnifying glass, an additional converging lens can be included to "premagnify" the object. The resulting optical instrument is called a **compound microscope**. The magnifying glass is called the **eyepiece** of the microscope, and the additional lens is called the **objective**. If f_o is the focal length of the objective, f_e the focal length of the eyepiece, and L the distance between the objective and eyepiece, the angular magnification of the compound microscope is given by

$$M \cong -\frac{(L - f_e)N}{f_o f_e} \qquad (26.11)$$

Example 15
The objective of a compound microscope has a focal length of 5.0 mm, while that of the eyepiece is 1.8 cm. If the two lenses are separated by 25.0 cm, determine the angular magnification if a person with normal vision is using the microscope (that is, N = 25 cm).

$$M \cong -\frac{(L - f_e)\ N}{f_o f_e} = -\frac{(25.0\ cm - 1.8\ cm)(25\ cm)}{(0.50\ cm)(1.8\ cm)} = -640$$

Note that when the eyepiece is used alone, the angular magnification would be $M = (N/f_e) + 1 = 15$. The compound microscope therefore represents an improvement in angular magnification by a factor of $640/15 = 43$ over that of the magnifying glass alone.

26.13 The Telescope

A telescope is an instrument for magnifying distant objects, such as stars and planets. A telescope consists of an objective lens and an eyepiece. The "first image" is formed by the objective lens and is real, inverted, and smaller than the object. If this first image falls inside the focal point of the eyepiece, the eyepiece acts like a magnifying glass. It forms a final image that is virtual, enlarged, and located near infinity. This final image can then be viewed with a fully relaxed eye.

The angular magnification of a telescope can be written as:

$$M = \frac{\theta'}{\theta} \cong -\frac{f_0}{f_e} \qquad\qquad (26.12)$$

Example 16

A refracting telescope has an objective and an eyepiece that have refractive powers of 1.25 diopters and 250 diopters, respectively. Find the angular magnification of the telescope.

To find the angular magnification we first need the focal lengths of the objective and eyepiece. Using equation (26.8) we can find the focal lengths. We have

$$\text{Refractive Power (diopters)} = \frac{1}{f \text{ (in meters)}}$$

therefore,

$$f_o = \frac{1}{1.25 \text{ diopters}} = 0.80 \text{ m} \qquad\qquad f_e = \frac{1}{250 \text{ diopters}} = 0.0040 \text{ m}$$

Finally, the angular magnification is given by

$$M = -\frac{f_o}{f_e} = -\frac{0.80 \text{ m}}{0.0040 \text{ m}}$$

$$M = -2.0 \times 10^2.$$

PRACTICE PROBLEMS

1. Light reflected from a fish strikes the surface of the water at an angle of 38° to the normal. What is the angle of refraction of the light into the air?

2. A clear spring has white shells on its bottom which appear to the occupants of a glass bottom boat to be only 45.0 ft below. What is the actual depth of the spring?

3. A laser beam is directed at an angle, θ, to the normal of one side of a rectangular block of fused quartz. Part of the light striking the side of the block passes into the block and strikes an adjacent face. For what angle θ will the laser beam be totally reflected in the block?

4. The light reflected from a solid of unknown refractive index is 100 % horizontally polarized when the light strikes the solid at an angle of 65° to the normal. What is the index of refraction of the solid?

5. Fill-in the following table for a single lens. Include the appropriate signs on ALL numbers.

Lens Type	focal length	object distance	image distance	magnification	real/virtual	inverted/upright
converging	25 cm	15 cm				
converging		22 cm	+22 cm			
diverging			12 cm	0.50		
diverging	11 cm		5 cm			
	8 cm		18 cm			inv

6. Two converging lenses with focal lengths 15 cm and 25 cm are placed 18 cm apart. An object is located 8.0 cm to the left of the 15 cm focal length lens. Where is the final image formed?

7. A converging lens is used to read the small print in a contract. The lens is held 9.0 cm from the print and produces a magnification of + 2.5. What is the focal length of the lens?

8. A camera is supplied with a 35.0 mm focal length lens. A 2.00 m tall man stands 16.0 m from the camera and has his picture taken. What is the size of the image of the man that the lens produces on the film?

9. A farsighted person cannot focus clearly on objects that are less than 145 cm from his eyes. To correct this problem, the person wears eyeglasses that are located 2.0 cm in front of his eyes. Determine the focal length that will permit this person to read a newspaper at a distance of 32.0 cm from his eyes.

10. A certain eye has a near point of 11.0 cm and a far point of 15.0 cm. (a) What is the refractive power of the lens that is required to place the far point at infinity? (b) What is the near point distance when the person uses the lens found in part (a)? (c) What is the refractive power of the lens required to yield a near point distance of 25 cm?

11. A jeweler whose near point is 65 cm from his eye uses a small magnifying glass to examine a watch held 5.0 cm from the lens. Find the angular magnification of the magnifying glass.

12. A microscope has an objective of 4.0-mm focal length and an eyepiece of 2.0-cm focal length. The distance between the two lenses, fixed by the microscope barrel, is 20.0 cm. Determine the magnification of this instrument for a person whose near point distance is 25 cm.

13. A microscope with a 22.0 cm barrel has an eyepiece whose focal length is 1.60 cm. What focal length of the objective lens will produce an angular magnification of - 450? Take the near point distance to be 25 cm.

14. A reflecting telescope uses a concave mirror whose radius of curvature is 4.00 m. If the angular magnification of the telescope is - 160.0, what is the focal length of the eyepiece?

15. A refracting telescope has an angular magnification of - 240. The length of the telescope barrel is 2.60 m. What are the focal lengths of the objective and eyepiece lenses?

16. The refracting telescope at the Yerkes Observatory has an objective whose focal length is 19.4 m. Its eyepiece has a focal length of 10.0 cm. (a) What is the angular magnification of the telescope? (b) If the telescope is used to look at a lunar crater (diameter 1500 m), what is the size of the first image, assuming that the moon is 3.85×10^8 m from the earth?

SOLUTIONS AND ANSWERS

Practice Problems

---

1. Snell's law gives

$$n_w \sin \theta_1 = n_a \sin \theta_2$$
$$\sin \theta_2 = (1.33) \sin 38°$$
$$\theta_2 = \mathbf{55°}$$

---

2. Equation (26.3) gives the actual depth to be

$$d = (n_1/n_2) \, d' = (1.33/1.00)(45.0 \text{ ft}) = \mathbf{59.9 \text{ ft}}.$$

---

3. Snell's law gives for the refraction into the block

$$(1.00) \sin \theta = n_2 \sin \theta_2$$

If the beam is totally internally reflected at the adjacent face, the incident angle on that face must be at least

$$\theta_c = \sin^{-1} (1.000/1.458) = 43.30°.$$

Since the block is rectangular, $\theta_2 + \theta_c = 90°$ so θ_2 is less than 46.70° for the internal reflection to occur.

$$\sin \theta = (1.458) \sin 46.70° = 1.06.$$

The sine function cannot be greater than one, so the angle, θ, which will produce the internal reflection must be less than or equal to 90°. That is, ALL incident angles will cause the light to be totally internally reflected at the adjacent face as long as the light strikes that face.

---

4. The incident angle of 65° must be Brewster's angle for the reflected light to be 100 % polarized. Equation (26.5)

$$n_2 = n_1 \tan \theta_B = (1.00) \tan 65° = \mathbf{2.1}.$$

---

5. See the table at the end of the Solutions and Answers.

---

6. Find the image location for the first lens using the thin lens equation.

$$1/d_{i1} = 1/f_1 - 1/d_{o1} = 1/(15 \text{ cm}) - 1/(8.0 \text{ cm})$$

which gives $d_{i1} = -17$ cm.

This image forms the object of the second lens located 17 cm + 18 cm = 35 cm away. The thin lens equation gives for the second lens

$$1/d_{i2} = 1/f_2 - 1/d_{o2} = 1/(25 \text{ cm}) - 1/(35 \text{ cm})$$

which gives $d_{i2} = 87.5$ cm.

The final image lies **87.5 cm to the right of the second lens** (25 cm focal length lens).

---

7. The fine print serves as the object for the lens, $d_O = 9.0$ cm. The image distance is then

$$d_i = -md_O = -(2.5 \text{ cm})(9.0 \text{ cm}) = -22 \text{ cm}$$

The thin lens equation gives

$$1/f = 1/d_O + 1/d_i = 1/(9.0 \text{ cm}) + 1/(-22 \text{ cm})$$

so f = **+ 15 cm.**

---

8. The angular size remains fixed so $\theta = h_i/d_i = h_O/d_O$. Solving for the image size h_i yields:

$$h_i = h_O(d_i/d_O) = (2.00 \text{ m})(35.0 \text{ mm})/(16.0 \text{ m}) = \textbf{4.38 mm}.$$

---

9. The near point is 145 cm and eyeglasses are 2.0 cm in front of the eyes. Therefore, $d_i = -143$ cm. The object is placed 32.0 cm from the eyes so $d_O = +30.0$ cm. The focal length is obtained from equation (26.6),

$$1/f = 1/d_O + 1/d_i = 1/(30.0 \text{ cm}) + 1/(-143 \text{ cm}) = 0.026 \text{ cm}^{-1}$$

hence, f = **38 cm.**

---

10. (a) We want to move the far point to infinity, so

$$1/f = 1/\infty + 1/(-0.15 \text{ m}) = \textbf{- 6.7 diopters}.$$

(b) We have
$$1/d_O = 1/f - 1/d_i = -6.7 \text{ m}^{-1} - 1/(-0.11 \text{ m}) = 2.4 \text{ m}^{-1}.$$

So d_O = **42 cm.**

(c) For the near point at 25 cm:

$$1/f = 1/(0.25 \text{ m}) + 1/(-0.11 \text{ m}) = \textbf{- 5.1 diopters}.$$

---

11. The angular magnification is

$$M = \theta'/\theta = (1/f - 1/d_i)N = N/d_O = (65 \text{ cm})/(5.0 \text{ cm}) = \textbf{13}.$$

---

12. Use equation (26.11):

$$M \cong -(L - f_e)N/f_O f_e = -(20.0 \text{ cm} - 2.0 \text{ cm})(25 \text{ cm})/[(0.40 \text{ cm})(2.0 \text{ cm})] = \textbf{- 560}.$$

---

13. Use equation (26.11) and solve for the objective focal length, f_O. We obtain:

$$f_O \cong -(L - f_e)N/f_e M = -(22.0 \text{ cm} - 1.60 \text{ cm})(25 \text{ cm})/[(1.60 \text{ cm})(- 450)] = \textbf{0.71 cm}.$$

---

14. The radius of curvature is 4.00 m, so the focal length is $f_o = R/2 = 2.00$ m. Using equation (26.12),

$$M \cong - f_o/f_e.$$

Solving for the eyepiece focal length yields:

$$f_e \cong - f_o/M = - (2.00 \text{ m})/(- 160.0) = \textbf{1.25 x 10}^{-2} \textbf{ m}.$$

15. The length of the barrel of a refracting telescope equals the sum of the focal lengths of the objective and eyepiece. Therefore, $f_o + f_e = 2.60$ m. From equation (26.12),

$$M = - f_o/f_e = - 240.$$

Thus, $f_o = 240 f_e$. Combining equations,

$$240 f_e + f_e = 2.60 \text{ m}.$$

The eyepiece focal length is $f_e = 0.011$ m $= \textbf{11 mm}$. The focal length of the objective is therefore,

$$f_o = 2.60 - 0.011 = \textbf{2.59 m}.$$

16. (a) The magnification is:
$$M \cong - f_o/f_e = - (19.4 \text{ m})/(0.100 \text{ m}) = \textbf{- 194}.$$

(b) The angular size of the crater is

$$\theta = h_o/d_o = (1500 \text{ m})/(3.85 \text{ x } 10^8 \text{ m}) = 3.9 \text{ x } 10^{-6} \text{ rad}.$$

The angular size of the first image is $\theta' = M\theta = (194)(3.9 \text{ x } 10^{-6} \text{ rad}) = 7.6 \text{ x } 10^{-4}$ rad. The size of the first image is therefore,

$$h = \theta' f_e = (7.6 \text{ x } 10^{-4} \text{ rad})(10.0 \text{ cm}) = \textbf{7.6 x 10}^{-3} \textbf{ cm}.$$

Practice Problem 5

Lens Type	focal length	object distance	image distance	magnification	real / virtual	inverted / upright
converging	+ 25 cm	+ 15 cm	-38 cm	+2.5	V	up
converging	+11 cm	+ 22 cm	+22 cm	-1.0	R	inv
diverging	-24 cm	+24 cm	- 12 cm	+ 0.50	V	up
diverging	- 11 cm	+9.2cm	-5 cm	+0.54	V	up
converging	+ 8 cm	+14 cm	+18 cm	-1.3	R	inv

CHAPTER 27
Interference and the Wave Nature of Light

PREVIEW

Light has wave properties. The geometrical optics presented up to now has not fully addressed this fact. In this chapter you will be introduced to wave optics which is the study of the wave nature of light. The wave properties of light include several familiar wave phenomena, such as superposition, interference, and diffraction.

QUICK REFERENCE

Important Terms

Geometrical optics
    The study of the straight line motion of light and the reflection and refraction of light.
Wave (Physical) Optics
    The study of the wave nature of light including interference and diffraction.
Superposition of light waves
    Adding the electric fields of two or more light waves to produce the electric field of a new wave.
Constructive interference
    The superposition of two or more light waves which results in a reinforcement of the waves.
Destructive interference
    The superposition of two or more light waves which results in a cancellation of the waves.
Coherent sources
    Sources of light whose waves maintain a constant phase relationship.
Diffraction of light
    The bending of light waves around obstacles or the edges of an opening.
Resolving power
    The ability of an optical instrument to distinguish between two closely spaced objects.
Diffraction grating
    A device with a large number of closely spaced parallel slits which is used to separate white light into its various colors.
Principal maxima of a grating
    The bright fringes produced by a diffraction grating.

Equations

Young's Double Slit Interference

For **bright fringes**

$$\sin \theta = m\frac{\lambda}{d} \qquad m = 0, 1, 2, 3, \ldots \qquad (27.1)$$

For **dark fringes**

$$\sin \theta = (m + \tfrac{1}{2})\frac{\lambda}{d} \qquad m = 0, 1, 2, 3, \ldots \qquad (27.2)$$

Thin Film Interference

The **wavelength which is important** for thin film interference is the wavelength of light in the film:

$$\lambda_{film} = \frac{\lambda_{vacuum}}{n} \qquad (27.3)$$

Diffraction

Single-slit dark fringes

$$\sin \theta = m\frac{\lambda}{W} \qquad m = 1, 2, 3, \ldots \qquad (27.4)$$

Circular opening, first dark fringe

$$\sin \theta = 1.22 \frac{\lambda}{D} \qquad (27.5)$$

Rayleigh criterion for resolution of two objects

$$\theta_{min} \cong 1.22 \frac{\lambda}{D} \qquad (27.6)$$

Principal maxima of a diffraction grating of spacing, d

$$\sin \theta = m\frac{\lambda}{d} \qquad m = 0, 1, 2, 3, \ldots \qquad (27.7)$$

Principles

Principle of **linear superposition of light** - when two or more light waves are at the same place at the same time the electric field of the waves combine to give the total electric field of the resulting wave.

Huygen's principle - every point on a wavefront acts as a source of wavelets which move forward with the same speed as the wave; the waveform at the later instant is the surface tangent to the wavelets.

DISCUSSION OF SELECTED SECTIONS

27.2 Young's Double-Slit Experiment

Coherent light falling on two closely spaced narrow slits is observed to produce a pattern of alternating bright and dark images of a slit (fringes) on a screen. The central fringe is the brightest with the brightness of the fringes decreasing on either side of the central fringe.

If the light falling on the slits consists of several different frequencies (colors), then overlapping interference patterns will be produced and the fringes will be less visible. If the light of each frequency is not coherent, then the pattern due to each frequency will change randomly in time. The average pattern seen by an observer is then "washed out". Most common sources of light produce such "washed out" patterns, hence much of the interference of light is not obvious in everyday life.

A good source of light for showing interference patterns is a laser, since the laser produces light that is coherent and practically of a single frequency. Young used sunlight which was passed through a small slit. Since a small sample of sunlight falling on the earth came from virtually one point on the sun, it is largely coherent.

Example 1

A helium-neon laser beam ($\lambda = 633$ nm) illuminates two narrow slits. The third bright fringe of the resulting pattern is 15° from the central maximum. How far apart are the slits?

Equation (27.1) applies to the position of a bright fringe in a double slit pattern. Solving for the separation of the slits, d, gives

$$d = \frac{m\lambda}{\sin\theta} = \frac{3(633 \times 10^{-9}\,\text{m})}{\sin 15°} = 7.3 \times 10^{-6}\,\text{m} = 7.3\ \mu\text{m}.$$

Example 2

The interference pattern of a double-slit ($d = 55.1\ \mu$m, $\lambda = 551$ nm) falls on a screen located 1.00 m from the slits. How far apart are the third and fourth bright fringes?

The location of a bright fringe is given by equation (27.1). The third bright fringe is located at an angle of

$$\theta = \sin^{-1}\left(\frac{m\lambda}{d}\right) = \sin^{-1}\left(\frac{3(551 \times 10^{-9}\,\text{m})}{55.1 \times 10^{-6}\,\text{m}}\right) = 1.72°$$

and a distance from the central fringe of

$$y_3 = L\tan\theta = (1.00\,\text{m})\tan(1.72°) = 3.00 \times 10^{-2}\,\text{m} = 3.00\,\text{cm}.$$

The fourth bright fringe is at an angle of

$$\theta = \sin^{-1}\left(\frac{m\lambda}{d}\right) = \sin^{-1}\left(\frac{4(551 \times 10^{-9}\,\text{m})}{55.1 \times 10^{-6}\,\text{m}}\right) = 2.29°$$

The position of the fourth bright fringe relative to the central fringe is

$$y_4 = L\tan\theta = (1.00\,\text{m})\tan(2.29°) = 4.00 \times 10^{-2}\,\text{m} = 4.00\,\text{cm}.$$

The distance between the fringes is then

$$y_4 - y_3 = 4.00\,\text{cm} - 3.00\,\text{cm} = 1.00\,\text{cm}.$$

An alternative way to find the distance between adjacent fringes if they correspond to small angular positions is to note that $\tan\theta$ and $\sin\theta$ are approximately the same for small angles ($< 10°$). The above analysis gives

$$\sin\theta = m\lambda/d \text{ and } \tan\theta = y_m/L$$

for the mth fringe. The position of the mth fringe is then

$$y_m = m\lambda L/d.$$

A similar expression for the $(m+1)^{th}$ fringe gives

$$y_{m+1} = (m+1)\lambda L/d$$

so that the distance between the adjacent fringes is

$$y_{m+1} - y_m = \lambda L/d.$$

Note that the distance between adjacent fringes is constant as long as the angle is small, that is, the fringes are close to the central fringe.

Using the numbers in example 2 we obtain

$$y_4 - y_3 = (551 \times 10^{-9} \text{ m})(1.00 \text{ m})/(55.1 \times 10^{-6} \text{ m}) = 1.00 \times 10^{-2} \text{ m} = 1.00 \text{ cm.}$$

This is the same result as before, so we can conclude that our approximation is valid.

Example 3
What is the largest number of dark fringes that could appear on a screen in a double slit arrangement (d = 0.100 mm) when the slits are illuminated with light of wavelength 455 nm?

If the screen were infinitely large, then all fringes with angles from 0° to 90° could fall on the screen. Equation (27.2) gives the m for the last fringe to be

$$m = (d/\lambda) \sin \theta - 1/2 = [(0.100 \times 10^{-3} \text{ m})/(455 \times 10^{-9} \text{ m})] \sin 90.0° - 1/2 = 219$$

There is a dark fringe for each m from 0 to 219, so there are 220 fringes which could appear on a sufficiently large screen.

27.3 Thin-Film Interference

All thin film problems involve at least three layers of materials as shown in the drawing. The region with n_2 is the actual film and the wavelength of the light in this material, λ_{film}, is the ONLY wavelength that is important in the analysis.

Any ray which reflects from a boundary of increasing index of refraction will suffer a 180° phase change upon reflection. A ray which reflects from a boundary of decreasing index of refraction undergoes no phase shift.

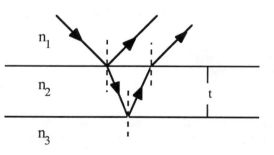

The following cases can be analyzed.

1. If $n_1 > n_2 > n_3$, then no phase shift occurs at either boundary. The only phase shift that can occur is due to the difference in distance traveled between the two reflected waves.

    Bright fringes occur for
    $$2t = m\lambda_{film}$$

    Dark fringes occur for
    $$2t = (m + 1/2)\lambda_{film}$$

2. If $n_1 < n_2 < n_3$, then a phase shift of 180° occurs at EACH boundary and the net phase shift of the reflected waves is again due solely to their path difference.

Bright fringes occur for

$$2t = m\lambda_{film}$$

Dark fringes occur for

$$2t = (m + 1/2)\lambda_{film}$$

3. If n_1 and n_3 are either both larger or smaller than n_2, then a 180° phase shift will occur at only one boundary and the reflected waves will be shifted $\lambda/2$ by the reflection.

Bright fringes occur for

$$2t = (m + 1/2)\lambda_{film}$$

Dark fringes occur for

$$2t = m\lambda_{film}$$

Example 4

A certain region of a soap bubble reflects red light of vacuum wavelength $\lambda = 650$ nm. What is the minimum thickness that this region of the soap bubble could have? Take the index of refraction of the soap film to be 1.41.

The soap film has air on both sides, so the reflections of the light produce a net 180° phase shift. The condition for bright fringes is $2t = (m + 1/2)\lambda_{film}$.

$$t = \frac{(m + 1/2)\lambda_{film}}{2} = \frac{(m + 1/2)\lambda}{2n} = \frac{(1/2)(650 \times 10^{-9} \, m)}{2(1.41)} = 1.2 \times 10^{-7} \, m$$

Example 5

A physics professor wants to find the diameter of a human hair by placing it between two flat glass plates, illuminating the plates with light of vacuum wavelength $\lambda = 552$ nm and counting the number of bright fringes produced along the plates. The Prof finds 125 bright fringes between the edge of the plates and the hair. What is the diameter of the hair?

The reflections from the boundaries will cause a net 180° phase shift. The condition for bright fringes is $2t = (m + 1/2)\lambda_{film}$. Now m = 124 since there is a bright fringe for m = 0 and $\lambda_{film} = \lambda/n$.

$$t = \frac{(m + \frac{1}{2})\lambda_{film}}{2} = \frac{(m + \frac{1}{2})\lambda}{2n} = \frac{(124 + \frac{1}{2})(552 \times 10^{-9} \, m)}{2(1.00)} = 3.44 \times 10^{-5} \, m$$

27.5 Diffraction

Light diffracts through an opening such as a slit and subsequently spreads. If a screen is placed in the path of the light, a point on the screen will receive light which has come from each point in the opening and interference of essentially an infinite number of waves occurs at the point. This interference is totally destructive at some points on the screen and gives dark fringes while the interference is totally constructive at other points and gives rise to bright fringes. Of course, intermediate interferences occur between the bright and dark fringes. The central fringe of the resulting diffraction pattern is wide and intense while the fringes to either side of the central fringe fall off in intensity and are narrower. Equation (27.4) locates the positions of the dark fringes. Note that this is the same equation which governed the diffraction of sound (see equation 17.1).

27.6 Resolving Power

The diffraction of light through an opening places limits on our, or an instrument's, ability to distinguish between closely spaced objects. Since the pupil of our eye is a circular opening, we see diffraction patterns which fall on the retina. This is the reason that stars which are so far away that they should be point sources of light appear to us to be circles of light. We see the diffraction pattern of a point source of light. If two such sources of light are close together, their individual diffraction patterns may overlap on the retina so much that they look like one source. When do two sources look like two sources, that is, when are they resolved? The Rayleigh criterion, equation (27.6), gives a useful, if not unique, definition. The assumption is that two objects are "just" resolved when the first minimum of the diffraction pattern of one falls on the central maximum of the other.

Example 6

The light from a red star which is essentially a point object is viewed by a child on a dark night when the child's pupil has a diameter of 7.00 mm. How wide does the star appear on the retina of the child? Assume that the index of refraction of the vitreous humor in the eye is 1.36, that light seen by the child has a wavelength of 625 nm, and that the distance from the pupil to the retina is 3.5 cm.

The child will see the star as the central fringe of the diffraction pattern which falls on the retina. The angular position of the first minimum of the pattern is given by

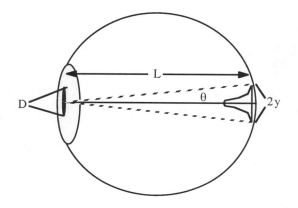

$$\sin \theta = 1.22 \, (\lambda/D)$$

where λ is the wavelength of light within the eye, $\lambda = (625 \text{ nm})/(1.36) = 4.60 \times 10^{-7}$ m and $D = 7.00 \times 10^{-3}$ m.

$$\theta = 4.59 \times 10^{-3} \, °.$$

The width of the central maximum is

$$2y = 2L \tan \theta = 2(3.5 \times 10^{-2} \text{ m}) \tan (4.59 \times 10^{-3} \, °) = 5.61 \times 10^{-6} \text{ m}.$$

Example 7

The headlights of a certain automobile are 1.50 m apart. How far away could the car be and the headlights just be resolved by a person's eye? Use the data for the eye given in the previous example and assume the wavelength of the light is 625 nm.

The headlines will be just resolved according to the Rayleigh criterion when the angle subtended by the headlights at the pupil is $\theta = 1.22 \, \lambda/D$. From the diagram (next page) we see that $\theta = d/x$ if the angle is small. Then, since $\lambda = (625 \text{ nm})/(1.36) = 4.60 \times 10^{-7}$ m,

$$x = \frac{dD}{1.22 \, \lambda} = \frac{(1.50 \text{ m})(7.00 \times 10^{-3} \text{ m})}{1.22(4.60 \times 10^{-7} \text{ m})} = 1.87 \times 10^4 \text{ m}$$

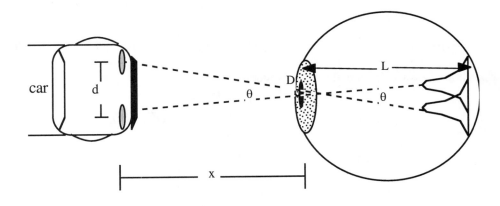

27.7 The Diffraction Grating

A set of closely spaced parallel slits is referred to as a diffraction grating. A grating with a large number of very closely spaced slits is capable of producing bright fringes which are much narrower and sharper than the bright fringes of a double slit. The bright fringes of a grating are often referred to as the principal maxima or orders. The positions of the principal maxima of a grating is given by equation (27.7). Here, d is the spacing between the slits of the grating. Usually a grating is specified by the number of slits or lines per inch or per cm. To find the spacing, d, simply take the reciprocal.

Example 8
How many orders (principal maxima) can be produced by a 1500 lines/cm grating when illuminated with green light whose wavelength is $\lambda = 510$ nm?

The largest angle that an order could possibly have is 90°. The order which is closest to having this angle is, according to equation (27.7)

$$m = \frac{d}{\lambda} = \frac{(1/1500) \times 10^{-2}\,m}{510 \times 10^{-9}\,m} = 13$$

The grating produces 13 orders along with the zero order, so it has a total of 14 orders.

The diffraction grating has a very important application in spectroscopes due to its ability to separate light into its various wavelengths (colors). Light of different wavelengths are diffracted at different angles, so the grating disperses light into its colors as a prism does. The grating produces a "rainbow" spectrum in each of its orders except the zero order. The grating in the above example can produce 13 rainbow spectra. The separation of the colors is larger in the higher orders which makes them especially useful for detailed analysis of the frequencies of light.

Example 9
A mixture of red (630 nm) and blue (490 nm) light passes through a 5500 slits/cm grating and falls on a screen located 1.0 m away. What are the separations of the colors on the screen in the first order and in the second order?

The angular position of a wavelength in the first order is

$$\theta = \sin^{-1} \frac{\lambda}{d} = \sin^{-1} [(5500 \times 10^2\,m^{-1})\lambda]$$

The position of this wavelength on the screen is $y = L \tan \theta$.

For $\lambda = 630$ nm, $\theta = 2.0 \times 10^1$ ° and $y = 0.37$ m.

For $\lambda = 490$ nm, $\theta = 16$ ° and $y = 0.29$ m.

The separation between the colors in the first order is 0.37 m - 0.29 m = 0.08 m = 8 cm.

The angular position of a wavelength in the second order is

$$\theta = \sin^{-1} \frac{2\lambda}{d} = \sin^{-1} [2(5500 \times 10^2 \text{ m}^{-1})\lambda]$$

For $\lambda = 630$ nm, $\theta = 44°$ and $y = 0.97$ m.

For $\lambda = 490$ nm, $\theta = 33°$ and $y = 0.65$ m.

The separation of the colors in the second order is 0.97 m - 0.65 m = 0.32 m = 32 cm.

PRACTICE PROBLEMS

1. Relative to the central fringe, what is the position on a screen located 2.0 m away of the third order fringe in a double slit arrangement (d = 0.058 mm) when the slits are illuminated with $\lambda = 480$ nm light?

2. Two parallel slits are illuminated by light composed of two wavelengths, one of which is 486 nm. On viewing the screen the light whose wavelength is known produces its third dark fringe at the same place where the light whose wavelength is unknown produces its fourth-order bright fringe. The fringes are counted relative to the central or zeroth-order bright fringe. What is the unknown wavelength?

3. A certain region of a soap bubble appears red ($\lambda = 630$ nm) due to reflection. What are the three smallest possible thicknesses of this region of the bubble? Take n = 1.36 for the soap film.

4. The minima of the diffraction pattern of light passing by a human hair lie in the same places as the minima of the diffraction pattern of a slit the same width as the hair. The distance between the two minima on either side of the central fringe is measured to be 2.2 cm on a screen located 2.5 m from a hair when $\lambda = 550$ nm light is used. What is the diameter of the hair?

5. Two stars, each located 580 light years from earth, are 2.4 light years apart. What is the smallest diameter telescope which could just resolve them using 630 nm light?

6. There are 5620 lines per centimeter in a grating that is used with light whose wavelength is 471 nm. A flat observation screen is located a distance of 0.750 m from the grating. What is the minimum width that the screen must have so the centers of all the principal maxima formed on either side of the central maximum fall on the screen?

7. A hologram is a photographic exposure of the interference between laser light reflected from an object and light coming directly from the laser. When the film is developed, there are ridges and valleys in the emulsion on the film which can act like a diffraction grating when light is shined through it. The light striking the hologram is diffracted into a three dimensional image of the object that was originally used. Assume that laser light of wavelength $\lambda = 630$ nm is incident on a small spot of the hologram and is diffracted in the first order through an angle of 22° to form an image. What is the effective lines per cm on this spot of the hologram?

8. You have a 4500 lines per cm grating available and need to separate light of wavelength 535 nm and of wavelength 585 nm by a distance of at least 15 cm on a screen 0.75 m away. What is the smallest order of diffraction you could use?

9. Light of wavelength 630 nm is incident on a small circular hole. The light travels to a wall located 2.0 m from the hole. What must be the diameter of the hole so that the first dark ring has a diameter of 2.5 cm?

SOLUTIONS AND ANSWERS

Practice Problems

---

1. The angular position of the third order fringe is

$$\theta = \sin^{-1} [3(480 \times 10^{-9} \text{ m})/(0.058 \times 10^{-3} \text{ m})] = 1.4°$$

The position of the fringe measured from the center of the screen is

$$y = L \tan \theta = (2.0 \text{ m}) \tan 1.4° = \textbf{5.0 cm.}$$

---

2. Equation (27.2) written for the third dark fringe of the known $\lambda = 486$ nm is:

$$\sin \theta = (2 + 1/2)\lambda/d$$

and, equation (27.1) written for the fourth order bright fringe of the unknown wavelength, λ' gives

$$\sin \theta = 4\lambda'/d.$$

The fringes are in the same position, so

$$\lambda' = (5/8)\lambda = (5/8)(486 \text{ nm}) = \textbf{3.04} \times \textbf{10}^2 \textbf{ nm.}$$

---

3. There is a 180° phase shift only at the soap-air interface, so the constructive interferences occur for

$$2t = (m + 1/2)\lambda_{film}$$

Thus, the three smallest thicknesses are:

$$2t_1 = (1/2)\lambda_{film}$$
$$2t_2 = (3/2)\lambda_{film}$$
$$2t_3 = (5/2)\lambda_{film}$$

$\lambda_{film} = \lambda/n$, so

$$t_1 = \textbf{115 nm}$$
$$t_2 = \textbf{350 nm}$$
$$t_3 = \textbf{580 nm.}$$

---

4. The first minima of the diffraction pattern of a slit, hence a hair of width, W, is found from equation (27.4). The position of the first minimum on the screen is $y = L \tan \theta$, so

$$\theta = \tan^{-1} (1.1 \times 10^{-2} \text{ m}/2.5 \text{ m}) = 0.25°.$$

Now, $W = \lambda/\sin \theta = (550 \times 10^{-9} \text{ m})/\sin 0.25° = 1.2 \times 10^{-4} \text{ m} = \textbf{0.12 mm.}$

---

5. The Rayleigh criterion for the resolution of two objects gives the needed diameter: $D = 1.22 \lambda/\theta_{min}$.

The angle subtended by the stars is

$$\theta_{min} = (2.4 \text{ ly})/(580 \text{ ly}) = 4.1 \times 10^{-3} \text{ rad.}$$

Then, from the Rayleigh criterion:

$$D = (1.22)(630 \times 10^{-9} \text{ m})/(4.1 \times 10^{-3} \text{ rad}) = \mathbf{1.9 \times 10^{-4} \text{ m}.}$$

6. The screen must have a total width,
$$2y = 2L \tan \theta \text{ where } \theta = \sin^{-1} m\lambda/d .$$

The order of the last maximum which can fall on one side of the screen is

$$m = (d/\lambda) \sin 90° = 3.77 \text{ or } 3$$

The angle locating the last maximum on the screen is

$$\theta = \sin^{-1} (3\lambda/d) = 52.7°$$

Now half the width of the screen is

$$y = (0.750) \tan 52.7° = 0.985 \text{ m}.$$

The required screen width is then **1.97 m.**

7. The distance between the "slits" is

$$d = \lambda/\sin \theta = (630 \times 10^{-9} \text{ m})/\sin 22° = 1.7 \times 10^{-6} \text{ m} = 1.7 \times 10^{-4} \text{ cm}$$

Then
$$1/d = \mathbf{5900 \text{ lines/cm}}$$

8. The angular position of the wavelengths on the screen is $\theta = y/L$ and $\theta' = y'/L$ where a small angle approximation has been used. Now $\theta = m\lambda/d$ and $\theta' = m\lambda'/d$, so

$$m = (y' - y)d/(\lambda' - \lambda)L = 8.9$$

The smallest order which could be used to see both lines is the **9th.**

9. The angular width of the first dark ring is
$$\theta \approx 1.22 \ \lambda/D.$$

The diameter of the first dark ring is
$$d \approx 2L\theta \text{ so}$$

$$D \approx 2(1.22)(2.0 \text{ m})(630 \times 10^{-9} \text{ m})/(2.5 \times 10^{-2} \text{ m}) = \mathbf{1.2 \times 10^{-4} \text{ m}.}$$

CHAPTER 28
Special Relativity

PREVIEW

In this chapter you will be introduced to some of the fascinating ideas contained in Einstein's theory of special relativity. You will learn how the theory has changed the way we view space and time, with the ideas of time dilation and length contraction. The relation between mass and energy will also be examined as well as some of the other seemingly strange predictions made by this famous theory.

QUICK REFERENCE

Important Terms

Event
A physical "happening" that occurs at a certain place and time.

Inertial reference frame
A frame of reference in which Newton's law of inertia is valid. That is, if the net force acting on a body is zero, the body either remains at rest or moves with constant velocity. In other words, the acceleration of a body is zero when measured in an inertial frame if there is no net force.

Relativity postulate
The laws of physics are the same in every inertial reference frame.

Speed of light postulate
The speed of light in a vacuum, measured in any inertial reference frame, always has the same value of c, no matter how fast the source of light and the observer are moving relative to each other.

Proper time interval
The time interval between two events measured by an observer who is at rest relative to the events and who views these events as occurring at the same place.

Time dilation
The lengthening of the time interval between two events, as measured by an observer who is in motion with respect to the events. It implies that "moving clocks run slow".

Proper length
The length of an object measured by an observer who is at rest relative to the object.

Length contraction
The shortening of distances as measured by a moving observer.

Mass-energy equivalence
Energy and mass are related through the equation $E_0 = mc^2$.

Equations

The **time dilation** equation is:

$$\Delta t = \frac{\Delta t_0}{\sqrt{1 - \frac{v^2}{c^2}}} \qquad (28.1)$$

The equation for **length contraction** is:

$$L = L_0 \sqrt{1 - \frac{v^2}{c^2}} \qquad (28.2)$$

The equation for the **relativistic momentum** is:

$$p = \frac{mv}{\sqrt{1 - \frac{v^2}{c^2}}} \qquad (28.3)$$

The **total energy** of an object is:

$$E = \frac{mc^2}{\sqrt{1 - \frac{v^2}{c^2}}} \qquad (28.4)$$

The **rest energy** of an object is:

$$E_0 = mc^2 \qquad (28.5)$$

The **relativistic kinetic energy** is:

$$KE = E - E_0 = mc^2 \left(\frac{1}{\sqrt{1 - \frac{v^2}{c^2}}} - 1 \right) \qquad (28.6)$$

The **velocity-addition** formula is:

$$v_{AB} = \frac{v_{AC} + v_{CB}}{1 + \frac{v_{AC}v_{CB}}{c^2}} \qquad (28.7)$$

DISCUSSION OF SELECTED SECTIONS

28.2 The Postulates of Special Relativity

The theory of special relativity deals with **inertial reference frames**. These are frames of reference in which Newton's law of inertia is valid. That is, if the net force acting on a body is zero, the body either remains at rest or moves at a constant velocity. The acceleration of such a body is zero when measured in an inertial reference frame. Rotating and otherwise accelerating reference frames are not inertial frames.

Einstein based his theory of special relativity on two fundamental assumptions or postulates:

1. **The Relativity Postulate**. The laws of physics are the same in every inertial reference frame.
2. **The Speed of Light Postulate**. The speed of light in a vacuum, measured in any inertial reference frame, always has the same value of c, no matter how fast the source of light and the observer are moving relative to each other.

We will find some very unusual predictions made by the theory of special relativity, most of which will not appeal to our "common sense". These seemingly strange predictions occur as a direct result of the application of the two postulates just stated. So keep in mind that these ideas, which may seem strange to our intuitive notions of space and time, are actually experimentally verified and accepted theories of physics.

28.3 The Relativity of Time: Time Dilation

Consider an observer on the earth and one on a spacecraft moving with respect to the earth. The theory of special relativity reveals that the earth-based observer sees time passing more slowly for the astronaut on the spacecraft than for himself. For example, suppose the astronaut has a clock on his spaceship. He measures the time interval between successive "ticks" on his clock (two **events**) to be Δt_0, which we will call the **proper time interval**. However, the earth-based observer measures a time interval Δt between these two events that is GREATER than the time interval Δt_0 measured by the astronaut. This effect is known as **time dilation** because the time interval Δt is dilated or "expanded" relative to Δt_0. In essence, the moving clock runs more slowly from the point of view of the earth-based observer. The relationship between the two time intervals is given by:

$$\Delta t = \frac{\Delta t_0}{\sqrt{1 - \frac{v^2}{c^2}}} \qquad (28.1)$$

where v is the relative speed between the two observers, and c is the speed of light in a vacuum.

Example 1
An astronaut travels away from the earth at v = 0.98c. If one hour ticks off on the astronaut's clock, what is the time interval that an earth observer measures on the astronaut's clock?

In equation (28.1) the proper time interval Δt_0 is taken to be 1.0 h, v = 0.98c, and we have

$$\Delta t = \frac{\Delta t_0}{\sqrt{1 - \frac{v^2}{c^2}}} = \frac{1.0 \text{ h}}{\sqrt{1 - \frac{(0.98c)^2}{c^2}}} = 5.0 \text{ h}$$

So, according to the earth observer, it takes 5.0 earth-hours for one hour to "tick off" on the astronaut's clock!

28.4 The Relativity of Length: Length Contraction

Because of time dilation, observers moving at a constant velocity relative to each other measure different time intervals between two events. Let's calculate the distance traveled by the astronaut in example 1, as measured by the earth observer and the astronaut. Suppose the spacecraft travels from point A to point B, the trip taking 5.0 h as measured by the earth observer. The astronaut measures a distance $L = v \, \Delta t_0 = (0.98c)(1.0 \text{ h}) = 0.98$ light-hours. (NOTE: a light-hour is the distance that light travels in one hour, i.e., $(3.0 \times 10^8 \text{ m/s})(3600 \text{ s/1 h}) = 1.1 \times 10^{12}$ m.) The earth observer measures the distance to be $L_0 = v \, \Delta t = (0.98c)(5.0 \text{ h}) = 4.9$ light-hours. For the astronaut, the distance has shortened, as compared to the earth observer's distance. This is an example of the phenomenon known as **length contraction**.

If L_0 is the length between two points as measured by an observer at rest with respect to these points (L_0 is referred to as the **proper length**), and L is the length measured by an observer moving at a constant speed v with respect to these points, the equation relating L and L_0 is given by:

$$L = L_0 \sqrt{1 - \frac{v^2}{c^2}} \qquad (28.2)$$

It is important to note that this length contraction occurs only along the direction of motion. Those dimensions that are perpendicular to the motion are not shortened.

Example 2

An astronaut carries a meter stick in his spaceship. The meter stick is aligned in the direction of motion of his spaceship, which is moving at a speed of 0.8c relative to the earth. What is the length of the meter stick, as measured by an observer back on earth?

The proper length of the meter stick is $L_0 = 1$ m. The speed of the spaceship is $v = 0.8c$. The length of the meter stick, as measured by the earth observer, is therefore

$$L = L_0 \sqrt{1 - \frac{v^2}{c^2}} = (1 \text{ m}) \sqrt{1 - \frac{(0.8c)^2}{c^2}} = 0.6 \text{ m}$$

So an observer on earth measures the astronaut's meter stick to be only 60 cm long!

28.5   Relativistic Momentum

When an object is moving at a speed v relative to an observer, the observer measures a momentum, p, called the **relativistic momentum** given by

$$p = \frac{mv}{\sqrt{1 - \frac{v^2}{c^2}}} \qquad (28.3)$$

As the velocity approaches the speed of light, the term in the denominator of the equation approaches zero and the relativistic momentum approaches infinity. When $v = c$, the relativistic momentum becomes infinite, and to accelerate the mass up to the speed of light would require an infinite amount of energy. An object with finite mass can therefore never have a speed equal to or greater than the speed of light. The speed of light therefore represents the "ultimate" speed for any such object.

Example 3

A proton whose mass is 1.67×10^{-27} kg is traveling at a speed $v = 0.999c$ relative to the earth. What is the relativistic momentum of the proton as measured by an earth based observer?

$$p = \frac{mv}{\sqrt{1 - \frac{v^2}{c^2}}} = \frac{(1.67 \times 10^{-27} \text{ kg})(0.999 \text{ c})}{\sqrt{1 - (0.999)^2}}$$

$$p = 1.12 \times 10^{-17} \text{ kg m / s.}$$

28.6 The Equivalence of Mass and Energy

An important result of special relativity is that mass and energy are equivalent. The equivalence between energy and mass is stated in

$$E = \frac{mc^2}{\sqrt{1 - \frac{v^2}{c^2}}}$$

where E represents the total energy of an object that is moving at speed v relative to an observer.

When an object is at rest relative to an observer, its energy is $E_0 = mc^2$, where E_0 represents the **rest energy** of the object. If the object is in motion, it has kinetic energy as well, so its total energy is a sum of its rest energy and its kinetic energy, $E = E_0 + KE$. Using equation (28.4), we can now express the kinetic energy of the object as

$$KE = E - E_0 = mc^2 \left(\frac{1}{\sqrt{1 - \frac{v^2}{c^2}}} - 1 \right) \qquad (28.6)$$

Example 4
For the proton of example 3 calculate (a) its rest energy, (b) its total energy, and (c) its kinetic energy.

(a) The rest energy is:

$$E_0 = mc^2 = (1.67 \times 10^{-27} \text{ kg})(3.00 \times 10^8 \text{ m/s})^2 = 1.50 \times 10^{-10} \text{ J}$$

(b) The total energy is:

$$E = \frac{mc^2}{\sqrt{1 - \frac{v^2}{c^2}}} = \frac{(1.67 \times 10^{-27} \text{ kg})(3.00 \times 10^8 \text{ m/s})^2}{\sqrt{1 - \frac{(0.999\,c)^2}{c^2}}} = 3.37 \times 10^{-9} \text{ J}.$$

(c) The kinetic energy is:

$$KE = E - E_0 = 3.37 \times 10^{-9} \text{ J} - 1.50 \times 10^{-10} \text{ kg} = 3.22 \times 10^{-9} \text{ J}.$$

28.7 The Relativistic Addition of Velocities

Consider the following situation shown in the drawing. A rocket is moving in the +x direction with respect to the earth. An earth observer and an astronaut on the rocket both see a flying saucer moving in the +x direction. However, each measures a different value for the velocity of the flying saucer. The relative velocities are:

v_{SE} = velocity of the flying **S**aucer relative to the **E**arth

v_{SR} = velocity of the flying **S**aucer relative to the **R**ocket

v_{RE} = velocity of the **R**ocket relative to the **E**arth

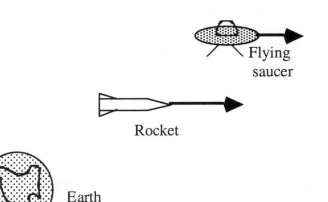

The theory of special relativity states that the velocities v_{SE}, v_{SR} and v_{RE} are related according to the **velocity-addition formula** (equation 28.7):

$$v_{SE} = \frac{v_{SR} + v_{RE}}{1 + \frac{v_{SR} v_{RE}}{c^2}}$$

When the velocities are one dimensional, they can have either positive or negative values, depending on whether they are directed along the positive or negative direction.

Example 5

An astronaut travels in a rocket along the +x direction with a velocity of +0.60c relative to the earth. She looks out her window and sees a flying saucer moving relative to her with a velocity of +0.90c in the +x direction. What is the velocity of the flying saucer relative to an observer on earth?

Using the velocity-addition formula with v_{SR} = +0.90c and v_{RE} = +0.60c, the velocity v_{SE} of the flying saucer relative to the earth is

$$v_{SE} = \frac{v_{SR} + v_{RE}}{1 + \frac{v_{SR} v_{RE}}{c^2}} = \frac{+0.90c + 0.60c}{1 + \frac{(+0.90c)(+0.60c)}{c^2}} = +0.97c$$

The earth observer measures the speed of the flying saucer to be v_{SE} = 0.97c, rather than the value of 1.50c (0.90c + 0.60c) which we would have obtained if we didn't know about relativity. Equation (28.7) ensures that relative speeds will never be greater than the speed of light.

Example 6

The rocket in Example 5 fires a photon torpedo (a burst of light waves traveling at the speed of light) at the flying saucer. (a) What is the velocity of the photon torpedo relative to an observer on earth? (b) What is the velocity of the photon torpedo relative to an observer on the flying saucer?

(a) The velocities are:

v_{PE} = velocity of the **P**hoton torpedo relative to the **E**arth

v_{PR} = velocity of the **P**hoton torpedo relative to the **R**ocket

v_{RE} = velocity of the **R**ocket relative to the **E**arth

The velocity v_{PE} of the photon torpedo relative to the earth is given by the velocity-addition formula, equation (28.7):

$$v_{PE} = \frac{v_{PR} + v_{RE}}{1 + \frac{v_{PR} v_{RE}}{c^2}}$$

According to the speed of light postulate, the speed of the photon torpedo relative to the rocket must be equal to the speed of light, so that $v_{PR} = +c$. We also know from Example 5 that $v_{RE} = +0.60c$, so v_{PE} is

$$v_{PE} = \frac{v_{PR} + v_{RE}}{1 + \frac{v_{PR} v_{RE}}{c^2}} = \frac{+c + 0.60c}{1 + \frac{(+c)(+0.60c)}{c^2}} = +c$$

(b) The velocities for this part of the problem are:

v_{PS} = velocity of the **P**hoton torpedo relative to the flying **S**aucer

v_{PE} = velocity of the **P**hoton torpedo relative to the **E**arth

v_{ES} = velocity of the **E**arth relative to the flying **S**aucer

The velocity v_{PS} of the photon torpedo relative to the flying saucer is given by the velocity-addition formula, equation (28.7), as

$$v_{PS} = \frac{v_{PE} + v_{ES}}{1 + \frac{v_{PE} v_{ES}}{c^2}}$$

We know from part (a) that $v_{PE} = +c$. The velocity v_{ES} of the earth relative to the flying saucer is the negative of the velocity of the flying saucer relative to the earth: $v_{ES} = - v_{SE} = -0.90c$. The velocity of the photon torpedo relative to the flying saucer is

$$v_{PS} = \frac{v_{PE} + v_{ES}}{1 + \frac{v_{PE} v_{ES}}{c^2}} = \frac{+c + (-0.90c)}{1 + \frac{(+c)(-0.90c)}{c^2}} = +c$$

So all observers measure the same value for the speed of the photon torpedo, namely, the speed of light. This is in keeping with the second postulate of special relativity, the speed of light postulate.

PRACTICE PROBLEMS

1. An unstable nuclear particle has a lifetime of 3.00×10^{-6} s when at rest on earth. If the particle is moving at 99.9% of the speed of light relative to earth, what is its lifetime as measured by an earth based observer?

2. If the lifetime of pions at rest in the laboratory is 2.6×10^{-8} s, at what speed must the pions travel with respect to the laboratory so that their lifetime is 7.8×10^{-8} s as measured by a laboratory observer?

3. As seen by an observer on earth, the Tau Ceti star system is 11.3 light years distant. A spaceship is sent to Tau Ceti at a speed of 0.95c.   (a) What is the distance to Tau Ceti as measured by an astronaut on the spaceship? (b) How much does the astronaut age according to the astronaut's clock?

4. At what rate must matter be converted to energy in order to supply the United States with the 1×10^{20} J of energy it consumes each year?

5. A proton has a rest energy of 938 MeV. If the proton is moving at a speed of 0.999 999c, what is its kinetic energy in MeV?

6. Electrons are accelerated in a linear accelerator so that their total energy is 3.0×10^4 times their rest energy. Traveling at this speed, what do the electrons measure for the length of the accelerator which is 2.0 km long in the laboratory?

7. An observer on the earth sees a spaceship approaching at a velocity of 0.50c. The spacecraft then launches an exploration vehicle that, according to the earth observer, approaches at 0.70c. What is the velocity of the exploration vehicle relative to the spaceship?

8. Consider the three blocks labeled 1, 2, and 3 in the figure below. Block 1 slides at a speed of 0.50c relative to the ground; block 2 slides at speed 0.50c relative to block 1; and block 3 slides at speed 0.50c relative to block 2. Find the speed of (a) block 2 relative to the ground; (b) block 3 relative to the ground.

SOLUTIONS AND ANSWERS

Practice Problems

1. The proper time is $\Delta t_0 = 3.00 \times 10^{-6}$ s. Using $v = 0.999c$,

$$\Delta t = \frac{\Delta t_0}{\sqrt{1 - v^2/c^2}} = 22.4 \ \Delta t_0$$

$$\Delta t = \mathbf{6.72 \times 10^{-5}} \text{ s.}$$

2. Using $\Delta t_0 = 2.6 \times 10^{-8}$ s and $\Delta t = 7.8 \times 10^{-8}$ s,

$$\Delta t_0/\Delta t = 0.33 = \sqrt{1 - v^2/c^2}.$$

Squaring this expression and solving for v yields,

$$1 - v^2/c^2 = 0.11$$

so that $v^2 = 0.89c^2$ and therefore, $v = \mathbf{0.94c}$.

3. (a) The proper length is $L_0 = 11.3$ light years. The distance to Tau Ceti as measured by the astronaut is

$$L = L_0 \sqrt{1 - v^2/c^2}.$$

Therefore,

$$L = (11.3 \text{ ly}) \sqrt{1 - (0.95)^2} = \mathbf{3.5 \text{ light years.}}$$

(b) Since $\Delta t = (11.3 \text{ ly})/(0.95c) = 11.9$ yr,

$$\Delta t_0 = \Delta t \sqrt{1 - \frac{v^2}{c^2}} = (11.9 \text{ yr}) \sqrt{1 - (0.95)^2} = \mathbf{3.7 \text{ years.}}$$

4. The power required to produce 1×10^{20} J in one year is

$$P = (1 \times 10^{20} \text{ J/yr})(1 \text{ yr}/365 \text{ d})(1 \text{ d}/86\,400 \text{ s}) = 3 \times 10^{12} \text{ J/s.}$$

So,

$$P = E_0/t = mc^2/t$$

which leads to

$$m/t = P/c^2 = (3 \times 10^{12} \text{ J/s})/(3.00 \times 10^8 \text{ m/s})^2 = \mathbf{3 \times 10^{-5} \text{ kg/s.}}$$

5. According to equation (28.6) the kinetic energy of the proton is

$$KE = (938 \text{ MeV}) \left[\frac{1}{\sqrt{1 - \frac{(0.999\,999\ c)^2}{c^2}}} - 1 \right] = \mathbf{6.62 \times 10^5 \text{ MeV.}}$$

6. According to equation (28.4) the energy is

$$E = 3.0 \times 10^4 \, E_0 = \frac{E_0}{\sqrt{1 - \dfrac{v^2}{c^2}}}$$

so that

$$\sqrt{1 - v^2/c^2} = 1/(3.0 \times 10^4).$$

Now, $L = L_0\sqrt{1 - v^2/c^2}$ and using $L_0 = 2.0$ km we obtain,

$$L = (2.0 \text{ km})/(3.0 \times 10^4) = 6.7 \times 10^{-5} \text{ km} = \textbf{6.7 cm}.$$

---

7. The following velocities are pertinent to this problem:

v_{VS} = velocity of the exploration **V**ehicle relative to the **S**paceship
v_{VE} = velocity of the exploration **V**ehicle relative to the **E**arth = +0.70c
v_{ES} = velocity of the **E**arth relative to the **S**paceship

Using the velocity-addition formula, equation (28.7), we can determine the velocity of the exploration vehicle relative to the spaceship:

$$v_{VS} = \frac{v_{VE} + v_{ES}}{1 + \dfrac{v_{VE}v_{ES}}{c^2}}$$

The velocity v_{ES} of the earth relative to the spaceship is the negative of the velocity of the spaceship relative to the earth: $v_{ES} = -v_{SE} = -0.50c$. The velocity of the exploration vehicle relative to the spaceship is

$$v_{VS} = \frac{v_{VE} + v_{ES}}{1 + \dfrac{v_{VE}v_{ES}}{c^2}} = \frac{+0.70c + (-0.50c)}{1 + \dfrac{(+0.70c)(-0.50c)}{c^2}} = \textbf{+0.31c}$$

---

8. (a) The velocities for this part of the problem are:

v_{2G} = velocity of block **2** relative to the **G**round
v_{21} = velocity of block **2** relative to block **1** = +0.50c
v_{1G} = velocity of block **1** relative to the **G**round = +0.50c

Using the velocity-addition formula, equation (28.7), the velocity of block 2 relative to the ground is

$$v_{2G} = \frac{v_{21} + v_{1G}}{1 + \dfrac{v_{21}v_{1G}}{c^2}} = \frac{+0.50c + 0.50c}{1 + \dfrac{(+0.50c)(+0.50c)}{c^2}} = \textbf{+0.80c}$$

(b) We will use the following velocities:

v_{3G} = velocity of block **3** relative to the **G**round
v_{32} = velocity of block **3** relative to block **2** = +0.50c
v_{2G} = velocity of block **2** relative to the **G**round = +0.80c

The velocity of block 3 relative to the ground is

$$v_{3G} = \frac{v_{32} + v_{2G}}{1 + \dfrac{v_{32}v_{2G}}{c^2}} = \frac{+0.50c + 0.80c}{1 + \dfrac{(+0.50c)(+0.80c)}{c^2}} = \mathbf{+0.93c}$$

CHAPTER 29
Particles and Waves

PREVIEW

This chapter begins the study of a branch of what is often called modern physics. Modern physics was developed mainly during this century and shows a new way of looking at nature which is both intriguing and often strange. One branch of modern physics called quantum mechanics views light as having both wave and particle properties and views matter as also having both particle and wave properties. Relativity is also part and parcel to modern physics and was discussed in an earlier chapter.

The main ideas of quantum mechanics grew out of the failure of classical physics to explain the results of two experiments: blackbody radiation and the photoelectric effect. Albert Einstein's explanation of the photoelectric effect led to the concept of light as being packets of energy called photons. Louis de Broglie, noticing that light appeared to behave both as a wave and a particle, suggested that electrons, protons, etc, which were normally thought of as particles may also behave like waves.

One of the foundation principles of quantum mechanics is the Heisenberg uncertainty principle which asserts that we cannot know some quantities such as the position and the momentum of an object accurately at the same time. In fact, the best we can do is to know the probability of the object having a certain momentum and a certain position.

QUICK REFERENCE

Important Terms

Wave-particle duality
   The idea that waves can exhibit particle-like characteristics and particles can exhibit wave-like characteristics.
Blackbody radiation
   The light emitted by a perfect emitter and absorber of light (a blackbody) due to its temperature.
Photoelectric effect
   The phenomenon in which light incident on a surface such as metal ejects electrons from the surface.
Photon
   A discrete packet of light energy postulated to exist by Albert Einstein to explain the photoelectric effect.
Work function
   The minimum energy needed to eject an electron from a metal surface.
Compton effect
   The phenomenon in which an x-ray photon is scattered from an electron with a change in frequency of the photon.
Compton wavelength of an electron
   Half the maximum wavelength change of a photon in a Compton scattering with an electron.
de Broglie wavelength
   The wavelength of the wave associated with a material particle. The de Broglie wavelength of a particle is inversely proportional to the momentum of the particle.
Wave function (Ψ)
   A function which contains all of the information which can be known about a particle or system of particles. The square of the wavefunction $(\Psi)^2$ is interpreted as being related to the probability of finding the particle at a certain position.
Quantum mechanics
   The theoretical framework for determining the wavefunction of a particle or system of particles.
Uncertainty principle
   A principle of quantum mechanics which asserts that it is impossible to know with 100% certainty the position and momentum of a particle at the same time.

Equations

Blackbody Radiation

The possible **energies** of the **atomic vibrators** which produce **blackbody radiation**.

$$E = nhf \qquad n = 0, 1, 2, 3, \ldots \qquad (29.1)$$

Photoelectric Effect

Energy of a photon

$$E = hf \qquad (29.2)$$

The **photoelectric** equation

$$hf = KE_{max} + W_0 \qquad (29.3)$$

Compton Effect

Conservation of energy for the **Compton effect**

$$hf = hf' + KE \qquad (29.4)$$

The **momentum of a photon**

$$p = \frac{hf}{c} = \frac{h}{\lambda} \qquad (29.6)$$

The **wavelength change** of the **scattered photon**

$$\lambda' - \lambda = \frac{h}{mc}(1 - \cos\theta) \qquad (29.7)$$

The **de Broglie wavelength** of a particle

$$\lambda = \frac{h}{p} \qquad (29.8)$$

The **Heisenberg uncertainty principle**

$$(\Delta p_y)(\Delta y) \geq \frac{h}{2\pi} \qquad (29.10)$$

$$(\Delta E)(\Delta t) \geq \frac{h}{2\pi} \qquad (29.11)$$

Planck's constant

$$h = 6.63 \times 10^{-34} \text{ J s}$$

$$h = 4.14 \times 10^{-15} \text{ (eV) s}$$

DISCUSSION OF SELECTED SECTIONS

29.3 Photons and the Photoelectric Effect

The explanation of the photoelectric effect by Albert Einstein led to the concept of the photon as a "particle" or quantum of light. The main results of the photoelectric effect experiment are:

1. Light incident on a clean metal surface can cause electrons to be ejected from the surface.
2. The time between the introduction of the light and the ejection of electrons is short.
3. Whether or not electrons were actually ejected from the metal depends on the frequency of the light. Above a certain frequency electrons are ejected, below this frequency no electrons are ejected.
4. If electrons are ejected, a higher frequency of light causes the electrons to have more kinetic energy (higher speeds).

Classical physics was useless in explaining the details of the photoelectric experiment because electromagnetic waves should:

1. give electrons in the metal energy a little at a time since the energy of a wave is spread out. Indeed it should take a long time (even days) for an electron to get enough energy to be ejected from the metal.
2. impart energy to the electron independent of the frequency of the wave. The energy carried by an electromagnetic wave depends on the square of its electric field.

Einstein assumed that light consisted of photons each of energy $E = hf$. This frequency dependence of the energy carried by light allowed Einstein to completely explain the photoelectric effect by an application of the conservation of energy (equation 29.3).

Example 1
What is the frequency of light which will cause electrons to be emitted from a magnesium surface whose work function is $W_0 = 3.68$ eV with a kinetic energy of 2.52 eV?

Equation (29.3) gives

$$f = \frac{KE_{max} + W_0}{h} = \frac{(1.60 \times 10^{-19} \text{ J}/ 1 \text{ eV})(2.52 \text{ eV} + 3.68 \text{ eV})}{6.63 \times 10^{-34} \text{ J s}}$$

$$f = 1.50 \times 10^{15} \text{ Hz}$$

Example 2
A laser beam has an average intensity of 1.5 W/m^2 and a wavelength of 515 nm. What is the density of photons in the beam?

The energy density of the beam is

$$u = \frac{\overline{S}}{c} = \frac{1.5 \text{ W/m}^2}{3.00 \times 10^8 \text{ m/s}} = 5.0 \times 10^{-9} \text{ J/m}^3$$

A single photon has energy

$$E = hf = hc/\lambda = (6.63 \times 10^{-34} \text{ J s})(3.00 \times 10^8 \text{ m/s})/(515 \times 10^{-9} \text{ m}) = 3.86 \times 10^{-19} \text{ J.}$$

The density of photons in the beam is then

$$\text{density} = u/E = (5.0 \times 10^{-9} \text{ J/m}^3)/(3.86 \times 10^{-19} \text{ J}) = 1.3 \times 10^{10} \text{ photons/m}^3.$$

29.4 The Momentum of a Photon and the Compton Effect

Evidence that light has particle properties comes from the collision of a photon with an electron. In this collision it can be demonstrated that BOTH energy and momentum are conserved if the momentum of the photon can be considered to be $p = h/\lambda$. The change in the wavelength of the scattered photons is given by equation (29.7).

Example 3
What is the momentum of a 0.055 nm photon? How fast must an electron travel to have this momentum (Assume the speed is much less than the speed of light)?

The momentum of the photon is

$$p = h/\lambda = (6.63 \times 10^{-34} \text{ J s})/(0.055 \times 10^{-9} \text{ m}) = 1.21 \times 10^{-23} \text{ kg m/s}.$$

An electron would have to travel with a speed of

$$v = p/m = (1.21 \times 10^{-23} \text{ kg m/s})/(9.11 \times 10^{-31} \text{ kg}) = 1.33 \times 10^{7} \text{ m/s}$$

to have the same momentum.

Example 4
A 3.10 MeV photon collides with a stationary electron and is scattered backwards. How much energy does the photon impart to the electron?

The photon gives the electron the same amount of energy that it loses since the collision is elastic. The change in the photon energy is

$$\Delta E = hf\,' - hf = hc(1/\lambda' - 1/\lambda)$$

The initial wavelength of the electron is

$$\lambda = c/f = hc/E$$
$$= (4.14 \times 10^{-15} \text{ eV s})(3.00 \times 10^{8} \text{ m/s})/(3.10 \times 10^{6} \text{ eV}) = 4.01 \times 10^{-13} \text{ m}.$$

The final wavelength of the photon is given by (29.7)

$$\lambda' = \lambda + (h/mc)(1 - \cos \theta) = 4.01 \times 10^{-13} \text{ m} + (2.43 \times 10^{-12} \text{ m})(1 - \cos 180°)$$

$$= 4.01 \times 10^{-13} \text{ m} + 4.86 \times 10^{-12} \text{ m} = 5.26 \times 10^{-12} \text{ m}.$$

Now

$$\Delta E = hc \left(\frac{1}{\lambda'} - \frac{1}{\lambda} \right) = (6.63 \times 10^{-34} \text{ J s})(3.00 \times 10^{8} \text{ m/s}) \left(\frac{1}{5.26 \times 10^{-12} \text{ m}} - \frac{1}{4.01 \times 10^{-13} \text{ m}} \right)$$

$$\Delta E = -4.58 \times 10^{-13} \text{ J}$$

The energy imparted to the electron is then 4.58×10^{-13} J.

29.5 The de Broglie Wavelength and the Wave Nature of Matter

Much of what you have already learned about waves applies to the wave nature of particles. The connection between the wave and particle nature of matter is in the de Broglie relationship (equation 29.8) which gives the wavelength of a particle. In principle, all matter has a wave nature, hence a de Broglie wavelength.

Example 5
A 75 kg person is running with a speed of 2.0 m/s. What is the de Broglie wavelength of the person? Comment on the potential importance of the wave nature of common objects. Equation (29.8) gives

$$\lambda = \frac{h}{mv} = \frac{6.63 \times 10^{-34} \text{ J s}}{(75 \text{ kg})(2.0 \text{ m/s})} = 4.4 \times 10^{-36} \text{ m}$$

This is an incredibly short wavelength compared to the size of objects in our everyday world. It's little wonder that we do not see the wave nature of people.

Example 6
An electron has been accelerated from rest through a potential difference of 2.5×10^3 V. What is the de Broglie wavelength of the electron? Comment on the potential importance of the wave nature of electrons.

The speed of the electron can be found by applying the conservation of mechanical energy

$$v = \sqrt{\frac{2eV}{m}} = \sqrt{\frac{2(1.6 \times 10^{-19} \text{ C})((2.5 \times 10^3 \text{ V})}{9.11 \times 10^{-31} \text{ kg}}} = 3.0 \times 10^7 \text{ m/s}$$

The de Broglie wavelength is then

$$\lambda = \frac{h}{mv} = \frac{6.63 \times 10^{-34} \text{ J s}}{(9.11 \times 10^{-31} \text{ kg})(3.0 \times 10^7 \text{ m/s})} = 2.4 \times 10^{-11} \text{ m}$$

The wavelength of the electron is small but comparable to the size of atoms and molecules. The wave nature of electrons is important, and we should expect them to interfere, diffract and generally behave like the other kinds of waves that you have studied.

29.6 The Heisenberg Uncertainty Principle

Since particles can behave like waves, it becomes difficult to tell exactly where a particle is located. It is very much like generating a wave on a pond and asking "Where is it?" Since the wave is spread out on the surface of the pond, its position is ambiguous. How much a water wave is spread out depends on the wavelength of the wave.

The Heisenberg uncertainty principle (equation 29.10) addresses this ambiguity for matter waves. The uncertainty of the knowledge of a particle's position is related to the uncertainty in the simultaneous knowledge of the particle's momentum. The more accurately you know the position, the less accurately you can know the momentum, and vice versa. Uncertainty also applies to the knowledge of the particle's energy and the time available for measuring the energy (equation 29.11). If you try to measure the energy in a short time the measurement will be very uncertain. However, if your measurement of the energy is made over a long time interval it can be quite accurate.

Example 7
A measurement of the position of an electron shows it to be within a region which is 0.15 nm wide. How much uncertainty is there in the electron's momentum? Compare this uncertainty with the momentum of an electron traveling at 5.5×10^6 m/s.

$$\Delta p \geq \frac{\frac{h}{2\pi}}{\Delta x} = \frac{\frac{6.63 \times 10^{-34} \text{ J s}}{2\pi}}{0.15 \times 10^{-9} \text{ m}} = 7.0 \times 10^{-25} \text{ kg m/s}$$

The momentum of an electron traveling at 5.5×10^6 m/s is

$$p = mv = (9.11 \times 10^{-31} \text{ kg})(5.5 \times 10^6 \text{ m/s}) = 5.0 \times 10^{-24} \text{ kg m/s}.$$

$$\Delta p/p = (7.0 \times 10^{-25} \text{ kg m/s})/(5.0 \times 10^{-24} \text{ kg m/s}) = 0.14.$$

The uncertainty is 14 % of the momentum.

Example 8
An atom loses energy by radiating away a photon. The process takes 1.5×10^{-8} s. What is the uncertainty in the energy of the photon? In its frequency?

Equation (29.11) gives the smallest uncertainty in the energy when the time interval is $\Delta t = 1.5 \times 10^{-8}$ s.

$$\Delta E = \frac{\frac{h}{2\pi}}{\Delta t} = \frac{\frac{6.63 \times 10^{-34} \text{ J s}}{2\pi}}{1.5 \times 10^{-8} \text{ s}} = 7.0 \times 10^{-27} \text{ J}$$

Now the frequency is $f = E/h$. If the energy of the photon is uncertain by at least ΔE, then the frequency of the photon is uncertain by at least

$$\Delta f = \Delta E/h = (7.0 \times 10^{-27} \text{ J})/(6.63 \times 10^{-34} \text{ J s}) = 1.1 \times 10^7 \text{ Hz}.$$

PRACTICE PROBLEMS

1. At night approximately 530 photons per second must enter an unaided human eye for an object to be seen, assuming the light is green. The light bulb in Example 1 in the text emits green light uniformly in all directions, and the diameter of the pupil of the eye is 7.0 mm. What is the maximum distance from which the bulb could be seen?

2. A photon scattered from an electron suffers a decrease in its wavelength of 25 % of the Compton wavelength of the electron. Through what angle was the electron scattered?

3. The repeating crystal structure of a solid can act like a diffraction grating for electrons. The atoms in the crystal correspond to the lines on a grating. Electrons traveling at a speed of 4.6×10^7 m/s are diffracted by a crystal whose atomic spacing is 150 nm. About how many orders of electron diffraction are produced?

4. Electrons traveling at a speed of 2.4×10^5 m/s are diffracted through a single slit and fall on a screen located 25 cm away. How wide must the slit be to produce a central fringe 0.50 mm wide?

5. Can any object ever be considered to be truly at rest? Explain why or why not on the basis of the Heisenberg uncertainty principle.

6. A vacuum in classical physics is considered to be empty space devoid of energy and matter. Is there such a thing as empty space devoid of energy or matter according to the uncertainty principle? HINT: Consider what could happen in a short time period, say 10^{-20} s.

7. In "Mr. Thompkins in Wonderland", by George Gamow, Mr. Thompkins, while listening to a stimulating lecture on quantum mechanics, doses off. He dreams that he is in a land where Planck's constant is a large value, say $h = 100.0$ J s. In this land a herd of giraffes moves through a row of evenly spaced trees with a speed of 0.12 m/s. If each giraffe has a mass of 360 kg, what must be the spacing of the trees for the giraffes to produce 3 complete orders of diffraction?

SOLUTIONS AND ANSWERS

Practice Problems

---

1. According to Example 1, 3.6×10^{18} photons/s are emitted by the bulb. These photons spread out over the surface of a sphere as they leave the bulb. After traveling a distance, R, to reach the eye, the light has an intensity

$$I = (3.6 \times 10^{18} \text{ photons/s})/(4\pi R^2).$$

In order for the eye to detect the light, the intensity entering the eye must be

$$I = (530 \text{ photons/s})/\pi r^2$$

The maximum distance that the light can travel and still be seen is

$$R = r\sqrt{(3.6 \times 10^{18} \text{ photons / s}) / (2120 \text{ photons / s})} = \mathbf{2.9 \times 10^5 \ m}$$

---

2. We must have $\lambda' - \lambda = (h/mc)(0.250)$. Comparing this with equation (29.7) reveals that

$$1 - \cos \theta = 0.250 \implies \theta = \mathbf{41.4} \ °.$$

---

3. The order number of the last order produced is

$$m = (d/\lambda) \sin 90° = d/\lambda.$$

The wavelength of the electron is

$$\lambda = h/p = (6.63 \times 10^{-34} \text{ J s})/[(9.11 \times 10^{-31} \text{ kg})(4.6 \times 10^7 \text{ m/s})] = 1.6 \times 10^{-11} \text{ m}.$$

The number of orders produced is then
$$m = \mathbf{9400.}$$

---

4. Using the small angle approximation and the same notation as used in the text, we find that the angle of diffraction is $\theta = \lambda/W$ and $\theta = y/L$ so
$$W = \lambda(L/y).$$

The de Broglie wavelength of the electrons is

$$\lambda = h/p = (6.63 \times 10^{-34} \text{ J s})/[(9.11 \times 10^{-31} \text{ kg})(2.4 \times 10^5 \text{ m/s})] = 3.0 \times 10^{-9} \text{ m}.$$

The slit width must be

$$W = (3.0 \times 10^{-9} \text{ m})[(2.5 \times 10^{-2} \text{ m})/(0.25 \times 10^{-3} \text{ m})] = \mathbf{3.0 \times 10^{-7} \ m.}$$

---

5. **No**. The Heisenberg uncertainty principle requires that the minimum uncertainty in the position of an object be

$$\Delta x = (h/2\pi)/\Delta p.$$

If an object is truly at rest, then Δp is zero and the object's position is completely uncertain. If an object is truly at rest though, it should be quite easy to find with great certainty. The contradiction leads us to the conclusion that the object cannot be at rest.

---

6. **No**. A measurement of the energy in the "vacuum" in a short time period will yield at least

$$\Delta E = (h/2\pi)/\Delta t.$$

For $\Delta t = 10^{-20}$ s the energy will be more than $\Delta E = 1.1 \times 10^{-14}$ J = 0.069 MeV.

---

7. The highest order of diffraction produced is m = 3. The separation needed between the trees is

$$d = (3\lambda/m) \sin 90° = 3\lambda.$$

The wavelength is

$$\lambda = h/p = (100.0 \text{ J s})/[(360 \text{ kg})(0.12 \text{ m/s})] = 2.3 \text{ m}.$$

Hence,

$$d = 3(2.3 \text{ m}) = \textbf{6.9 m}.$$

CHAPTER 30
The Nature of the Atom

PREVIEW

In this chapter you will study the nature of the atom. You will learn about line spectra, the Bohr model of the hydrogen atom, energy level diagrams, the quantum mechanical picture of the hydrogen atom, the Pauli exclusion principle, X-rays, and lasers.

QUICK REFERENCE

Important Terms

Line spectrum
    A series of discrete electromagnetic wavelengths emitted by the atoms of a low-pressure gas that is subjected to a sufficiently high potential difference. Certain groups of discrete wavelengths are referred to as a series.

Bohr model
    The electron in a single electron atom exists in circular orbits called stationary orbits. A photon is emitted when an electron changes from a higher energy orbit to a lower energy orbit.

Ionization energy
    The energy needed to remove an electron completely from an atom.

Principal quantum number
    An integer number, n, which determines the total energy of the hydrogen atom.

Orbital quantum number
    An integer number, l, which determines the angular momentum of the electron due to its orbital motion.

Magnetic quantum number
    An integer number, m_l, which determines the component of the angular momentum along a specific direction.

Spin quantum number
    A number, m_s, which is used to describe the intrinsic spin angular momentum of an electron.

Pauli exclusion principle
    No two electrons in an atom can have the same set values for the four quantum numbers n, l, m_l, m_s. This determines the way in which the electrons in multiple-electron atoms are distributed into shells and subshells, thus explaining the periodic table.

X-rays
    Electromagnetic waves emitted when high-energy electrons strike a metal target contained within an evacuated glass tube.

Laser
    A device that generates electromagnetic waves via a process known as "stimulated emission". The process produces waves that are coherent and may be confined to a very narrow beam.

Equations

The wavelengths of the **spectral line series of hydrogen** are given by:

$$\text{Lyman series:} \quad \frac{1}{\lambda} = R\left(\frac{1}{1^2} - \frac{1}{n^2}\right) \qquad n = 2, 3, 4, \ldots \qquad (30.1)$$

$$\text{Balmer series:} \quad \frac{1}{\lambda} = R\left(\frac{1}{2^2} - \frac{1}{n^2}\right) \qquad n = 3, 4, 5, \ldots \qquad (30.2)$$

$$\text{Paschen series:} \quad \frac{1}{\lambda} = R\left(\frac{1}{3^2} - \frac{1}{n^2}\right) \qquad n = 4, 5, 6, \ldots \qquad (30.3)$$

The **energy of the photon** emitted when an electron changes from a higher to a lower energy level is:

$$E_i - E_f = hf \qquad (30.4)$$

The **total energy of the atom** (Bohr theory) is given by:

$$E = KE + EPE = \frac{1}{2}mv^2 - \frac{kZe^2}{r} \qquad (30.5)$$

Equating the centripetal and Coulomb forces yields:

$$mv^2 = \frac{kZe^2}{r} \qquad (30.6)$$

The total energy can therefore be written:

$$E = \frac{1}{2}\left(\frac{kZe^2}{r}\right) - \frac{kZe^2}{r} = -\frac{kZe^2}{2r} \qquad (30.7)$$

According to Bohr, the **angular momentum L_n** is given by:

$$L_n = mv_n r_n = n\frac{h}{2\pi} \qquad n = 1, 2, 3, \ldots \qquad (30.8)$$

The **radius of the nth Bohr orbit** is:

$$r_n = \left(\frac{h^2}{4\pi^2 mke^2}\right)\frac{n^2}{Z} \qquad n = 1, 2, 3, \ldots \qquad (30.9)$$

Substituting the constants into equation (30.9) yields:

$$r_n = (5.29 \times 10^{-11}\,\text{m})\frac{n^2}{Z} \qquad n = 1, 2, 3, \ldots \qquad (30.10)$$

The **energy of the nth Bohr orbit** is:

$$E_n = -\left(\frac{2\pi^2 mk^2 e^4}{h^2}\right)\frac{Z^2}{n^2} \qquad n = 1, 2, 3, \ldots \quad (30.11)$$

The **Bohr energy levels** can then be written as:

$$E_n = -(2.18 \times 10^{-18} \text{ J})\frac{Z^2}{n^2} \qquad n = 1, 2, 3, \ldots \quad (30.12)$$

The **Bohr energy levels** expressed **in electron volts**:

$$E_n = -(13.6 \text{ eV})\frac{Z^2}{n^2} \qquad n = 1, 2, 3, \ldots \quad (30.13)$$

The **wavelengths radiated by the hydrogen atom** are given by:

$$\frac{1}{\lambda} = \frac{2\pi^2 mk^2 e^4}{h^3 c}\,(Z^2)\left(\frac{1}{n_f^2} - \frac{1}{n_i^2}\right) \qquad (30.14)$$

$$n_i, n_f = 1, 2, 3, \ldots \text{ and } n_i > n_f$$

The **magnitude of the angular momentum of an electron** is:

$$L = \sqrt{l\,(l+1)}\ \frac{h}{2\pi} \qquad (30.15)$$

The **angular momentum in the z direction** is:

$$L_z = m_l\,\frac{h}{2\pi} \qquad (30.16)$$

The **minimum wavelength of the photon emitted by an X-ray tube** is:

$$\lambda_0 = \frac{hc}{eV} \qquad (30.17)$$

DISCUSSION OF SELECTED SECTIONS

30.2 Line Spectra

When a low-pressure gas contained within an evacuated tube is subjected to a sufficiently large potential difference, the gas will emit electromagnetic waves. With a grating spectrometer the individual wavelengths emitted by the gas can be separated and identified as a series of bright fringes or lines. The series of lines is called a **line spectrum**. The simplest line spectrum is that of the hydrogen atom. In schematic form, the following figure illustrates one series of lines for atomic hydrogen, the **Balmer series**.

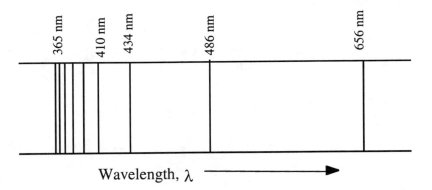

Balmer found an empirical equation that gave the values for the observed wavelengths. This equation is given below, along with similar equations that apply to the **Lyman** and **Paschen** series of hydrogen lines.

$$\text{Lyman series:} \qquad \frac{1}{\lambda} = R\left(\frac{1}{1^2} - \frac{1}{n^2}\right) \qquad n = 2, 3, 4, \ldots$$

$$\text{Balmer series:} \qquad \frac{1}{\lambda} = R\left(\frac{1}{2^2} - \frac{1}{n^2}\right) \qquad n = 3, 4, 5, \ldots$$

$$\text{Paschen series:} \qquad \frac{1}{\lambda} = R\left(\frac{1}{3^2} - \frac{1}{n^2}\right) \qquad n = 4, 5, 6, \ldots$$

In these equations, the constant term R, known as the **Rydberg constant**, has the value $R = 1.097 \times 10^7 \text{ m}^{-1}$.

Example 1
Find the wavelength of the $n = 4$ Balmer line. Using equation (30.2) we obtain,

$$\frac{1}{\lambda} = R\left(\frac{1}{2^2} - \frac{1}{n^2}\right) = (1.097 \times 10^7 \text{ m}^{-1})\left(\frac{1}{2^2} - \frac{1}{4^2}\right) = 2.057 \times 10^6 \text{ m}^{-1} \Rightarrow \lambda = 486 \text{ nm}$$

30.3 The Bohr Model of the Hydrogen Atom

Bohr's theory of the hydrogen atom has a positively charged nucleus and a negatively charged electron moving around it in a circular orbit. Bohr hypothesized that in a hydrogen atom there can only be certain values of the total energy (electron kinetic energy plus potential energy). These allowed energy levels correspond to different orbits for the electron as it moves around the nucleus, the larger orbits being associated with larger total energies. An electron in one of these orbits does not radiate electromagnetic waves. These orbits are sometimes referred to as **stationary orbits** or **stationary states**.

Bohr theorized that a photon is emitted only when the electron changes orbits from a larger one with higher energy to a smaller one with a lower energy. If E_i is the energy of the initial orbit (higher energy orbit), E_f the energy of the final orbit (lower energy), and the energy of the photon is hf, where f is the frequency and h is Planck's constant, then

$$E_i - E_f = hf$$

For an electron of mass m and speed v in an orbit of radius r, the total energy is the kinetic energy of the electron plus the electric potential energy. Therefore, we can write

$$E = KE + PE = \frac{1}{2}mv^2 - \frac{kZe^2}{r} \qquad (30.5)$$

Equation (30.5) can be modified to yield the radii of the various orbits r_n and the energies of these orbits E_n. As detailed in the textbook, by using the Coulomb electrostatic force as the centripetal force (equation 30.6), and by assuming that the allowed energy levels have angular momenta that are integral multiples of Planck's constant (equation 30.8), the radii for the **Bohr orbits** are shown to be

$$r_n = (5.29 \times 10^{-11} \text{ m}) \frac{n^2}{Z} \qquad n = 1, 2, 3, \ldots \qquad (30.10)$$

For the hydrogen atom (Z = 1) and the smallest Bohr orbit (n = 1), $r_1 = 5.29 \times 10^{-11}$ m. This particular value is often referred to as the **Bohr radius**. Equation (30.10) can now be used in equation (30.7) to show that the energy of the nth Bohr orbit is given by

$$E_n = -(2.18 \times 10^{-18} \text{ J}) \frac{Z^2}{n^2} \qquad n = 1, 2, 3, \ldots$$

This energy is often expressed in electron volts, where 1 eV = 1.60×10^{-19} J. Therefore, we can write

$$E_n = -(13.6 \text{ eV}) \frac{Z^2}{n^2} \qquad n = 1, 2, 3, \ldots \qquad (30.13)$$

It is useful to represent the energy values given by equation (30.13) on an **energy level diagram**. For the hydrogen atom (Z = 1), the highest energy level corresponds to n = ∞ in equation (30.13) and has an energy of 0 eV. This is the energy of the atom when the electron is completely removed (r = ∞) from the nucleus and is at rest. The lowest energy level corresponds to n = 1 and has a value of - 13.6 eV. The lowest energy level is called the **ground state** while higher energy levels are called **excited levels**. To determine the wavelengths radiated by the atom, equation (30.13) for the energies can be used in equation (30.4) with f = c/λ to yield equation (30.14). The Lyman, Balmer, and Paschen series of lines in the hydrogen atom spectrum correspond to transitions that the electron makes between higher and lower energy levels, as shown in the diagram below.

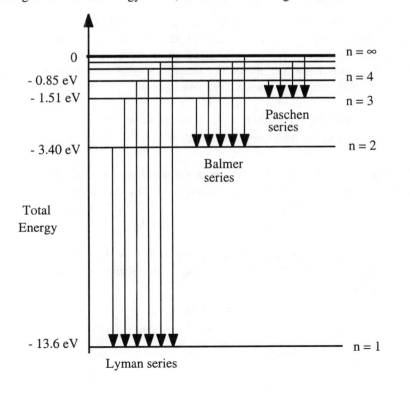

The wavelengths of the line spectra are given by:

$$\frac{1}{\lambda} = \frac{2\pi^2 m k^2 e^4}{h^3 c} (Z^2)\left(\frac{1}{n_f^2} - \frac{1}{n_i^2}\right) \qquad (30.14)$$

$$n_i, n_f = 1, 2, 3, \ldots \text{ and } n_i > n_f$$

Example 2
Lines in the Paschen series are produced when electrons, excited to higher levels, make transitions to the n = 3 level. Determine (a) the longest wavelength in the series, (b) the wavelength that corresponds to the transition from $n_i = 7$ to $n_f = 3$. (c) What spectral region are these lines found in?

(a) The longest wavelength corresponds to the transition which has the smallest energy change. This occurs between levels $n_i = 4$ to $n_f = 3$. Using equation (30.14) with Z = 1, we find that

$$\frac{1}{\lambda} = (1.097 \times 10^7 \, m^{-1})\left(\frac{1}{3^2} - \frac{1}{4^2}\right) = 5.33 \times 10^5 \, m^{-1} \Rightarrow \lambda = 1880 \text{ nm}$$

(b) For the transition from $n_i = 7$ to $n_f = 3$:

$$\frac{1}{\lambda} = (1.097 \times 10^7 \, m^{-1})\left(\frac{1}{3^2} - \frac{1}{7^2}\right) = 9.95 \times 10^5 \, m^{-1} \Rightarrow \lambda = 1010 \text{ nm}$$

(c) According to Figure 24.9, these lines lie in the **infrared** region of the electromagnetic spectrum.

30.5 The Quantum Mechanical Picture of the Hydrogen Atom

The picture of the hydrogen atom that quantum mechanics provides differs in a number of ways from the Bohr model. The Bohr model uses a single integer number n to identify the various electron orbits and the associated energies. Because this number can have only discrete values, rather than a continuous range of values, n is called a **quantum number**. Quantum mechanics, on the other hand, reveals that four different quantum numbers are needed to describe each state of the hydrogen atom. They are described below.

1. **The principal quantum number n**. This is the same number used by Bohr to describe the total energy of the atom and can have only integer values: n = 1, 2, 3, ... $E_n = - (13.6 \text{ eV}) Z^2/n^2$.

2. **The orbital quantum number l**. This number determines the angular momentum of the electron due to its orbital motion. Only the following integer values of l are allowed:

$$l = 0, 1, 2, \ldots, (n - 1)$$

For instance, if n = 1, the orbital quantum number can only have the value $l = 0$. For n = 5, possible values for l are $l = 0$, 1, 2, 3, and 4. According to the Schrodinger equation, the magnitude L of the angular momentum of the electron is

$$L = \sqrt{l (l + 1)} \, \frac{h}{2\pi}$$

3. **The magnetic quantum number m_l**. This number determines the component of the angular momentum along a specific direction, which is called the z direction by convention. The values that m_l can have depend on the value of l , with only the following positive and negative integers being permitted:

$$m_l = 0, \pm 1, \pm 2, \ldots, \pm l$$

For example, if the angular momentum quantum number is $l = 2$, then the magnetic quantum number can have values $m_l = -2, -1, 0, +1, +2$. The Schrodinger equations shows that the component L_z of the angular momentum in the z direction is

$$L_z = m_l \frac{h}{2\pi}$$

4. **The spin quantum number m_s**. This number is needed because the electron has an intrinsic property called spin angular momentum. There are only two possible values for the spin quantum number. They are:

$$m_s = +\frac{1}{2} \quad \text{(spin up)} \qquad \text{or} \qquad m_s = -\frac{1}{2} \quad \text{(spin down)}$$

Example 3
Write down the eighteen possible sets of four quantum numbers that exist when the principal quantum number is 3.

Using the rules listed above, for n = 3 we have:

n	l	m_l	m_s
3	0	0	1/2
3	0	0	-1/2
3	1	1	1/2
3	1	1	-1/2
3	1	0	1/2
3	1	0	-1/2
3	1	-1	1/2
3	1	-1	-1/2
3	2	2	1/2
3	2	2	-1/2
3	2	1	1/2
3	2	1	-1/2
3	2	0	1/2
3	2	0	-1/2
3	2	-1	1/2
3	2	-1	-1/2
3	2	-2	1/2
3	2	-2	-1/2

30.6   The Pauli Exclusion Principle and the Periodic Table of the Elements

In multiple-electron atoms, all electrons with the same value of n are said to be in the same **shell**. For example, electrons with n = 1 are in a single shell (sometimes called the K shell), those with n = 2 are in another shell (the L shell), those with n = 3 are in another (the M shell), and so on. Those electrons with the same values for both n and l are said to be in the same **subshell**. So the n = 1 shell has one subshell, $l = 0$. The n = 2 shell has two subshells, $l = 0$ and $l = 1$, and so forth.

In most atoms, the electron spends most of its time in the lowest energy level n = 1 called the **ground state**. However, when a multiple-electron atom is in the ground state, not every electron is crowded into the n = 1 shell. The reason for this is that electrons obey the **Pauli exclusion principle**. This principle states that no two electrons in an atom can have the same set of values for the four quantum numbers n, l, m_l, and m_s.

Because of the Pauli exclusion principle, there is a maximum number of electrons that can fit into an energy level or subshell. For example, the n = 1, l = 0 subshell can hold at most two electrons. The n = 2, l = 1 subshell, however, can hold six electrons, because with l = 1 there are three possibilities for m_l (-1, 0, 1), each of which have two values of m_s (+1/2, -1/2). The maximum number of electrons the l subshell can hold is $2(2l + 1)$. So, for instance, the n = 5, l = 4 subshell can hold 2[2(4) + 1] = 18 electrons.

Example 4
How many electrons can be put into (a) the n = 4 shell and (b) the n = 5 shell?

(a) Since there are $2(2l + 1)$ possible electrons in each subshell, and since l can have values up to (n - 1), for n = 4, we have l = 0, 1, 2, and 3. So for l = 0 there are two electrons, for l = 1 there are six electrons, for l = 2 there are ten electrons and for l = 3 there are fourteen electrons. For the n = 4 shell, therefore, there are a maximum of N = 2 + 6 + 10 + 14 = 32 electrons.

(b) For the n = 5 shell, the l = 4 subshell can have 18 possible electrons. Therefore, in the entire n = 5 shell there can be as many as N = 2 + 6 + 10 + 14 + 18 = 50 electrons.

A shorthand notation is often used for the electronic configuration of the atom. For instance, the l = 0 subshell is called the s subshell. The l = 1 and 2 subshells are known as the p and d subshells, respectively. Higher values of l are referred to as f, g, h, etc. This convention of letters is used in a shorthand notation that indicates the principal quantum number n, the orbital quantum number l, and the number of electrons in the n, l, subshell. We have

The ground state configurations for a few elements in the periodic table are listed below.

Element	Number of Electrons	Electronic Configuration
Hydrogen	1	$1s^1$
Helium	2	$1s^2$
Lithium	3	$1s^2 2s^1$
Boron	5	$1s^2 2s^2 2p^1$
Neon	10	$1s^2 2s^2 2p^6$
Sodium	11	$1s^2 2s^2 2p^6 3s^1$
Aluminum	13	$1s^2 2s^2 2p^6 3s^2 3p^1$
Argon	18	$1s^2 2s^2 2p^6 3s^2 3p^6$
Potassium	19	$1s^2 2s^2 2p^6 3s^2 3p^6 4s^1$

PRACTICE PROBLEMS

1. Determine the radius and energy of the hydrogen atom in the n = 3 state of the Bohr model.

2. What are the wavelengths of the longest and shortest wavelength lines in the Brackett series which has $n_f = 4$.

3. Compute the ionization energy of doubly ionized lithium Li^{2+} (Z = 3).

4. Light of 1090-nm wavelength is emitted from a hydrogen discharge tube. If the transition that produces this emission occurs to the $n_f = 3$ level, from which level n_i did it originate?

5. Assume that 1200 hydrogen atoms are initially in the n = 4 state. The electrons then make transitions to lower energy levels. (a) How many distinct spectral lines will be emitted? (b) Assume, for simplicity, that all possible transitions to lower levels are equally probable from a given excited state. Determine the total number of photons emitted.

6. Write down the fourteen sets of the four quantum numbers that correspond to the electrons in a completely filled 4f shell.

7. Write down the ground state electronic configuration for iron ($Z = 26$).

8. Suppose tungsten ($Z = 74$) has all but one of its electrons removed. (a) How much energy (in eV) is required to remove the last electron completely if the ionized atom is initially in the ground state? (b) What would be the wavelength of the photon emitted if the remaining electron were to make the transition $n = 2$ to $n = 1$? (c) In what region of the spectrum is such a photon found?

9. Suppose the X-ray machine in a dentist's office uses a potential difference of 25 kV to operate the X-ray tube. What is the shortest X-ray wavelength emitted by the machine?

10. Calculate the approximate minimum energy necessary to observe the $n = 2$ to $n = 1$ characteristic X-ray of gold.

SOLUTIONS AND ANSWERS

Practice Problems

1. Using equation (30.10) we have,

$$r_3 = (5.29 \times 10^{-11} \text{ m})(3)^2/1 = \textbf{4.76} \times \textbf{10}^{\textbf{-10}} \textbf{ m}.$$

   Equation (30.13) gives,

$$E_3 = -(13.6 \text{ eV})(1)^2/(3)^2 = \textbf{- 1.51 eV}.$$

2. The longest wavelength occurs from $n_i = 5$ to $n_f = 4$. Equation (30.14) gives,

$$(1/\lambda) = (1.097 \times 10^7 \text{ m}^{-1})(1/4^2 - 1/5^2) = 2.47 \times 10^5 \text{ m}^{-1},$$

   so that $\lambda = \textbf{4050 nm}$.
   The shortest wavelength occurs when $n_i = \infty$. Therefore,

$$(1/\lambda) = (1.097 \times 10^7 \text{ m}^{-1})(1/4^2 - 1/\infty^2) = 6.86 \times 10^5 \text{ m}^{-1},$$

   so that $\lambda = \textbf{1460 nm}$.

3. The ionization energy for doubly ionized lithium ($Z = 3$) can be found using equation (30.13).

$$E_n = -(13.6 \text{ eV}) Z^2/n^2 = -(13.6 \text{ eV})(3)^2/(1)^2 = \textbf{- 122 eV}.$$

4. Using equation (30.14),

$$(1/\lambda) = (1.097 \times 10^7 \text{ m}^{-1})(1/n_f^2 - 1/n_i^2).$$

   With $\lambda = 1090 \text{ nm} = 1.09 \times 10^{-6}$ m and $n_f = 3$ we obtain $n_i = \textbf{6}$.

5. (a) Transitions beginning on $n = 4$ and ending on $n = 1$ can take the following paths:

$$n = 4 \rightarrow n = 3 \rightarrow n = 2 \rightarrow n = 1 \text{ (3 spectral lines)},$$

   or

$$n = 4 \rightarrow n = 2 \rightarrow n = 1 \text{ (one new line)},$$

   or

$$n = 4 \rightarrow n = 1 \text{ (one new line)},$$

   and also

$$n = 4 \rightarrow n = 3 \rightarrow n = 1 \text{ (one new line)}.$$

   The total number of distinct lines = **6**.

   (b) If all transitions are equally probable, begin with 1200 in the $n = 4$ state. One-third goes to $n = 3$, one third goes to $n = 2$ and one third goes to $n = 1$ (for a total of $400 + 400 + 400 = 1200$ photons).
   In the $n = 3$ state, half goes to $n = 2$ and half to $n = 1$ ($200 + 200 = 400$ photons). In the $n = 2$ state, where there are now 600 electrons, all 600 produce photons in going to the $n = 1$ state. The total number of photons produced is therefore,

$$1200 + 400 + 600 = \textbf{2200 photons}.$$

6.

n	l	m_l	m_s
4	3	3	1/2
4	3	3	-1/2
4	3	2	1/2
4	3	2	-1/2
4	3	1	1/2
4	3	1	-1/2
4	3	0	1/2
4	3	0	-1/2
4	3	-1	1/2
4	3	-1	-1/2
4	3	-2	1/2
4	3	-2	-1/2
4	3	-3	1/2
4	3	-3	-1/2

7. The ground state configuration for iron (Z = 26) is

$$1s^2 2s^2 2p^6 3s^2 3p^6 4s^2 3d^6$$

8. (a) To find the ionization energy use equation (30.13),

$$E_I = (13.6 \text{ eV})(74)^2 = \textbf{74 keV}.$$

(b) Use equation (30.14),

$$(1/\lambda) = (1.097 \times 10^7 \text{ m}^{-1})(74)^2 (1/1^2 - 1/2^2) = 4.5 \times 10^{10} \text{ m}^{-1}$$

which gives $\lambda = \textbf{2.2} \times \textbf{10}^{\textbf{-11}}$ **m**.

(c) This wavelength corresponds to a photon in the **X-ray** region of the spectrum.

9. Use equation (30.17) to obtain

$$\lambda_0 = hc/eV = (6.63 \times 10^{-34} \text{ J s})(3.00 \times 10^8 \text{ m/s})/[(1.60 \times 10^{-19} \text{ C})(25\ 000 \text{ V})]$$
$$\lambda_0 = \textbf{5.0} \times \textbf{10}^{\textbf{-11}} \textbf{ m}.$$

10. To account for the shielding effect, Z = 78 rather than 79 should be used in equation (30.13). We have,

$$E_n = - (13.6 \text{ eV}) Z^2/n^2 = - (13.6 \text{ eV}) (78)^2 [1/(1)^2 - 1/(2)^2] = \textbf{- 6.2} \times \textbf{10}^{\textbf{4}} \textbf{ eV}.$$

CHAPTER 31
Nuclear Physics
and Radioactivity

PREVIEW

One of the great successes of physics during the twentieth century was the development of the model of the atom and the subsequent description of its nucleus. In this chapter you will learn about the nucleon inhabitants of the nucleus: the neutron and proton. You will learn how the size of the nucleus depends on the number of nucleons.

Also, you will learn about the force which binds the nucleons together and the energy associated with this binding. If the nucleus is "split", some of this energy can be released. This fact provides the basis for nuclear power.

One-half of radioactive nuclei spontaneously disintegrate after a certain amount of time. You will be able to calculate this half-life in some cases and to determine how much radioactive material has disintegrated after a given time. This idea is used for dating archeological artifacts.

QUICK REFERENCE

Important Terms

Nucleons
Protons and neutrons.

Atomic number (Z)
The total number of protons in a nucleus.

Atomic mass number (A)
The total number of nucleons in a nucleus.

Isotopes
Nuclei which have the same number of protons but different number of neutrons.

Strong nuclear force
One of the four known fundamental forces in nature. The strong force is the force exerted between nucleons which binds them together in a nucleus.

Radioactivity
The spontaneous disintegration or rearrangement of the internal structure of nuclei.

Binding energy
The energy required to break apart a nucleus into separated nucleons.

Mass defect
The difference in the mass of an intact nucleus and total mass of the separated nucleons. This is the mass equivalent of the binding energy of the nucleus.

α particle
A helium nucleus consisting of two protons and two neutrons.

α decay
The radioactive emission of an α particle from a nucleus.

Transmutation
The changing of one element to another by a loss or gain of one or more protons. Often transmutation is caused by α decay.

β decay
The radioactive decay of a nucleus by the emission of an electron or positron.

γ decay

The radioactive decay of a nucleus by emission of a high frequency "gamma" photon.

Neutrino

A particle created during the β decay of a nucleus.

Half-life

The time necessary for one-half of a particular radioactive substance to undergo spontaneous decay.

Radioactive decay series

The sequential decay of one nuclear species after another.

Geiger counter

A common gas filled device used for detecting α, β, and γ emissions of radioactive nuclei.

Scintillation counter

A device consisting of a material which emits light when struck by ionizing radiation. The material is mounted on a photomultiplier tube.

Semiconductor detector

A device which detects ionizing radiation by the electrons and holes produced in a semiconductor material.

Cloud chamber

A device which detects ionizing radiation by its condensation track left in a supercooled gas.

Bubble chamber

A device which detects ionizing radiation by its vapor track left in a superheated liquid.

Equations

The **relationship** between the **atomic number**, Z, number of neutrons, N, and the **atomic mass number**, A, of a nucleus

$$A = Z + N \tag{31.1}$$

The approximate **radius of a nucleus**

$$r \cong (1.2 \times 10^{-15} \, m) A^{1/3} \tag{31.2}$$

The **binding energy** of a nucleus

$$\text{Binding energy} = (\Delta m) c^2 \tag{31.3}$$

The **activity** of a radioactive sample is the magnitude of

$$\frac{\Delta N}{\Delta t} = -\lambda N \tag{31.4}$$

The **number of radioactive nuclei** present at time, t, in a sample

$$N = N_0 \, e^{-\lambda t} \tag{31.5}$$

The **half-life** of a radioactive material

$$T_{1/2} = \frac{\ln 2}{\lambda} = \frac{0.693}{\lambda} \tag{31.6}$$

DISCUSSION OF SELECTED SECTIONS

31.1 Nuclear Structure

Example 1

Two isotopes of chlorine occur in nature. The $^{35}_{17}$Cl isotope has an atomic mass of 34.968 85 u and a natural abundance of 75.77%. The $^{37}_{17}$Cl has an atomic mass of 36.965 90 u and a natural abundance of 24.23%. By a calculation of your own, verify that the value of 35.45 u listed in the periodic table is a weighted average of these individual atomic masses.

The weighted average of the masses is

$$(0.7577)(34.968\ 85\ u) + (0.2423)(36.965\ 90\ u) = 35.45\ u$$

Example 2

What is the approximate radius of lead $^{207}_{82}$Pb ?

The atomic mass number of the lead is A = 207. Equation (31.2) gives the approximate radius to be

$$r \approx (1.2 \times 10^{-15}\ m)\ A^{1/3} = (1.2 \times 10^{-15}\ m)(207)^{1/3} = 7.1 \times 10^{-15}\ m$$

31.3 The Mass Defect of the Nucleus and Nuclear Binding Energy

Relativity tells us that mass and energy are equivalent and related by $E = mc^2$. A nucleus has less energy than the total energy of its separated constituents since some energy is released in the formation of the nucleus (binding energy). Hence, the mass of the nucleus is less than the total mass of its separated nucleons. The difference in mass is called the mass defect and is the mass equivalent of the binding energy of the nucleus. The specific relationship between the binding energy and mass defect is given by equation (31.3).

Note: The total mass of an atom includes the mass of its complement of electrons. The mass of the electrons must be taken into account when calculating mass defects.

Example 3

What is the mass defect and the binding energy for $^{239}_{93}$Np ?

The total mass of the atom including electrons is found in Appendix F to be 239.052 933 u. The atom has 93 protons, 93 electrons and 146 neutrons. The total mass of these separated particles is

$$93(1.007\ 276\ u) + 93(5.485\ 799 \times 10^{-4}\ u) + 146(1.008\ 665\ u) = 240.992\ 776\ u.$$

The mass defect of the nucleus is then

$$\Delta m = 240.992\ 776\ u - 239.052\ 933\ u = 1.939\ 843\ u$$

The corresponding binding energy is

$$(\Delta m)c^2 = (1.939\ 843\ u)((931.5\ MeV/1\ u) = 1807\ MeV$$

31.4 Radioactivity

During radioactive decay a nucleus emits an α particle, a β particle or a γ ray. During any of these decay processes, the mass/energy, charge, linear momentum, angular momentum and nucleon number are conserved.

Example 4

What daughter nucleus would be formed if $^{238}_{92}U$ could undergo (a) α decay? (b) β decay? (c) γ decay?

(a) Write the reaction as

$$^{238}_{92}U \rightarrow {}^{A}_{Z}X + {}^{4}_{2}He$$

The conservation of nucleon number requires

$$238 = A + 4 \rightarrow A = 234$$

The conservation of charge requires

$$92 = Z + 2 \rightarrow Z = 90$$

We find that the daughter nucleus is $^{234}_{90}Th$.

(b) The reaction is

$$^{238}_{92}U \rightarrow {}^{A}_{Z}X + {}^{0}_{-1}e$$

The conservation of nucleon number gives

$$238 = A + 0 \rightarrow A = 238$$

The conservation of charge requires

$$92 = Z + (-1) \rightarrow Z = 93$$

The periodic chart shows that the daughter nucleus is $^{238}_{93}Np$.

(c) The reaction in this case is

$$^{238}_{92}U \rightarrow {}^{A}_{Z}X + \gamma$$

The γ ray does not have a nucleon number or a charge, hence A = 238 and Z = 92. The daughter nucleus is $^{238}_{92}U$.

31.6 Radioactive Decay and Activity

Example 5
The radioactive isotopes of strontium are particularly dangerous to animals since strontium behaves chemically like calcium and is taken up by the body and deposited in bone. Much of the radioactive strontium in the atmosphere is $^{90}_{38}$Sr which was released by atmospheric nuclear bomb testing during the 1950's and early 1960's. Assume that no strontium has been introduced into the atmosphere for 28.0 years. What percentage of the original radioactive nuclei is still present in the atmosphere? Strontium has a half-life of 28.5 years.

In any given volume of air there are still

$$N = N_0 e^{-\lambda t}$$

radioactive nuclei left. The percentage left is then

$$\frac{N}{N_0} \times 100\% = 100\% \times e^{-\lambda t}$$

The decay constant, λ, of the strontium isotope is

$$\lambda = (0.693)/T_{1/2} = (0.693)/(28.5 \text{ yr}) = 0.02432 \text{ yr}^{-1}$$

Note that we have carried an extra figure for additional accuracy. The percent of strontium left in the atmosphere is

$$e^{-(0.02432 \text{ yr}^{-1})(28.0 \text{ yr})} \times 100\% = 50.6\%$$

31.7 Radioactive Dating

Example 6
The shroud of Turin was originally believed to be about two thousand years old, but was determined by dating techniques to be about 700 years old. Assuming that the shroud had a carbon-14 activity of 0.23 Bq per gram when it was made, what is its activity now?

The number of nuclei left in a given amount of the shroud is

$$N = N_0 e^{-\lambda t}$$

The activity, $A = \lambda N$ so

$$A = A_0 e^{-\lambda t}$$

The decay constant for carbon-14 is

$$\lambda = (0.693)/T_{1/2} = (0.693)/(5730 \text{ yr}) = 1.21 \times 10^{-4} \text{ yr}^{-1}$$

Now the per gram activity is

$$A = (0.23 \text{ Bq})e^{-(1.21\times10^{-4} \text{ yr}^{-1})(700 \text{ yr})} = 0.21 \text{ Bq}$$

PRACTICE PROBLEMS

1. What is the radius of a nucleus of $^{28}_{14}\text{Si}$?

2. What is the binding energy per nucleon of $^{28}_{14}\text{Si}$?

3. Polonium $^{214}_{84}\text{Po}$ undergoes both α and γ decay. What daughter nucleus is produced in each case?

4. An intermediate nucleus in the decay series of $^{238}_{92}\text{U}$ is $^{230}_{90}\text{Th}$. After several successive α decays, this intermediate nucleus leads to the isotope $^{214}_{82}\text{Pb}$. How many α decays have occurred?

5. The isotope $^{89}_{36}$Kr has a lifetime of 3.16 min. What percentage of Krypton nuclei are left after 15.0 min?

6. The activity of a sample of a radioactive material is 1.50 Ci and the half-life is 1.6 X 10^3 yr. How many of the radioactive nuclei are left after 250 yr?

7. Plutonium $^{239}_{94}$Pu (atomic mass = 239.052 157 u) undergoes α decay. Assuming all the released energy is in the form of kinetic energy of the α particle and ignoring the recoil of the daughter nucleus, find the speed of the α particle.

8. One gram of charcoal from an ancient campfire is analyzed for $^{14}_{6}$C and found to have an activity of 0.11 Bq. How old is the charcoal?

SOLUTIONS AND ANSWERS

Practice Problems

---

1. Equation (31.2) gives the approximate radius to be

$$r \approx (1.2 \times 10^{-15} \text{ m})(28)^{1/3} = \textbf{3.6} \times \textbf{10}^{-15} \text{ m.}$$

---

2. The mass defect for the nucleus is

$$\Delta m = 14(1.007\ 825 \text{ u}) + 14(1.008\ 665 \text{ u}) - 27.976\ 927 \text{ u} = 0.253\ 933 \text{ u}$$

The energy equivalent of this mass defect is

$$\text{Binding energy} = (931.5 \text{ MeV/1 u})(0.253\ 933 \text{ u}) = 236.5 \text{ MeV.}$$

There are 28 nucleons, so the energy corresponds to an energy/nucleon of **8.448 MeV/nucleon**.

---

3. For α decay the atomic mass number is conserved so the daughter nucleus has A = 214 - 4 = 210. The charge is also conserved so the daughter nucleus has an atomic number of Z = 84 - 2 = 82. Thus we see that the daughter nucleus is $^{210}_{82}\textbf{Pb}$.

In a gamma decay the nucleus loses only energy, hence the daughter is a de-excited $^{214}_{84}\textbf{Po}$.

---

4. The reaction is

$$^{230}_{90}\text{Th} \quad \rightarrow \quad ^{214}_{82}\text{Pb} + \text{N} \, ^{4}_{2}\text{He}$$

where N is the number of α decays which have occurred.

The conservation of charge gives
$$90 = 82 + 2\text{N} \implies \text{N} = 4.$$

The conservation of atomic mass number gives

$$230 = 214 + 4\text{N} \implies \text{N} = 4.$$
The number of α decays needed is **4.**

---

5. Equation (31.5) gives the percentage of nuclei left.

$$(\text{N/N}_0) \times 100 \% = e^{-\lambda t} \times 100 \%$$

where $\lambda = (0.693)/\text{T}_{1/2} = 0.2193 \text{ min}^{-1}$.

The percentage of nuclei left is **3.73 %**.

---

6. Equation (31.4) gives for the initial case: Activity = $-\lambda N_0$, where $\lambda = (0.693)/T_{1/2}$. The number of nuclei present initially is

$$N_0 = (1.5 \text{ Ci})(3.70 \times 10^{10} \text{ Bq}/ 1 \text{ Ci})/[(0.693)/(1.6 \times 10^3 \text{ yr})] = 1.3 \times 10^{14}.$$

The number of nuclei remaining after 250 yr is

$$N = (1.3 \times 10^{14}) e^{-(0.000433)(250)} = \mathbf{1.2 \times 10^{14}}.$$

---

7. The conservation of atomic mass number requires that the daughter have a mass number of 235 and the conservation of charge requires that it have an atomic number of 92. The daughter is uranium. The kinetic energy of the α particle is the equivalent of the total mass defect of the process.

$$M_{lost} = 239.052\ 157 \text{ u} - 235.043\ 924 \text{ u} - 4.002\ 603 \text{ u} = 0.005\ 630 \text{ u}.$$

The kinetic energy of the α particle is $KE = M_{lost} c^2$

The speed of the α particle is then

$$v = \sqrt{\frac{2M_{lost}}{M_\alpha}}\ c = \sqrt{\frac{2(0.005\ 630 \text{ u})}{4.002\ 603 \text{ u}}}\ (3.00 \times 10^8 \text{ m/s}) = \mathbf{1.59 \times 10^7 \ m/s}$$

---

8. Assume that the original activity of the radioactive carbon in the wood was 0.23 Bq. The activity at any time is

$$A = A_0 e^{-\lambda t} \text{ where } \lambda = (0.693)/(5730 \text{ yr}) = 1.21 \times 10^{-4} \text{ yr}^{-1}$$

Taking the natural logarithm of the above equation and solving for the time yields

$$t = -(1/\lambda) \ln (A/A_0) = -(1/1.21 \times 10^{-4} \text{ yr}^{-1}) \ln (0.11 \text{ Bq}/0.23 \text{ Bq}) = \mathbf{6100 \ yr.}$$

CHAPTER 32
Ionizing Radiation, Nuclear Energy, and Elementary Particles

PREVIEW

In this final chapter you will learn about the effects of ionizing radiation on living cells, about nuclear fission and fusion reactions, about nuclear reactors, and about elementary particles including the latest theory of "quarks".

QUICK REFERENCE

Important Terms

Ionizing radiation
Radiation consisting of photons and/or moving particles that have enough energy to remove an electron from an atom or molecule.

Exposure
A measure of the ionization produced in air by X-rays or γ rays. Exposure is measured in roentgens.

Absorbed dose
The amount of energy absorbed from the radiation per unit mass of absorbing material. The SI unit of absorbed dose is the gray (Gy); 1 Gy = 1 J/kg.

Relative biological effectiveness (RBE)
The ratio of the dose of 200-keV X-rays needed to produce a certain biological effect to the dose of a given type of radiation required to produce the same effect.

Biologically equivalent dose
The product of the absorbed dose (in rad) and the RBE. The unit is the rem.

Induced nuclear transmutation
The process whereby an incident particle strikes a nucleus and causes the production of a new element.

Nuclear fission
A massive nucleus splits into two less-massive fragments.

Chain reaction
Neutrons released from a fission process induce other nuclei to fission.

Fission reactor
A device that generates energy by a controlled chain reaction.

Nuclear fusion
Two smaller masses combine to form a single nucleus with a larger mass. Energy is released when the binding energy per nucleon is greater for the larger nucleus than for the smaller nuclei.

Elementary particles
The basic building blocks of all matter.

Quarks
Elementary particles which are believed to make up the hadrons.

Equations

$$\text{Exposure (in roentgens)} = \left(\frac{1}{2.58 \times 10^{-4}}\right)\frac{q}{m} \qquad (32.1)$$

$$\text{Absorbed dose} = \frac{\text{Energy absorbed}}{\text{Mass of absorbing material}} \qquad (32.2)$$

$$\begin{matrix}\text{Relative biological} \\ \text{effectiveness (RBE)}\end{matrix} = \frac{\begin{matrix}\text{The dose of 200-keV X-rays that} \\ \text{produces a certain biological effect}\end{matrix}}{\begin{matrix}\text{The dose of radiation that} \\ \text{produces the same effect}\end{matrix}} \qquad (32.3)$$

$$\begin{matrix}\text{Biologically equivalent dose} \\ \text{(in rem)}\end{matrix} = \begin{matrix}\text{Absorbed dose} \\ \text{(in rad)}\end{matrix} \times \text{RBE} \qquad (32.4)$$

DISCUSSION OF SELECTED SECTIONS

32.1 Biological Effects of Ionizing Radiation

Ionizing radiation consists of photons and/or moving particles that have sufficient energy to knock an electron out of an atom or molecule, thus forming an ion. The photons usually lie in the ultraviolet, X-ray, and γ-ray regions of the electromagnetic spectrum, while the moving particles can be the α and β particles emitted during radioactive decay. A single high-energy photon or particle can ionize thousands of molecules.

Nuclear radiation is potentially harmful to humans, because the ionization it produces can significantly alter the structure of molecules within a living cell. The alterations cause the cell to malfunction and, if severe enough, can lead to the death of the cell and even the organism itself. Because of these hazards, it is important to learn the fundamentals of radiation exposure, including dose units, and the biological effects of radiation.

Exposure is a measure of the ionization produced in air by X-rays or γ rays, and it is defined as follows: A beam of X-rays or γ rays sent through a mass m of dry air at STP will produce ions whose total charge is q. The SI unit of exposure is C/kg. However, for historical reasons, a radiation unit known as the **roentgen (R)**, is still used. The exposure in roentgens is given by

$$\text{Exposure (in roentgens)} = \left(\frac{1}{2.58 \times 10^{-4}}\right)\frac{q}{m} \qquad (32.1)$$

Thus, an exposure of one roentgen produces $q = 2.58 \times 10^{-4}$ coulombs of positive charge in m = 1 kg of dry air, so $1\text{ R} = 2.58 \times 10^{-4}$ C/kg (dry air, STP).

For biological purposes, the **absorbed dose** is a more suitable quantity. It is defined as

$$\text{Absorbed dose} = \frac{\text{Energy absorbed}}{\text{Mass of absorbing material}} \qquad (32.2)$$

The SI unit of absorbed dose is the **gray (Gy)**, which is 1 Gy = 1 J/kg. The absorbed dose is a concept applicable to all types of radiation and absorbing media. A unit also used for absorbed dose is the **rad** (an acronym for radiation absorbed dose). The gray and rad are related by 1 rad = 0.01 Gy.

The amount of biological damage produced by ionizing radiation is different for different types of radiation. For instance, a 1-rad dose of protons is about 10 times as effective in producing biological damage as a 1-rad dose of X-rays. To compare the damage done by different types of radiation, the **relative biological effectiveness (RBE)** is used. It is defined by the following equation.

$$\text{Relative biological effectiveness (RBE)} = \frac{\text{The dose of 200-keV X-rays that produces a certain biological effect}}{\text{The dose of radiation that produces the same effect}} \qquad (32.3)$$

The RBE depends on the nature of the ionizing radiation and its energy, as well as the type of tissue being irradiated. As seen from Table 32.1, 200-keV X-rays, γ rays, and β^- particles produce less damage than protons, neutrons, and α particles. The RBE is often used in conjunction with the absorbed dose to reflect the damage-producing character of the radiation on tissue. The **biologically equivalent dose** is defined as:

$$\text{Biologically equivalent dose (in rem)} = \text{Absorbed dose (in rad)} \times \text{RBE} \qquad (32.4)$$

Example 1

What absorbed dose of α particles (RBE = 20) causes as much biological damage as a 60-rad dose of protons (RBE = 10)?

Using equation (32.4) we can obtain the biologically equivalent dose for the protons,

$$\text{Biologically equivalent dose (in rem)} = \text{Absorbed dose (in rad)} \times \text{RBE} = (60 \text{ rad})(10) = 600 \text{ rem}$$

For the alpha particles, the biologically equivalent dose is the same (600 rem). Therefore,

$$\text{Absorbed dose} = \frac{\text{Biologically equivalent dose}}{\text{RBE}} = \frac{600 \text{ rem}}{20} = 30 \text{ rad}$$

32.2 Induced Nuclear Reactions

A **nuclear reaction** occurs when a nucleus, particle, or photon strikes a target nucleus and causes a change in the target nucleus. For example, an α particle (a helium nucleus) strikes a target nitrogen nucleus producing a nucleus of oxygen and a proton. The reaction may be written as

$$_2^4\text{He} + {}_7^{14}\text{N} \rightarrow {}_8^{17}\text{O} + {}_1^1\text{H}$$

A shorthand notation is often used in writing reactions like the one shown above. The above reaction can be written

$$_7^{14}\text{N} \, (\alpha, \, p)_8^{17}\text{O}$$

The first and last symbols represent the initial and final nuclei, respectively. The symbols inside the parentheses denote the incident particle (on the left) and the small emitted particle (on the right).

In any nuclear reaction, both the total electric charge of the nucleons and the total number of nucleons are conserved during the process. The fact that these quantities are conserved makes it possible to identify the nucleus produced in a nuclear reaction.

Example 2

Determine the values of the atomic number Z (number of protons) and atomic mass number A (number of nucleons) for the following reactions. Identify the final nucleus and write each reaction in shorthand notation.

$$^{32}_{16}S + ^{1}_{0}n \rightarrow ^{A}_{Z}X + ^{1}_{1}H$$

$$^{40}_{18}Ar + ^{4}_{2}He \rightarrow ^{A}_{Z}X + ^{1}_{1}H$$

$$^{9}_{4}Be + ^{1}_{1}H \rightarrow ^{A}_{Z}X + ^{4}_{2}He$$

The atomic number Z and mass number A must be conserved. Therefore, the left and right hand sides of each reaction must balance. In the first reaction A = 32 + 1 - 1 = 32, and Z = 16 + 0 - 1 = 15. The element with atomic number Z = 15 is P. The reaction may therefore be written in shorthand as

$$^{32}_{16}S\,(n,\,p)^{32}_{15}P$$

Similarly, for the other two reactions we have

$$^{40}_{18}Ar\,(\alpha,\,p)^{43}_{19}K \qquad \text{and} \qquad ^{9}_{4}Be\,(p,\,\alpha)^{6}_{3}Li$$

32.3 Nuclear Fission

In 1939 four German physicists found that a uranium nucleus, after absorbing a neutron, splits into two fragments, each with a smaller mass than the original nucleus. The splitting of a massive nucleus into two less-massive fragments is known as **nuclear fission**. The following reaction shows one possible fission reaction.

$$^{1}_{0}n + ^{235}_{92}U \rightarrow ^{236}_{92}U \rightarrow ^{141}_{56}Ba + ^{92}_{36}Kr + 3^{1}_{0}n$$

In this reaction, a uranium-235 nucleus absorbs a neutron, creating a "compound nucleus", uranium-236. The compound nucleus disintegrates quickly into barium-141, krypton-92, and three neutrons.

The fission of uranium is accompanied by the release of an enormous amount of energy, which appears primarily as kinetic energy of the fission products. An average of 200 MeV of energy is released per fission.

The fact that the uranium fission reaction releases 2 or 3 neutrons makes it possible for a self-sustaining series of fissions to occur. Each neutron released by the fission process can initiate another fission reaction, resulting in the release of still more neutrons, followed by more fissions, and so on. This series of fission reactions is referred to as a **chain reaction**. If the reaction is uncontrolled, it would not be unusual for the number of fissions to increase a thousand fold within a few millionths of a second. With an average energy of 200 MeV per fission being released, an incredible amount of energy can be released in a very short time. This is the basis of the atomic bomb.

By limiting the number of neutrons allowed to participate in further fission reactions, it is possible to control the rate of fissions, and hence the amount of energy released. This controlled nuclear fission chain reaction is the principle behind the nuclear reactors used in the commercial generation of electric power.

32.5   Nuclear Fusion

Another way of generating nuclear energy is to combine two very low-mass nuclei with relatively small binding energies per nucleon, and "fuse" them together into a single, more massive nucleus that has a greater binding energy per nucleon. A substantial amount of energy can be released in such a reaction called a **nuclear fusion** reaction. A typical fusion reaction is shown below

$$^2_1H + {}^3_1H \rightarrow {}^4_2He + {}^1_0n$$

This reaction involves two isotopes of hydrogen, ^2_1H (deuterium) and ^3_1H (tritium), which fuse to form ^4_2He with the release of a neutron. The amount of energy released in the process can be determined by looking at the initial and final masses of the reacting nuclei. We have

	Initial Masses		Final Masses
^2_1H	2.014 102 u	^4_2He	4.002 603 u
^3_1H	3.016 050 u	^1_0n	1.008 665 u
Total	5.030 152 u		5.011 268 u

The **mass defect** is

$$\Delta m = 5.030\ 152\ u - 5.011\ 268\ u = 0.018\ 884\ u.$$

Since 1 u is equivalent to 931.5 MeV, the energy released by the fusion reaction is

$$(931.5\ MeV/1\ u)(0.018\ 884\ u) = 17.59\ MeV.$$

Example 3

In one type of fusion reaction a proton fuses with a neutron to form a deuterium nucleus:

$$^1_1H + {}^1_0n \rightarrow {}^2_1H.$$

The masses are ^1_1H (1.007 825 u), ^1_0n (1.008 665 u), and ^2_1H (2.014 102 u). How much energy (in MeV) is released by this reaction?

We just need to find the mass deficit between the proton plus neutron, and the deuterium. We have

$$\Delta m = 1.007\ 825\ u + 1.008\ 665\ u - 2.014\ 102\ u = 0.002\ 388\ u.$$

The energy released is therefore,

$$\Delta E = (0.002\ 388\ u)(931.5\ MeV/1\ u) = 2.224\ MeV.$$

PRACTICE PROBLEMS

1. A beam of X-rays passes through 7.5×10^{-4} kg of dry air and generates 1.2×10^{13} singly charged ions. What is the exposure in roentgens?

2. A person stands near a radioactive source and receives doses of the following radiation: β^- particles (35 mrad, RBE = 1), protons (7.5 mrad, RBE = 10), and α particles (3.0 mrad, RBE = 15). What is the total biologically equivalent dose?

3. During an X-ray examination, a person is exposed to radiation at a rate of 110 milligrays per hour. The exposure time is 0.10 s, and the mass of the exposed tissue is 1.2 kg. Determine the energy absorbed.

4. Write the following reactions in shorthand form.

$$^{14}_{7}\text{N} + ^{1}_{0}\text{n} \rightarrow ^{14}_{6}\text{C} + ^{1}_{1}\text{H}$$

$$^{43}_{20}\text{Ca} + ^{4}_{2}\text{He} \rightarrow ^{46}_{21}\text{Sc} + ^{1}_{1}\text{H}$$

$$^{55}_{25}\text{Mn} + ^{1}_{1}\text{H} \rightarrow ^{52}_{24}\text{Cr} + ^{4}_{2}\text{He}$$

5. Complete the following nuclear reactions, assuming the unknown quantity signified by a question mark is a single entity:

$$^{16}_{8}O\,(\alpha,\,n)\,?$$

$$?(p,\,\gamma\,)\,^{29}_{15}P$$

$$^{48}_{22}Ti\,(?,\,\alpha)^{45}_{21}Sc$$

$$^{35}_{17}Cl\,(?,\,n)^{38}_{19}K$$

6. $^{235}_{92}U$ absorbs a thermal neutron and fissions into rubidium $^{93}_{37}Rb$ and cesium $^{141}_{55}Cs$. What other nucleons are produced by the fission, and how many are there?

7. A nuclear reactor which uses $^{235}_{92}U$ as fuel has an output of 35 MW. If the energy released in the fission of one uranium-235 nucleus is 210 MeV, how many grams of $^{235}_{92}U$ are consumed in one day?

8. What energy (in MeV) is liberated by the following fission reaction?

$$^{235}_{92}U\,+\,^{1}_{0}n\,\rightarrow\,^{93}_{37}Rb\,+\,^{141}_{55}Cs\,+\,2^{1}_{0}n$$

Use the following values for the masses: n = 1.0087 u; U = 235.0439 u; Rb = 92.9217 u; Cs = 140.9195 u.

SOLUTIONS AND ANSWERS

Practice Problems

1. The exposure is obtained using equation (32.1),

   Exposure = $[1/(2.58 \times 10^{-4})](q/m) = [1/(2.58 \times 10^{-4})][(1.2 \times 10^{13})(1.6 \times 10^{-19} \text{ C})/(7.5 \times 10^{-4} \text{ kg})]$
   Exposure = **9.9 R**.

2. The biologically equivalent dose (BED) will be the sum of the individual doses. Using equation (32.4) we have

   $$\text{Biol.Equiv.Dose} = \text{BED}(\beta) + \text{BED}(\text{prot}) + \text{BED}(\alpha)$$
   $$\text{Biol.Equiv.Dose} = (35 \times 10^{-3} \text{ rd})(1) + (7.5 \times 10^{-3} \text{ rd})(10) + (3.0 \times 10^{-3} \text{ rd})(15)$$

   The biologically equivalent dose is therefore, BED = **160 mrem**.

3. The absorbed dose is,
   $$\text{AD} = (110 \times 10^{-3} \text{ Gy/hr})(1 \text{ hr}/3600 \text{ s})(0.10 \text{ s}) = 3.1 \times 10^{-6} \text{ Gy}.$$

   The energy absorbed is, by equation (32.2),

   $$\text{Energy absorbed} = \text{AD} \times \text{mass of absorbed matter} = (3.1 \times 10^{-6} \text{ Gy})(1.2 \text{ kg}) = \mathbf{3.7 \times 10^{-6} \text{ J}}$$

4. The shorthand notation is:

   $$^{14}_{7}\text{N (n, p) }^{14}_{6}\text{C}$$

   $$^{43}_{20}\text{Ca } (\alpha, \text{p}) \, ^{46}_{21}\text{Sc}$$

   $$^{55}_{25}\text{Mn (p, }\alpha)^{52}_{24}\text{Cr}$$

5. Adding up the atomic numbers and atomic masses on each side of the equations yields:

   a. $A = 16 + 4 - 1 = 19$, $Z = 8 + 2 - 0 = 10$. The unknown quantity has 10 protons and is $^{19}_{10}\text{Ne}$.

   b. $A = 29 + 0 - 1 = 28$, $Z = 15 + 0 - 1 = 14$. The unknown quantity has 14 protons and is $^{28}_{14}\text{Si}$.

   c. $A = 45 + 4 - 48 = 1$, $Z = 21 + 2 - 22 = 1$. The unknown quantity is therefore $^{1}_{1}\text{H}$.

   d. $A = 38 + 1 - 35 = 4$, $Z = 19 + 0 - 17 = 2$. The unknown quantity is therefore $^{4}_{2}\text{He}$.

6. If we balance the equation we have

$$A = 235 + 1 - 93 - 141 = 2, Z = 92 + 0 - 37 - 55 = 0.$$

Therefore, since $Z = 0$ and $A = 2$, the only possible nucleons are **two neutrons**.

---

7. The power is 35 MW = 35×10^6 J/s. In one day the energy requirement is,

$$E = (3.5 \times 10^7 \text{ J/s})(86\ 400 \text{ s}) = 3.0 \times 10^{12} \text{ J.}$$

We know that one gram of uranium-235 has

$$(6.02 \times 10^{23} \text{ mol}^{-1})(1 \text{ g})/(235 \text{ g mol}^{-1}) = 2.56 \times 10^{21} \text{ atoms.}$$

If each atom yields 210 MeV, then 1 g of uranium-235 yields

$$\Delta E = (2.56 \times 10^{21} \text{ atoms})(210 \times 10^6 \text{ eV})(1.60 \times 10^{-19} \text{ J/1 eV}) = 8.6 \times 10^{10} \text{ J.}$$

Finally, the amount of uranium-235 needed in one day is

$$m = (1 \text{ g})(E/\Delta E) = (1 \text{ g})(3.0 \times 10^{12} \text{ J})/(8.6 \times 10^{10} \text{ J}) = \textbf{35 g.}$$

---

8. Find the mass deficit:

$$\Delta m = 235.0439 \text{ u} + 1.0087 \text{ u} - 92.9217 \text{ u} - 140.9195 \text{ u} - 2(1.0087 \text{ u}) = 0.194 \text{ u.}$$

Since 1 u = 931.5 MeV the energy is

$$(931.5 \text{ MeV/1 u})(0.194 \text{ u}) = \textbf{181 MeV.}$$
